Finding Our Place in the Solar System

The Scientific Story of the Copernican Revolution

Finding Our Place in the Solar System gives a detailed account of how the Earth was displaced from its traditional position at the center of the universe to be recognized as one of several planets orbiting the Sun under the influence of a universal gravitational force. The transition from the ancient geocentric worldview to a modern understanding of planetary motion, often called the Copernican Revolution, is one of the great intellectual achievements of humankind. This book provides a deep yet accessible explanation of the scientific disputes over our place in the solar system and the work of the great scientists who helped settle them. Readers will come away knowing not just that the Earth orbits the Sun, but why we believe that it does so. The Copernican Revolution also provides an excellent case study of what science is and how it works.

TODD TIMBERLAKE has taught physics and astronomy at Berry College near Rome, Georgia, USA, since 2001. He teaches courses on the Copernican Revolution, the history of galactic astronomy and cosmology, and extra-terrestrial life, among others. Although he usually teaches college students, he has taught audiences ranging from kindergarten children in the USA to Tibetan Buddhist monks in India. He is passionate about incorporating the history and philosophy of science into the teaching of science and he is an expert at using computers as a tool for teaching physics and astronomy.

PAUL WALLACE teaches physics and astronomy at Agnes Scott College in Decatur, Georgia, USA. His scientific work was in experimental nuclear physics and gamma-ray astrophysics, but his love of the history of astronomy drew him away from technical research. A book lover, seminary graduate, and ordained minister, Paul now works at the intersection of literature, science, and theology. His unusual background and informal approach make him a sought-after speaker at churches, conferences, colleges, and retreats. His first book, *Stars Beneath Us: Finding God in the Evolving Cosmos*, was published by Fortress Press in 2016.

Finding Our Place in the Solar System

The Scientific Story of the Copernican Revolution

TODD TIMBERLAKE
Berry College, Mount Berry, GA, USA

PAUL WALLACE
Agnes Scott College, Decatur, GA, USA

CAMBRIDGE
UNIVERSITY PRESS

CAMBRIDGE
UNIVERSITY PRESS

University Printing House, Cambridge CB2 8BS, United Kingdom

One Liberty Plaza, 20th Floor, New York, NY 10006, USA

477 Williamstown Road, Port Melbourne, VIC 3207, Australia

314–321, 3rd Floor, Plot 3, Splendor Forum, Jasola District Centre, New Delhi – 110025, India

79 Anson Road, #06–04/06, Singapore 079906

Cambridge University Press is part of the University of Cambridge.

It furthers the University's mission by disseminating knowledge in the pursuit of education, learning, and research at the highest international levels of excellence.

www.cambridge.org
Information on this title: www.cambridge.org/9781107182295
DOI: 10.1017/9781316856208

First published 2019

Printed in the United Kingdom by TJ International Ltd. Padstow Cornwall

A catalogue record for this publication is available from the British Library.

Library of Congress Cataloging-in-Publication Data
Names: Timberlake, Todd Keene, 1973– author. | Wallace, Paul, 1968– author.
Title: Finding our place in the solar system : the scientific story of the
    Copernican revolution / Todd Timberlake (Berry College, [Mount Berry], Georgia),
    Paul Wallace (Agnes Scott College, Decatur [Georgia]).
Description: Cambridge : Cambridge University Press, 2019. |
    Includes bibliographical references and index.
Identifiers: LCCN 2018040457 | ISBN 9781107182295 (hardback : alk. paper)
Subjects: LCSH: Celestial mechanics–Study and teaching (Higher) | Astronomy–Study
    and teaching (Higher) | Planetary theory–Study and teaching (Higher)
Classification: LCC QB351 .T56 2019 | DDC 521–dc23
    LC record available at https://lccn.loc.gov/2018040457

ISBN 978-1-107-18229-5 Hardback

For my mother, Susan Timberlake (1944–2018), who always encouraged me to pursue my dreams. I wish she could have seen this particular dream come to fruition.

Contents

Preface

This book began as a college course. Astronomy 120: The Copernican Revolution was originally developed by my colleague Paul Wallace. The course was intended to fulfill a science requirement for students majoring in non-scientific fields at Berry College in Georgia (USA), where Paul and I taught. Paul's course was a great success, particularly when he taught it as a summer international course with stops in Poland, the Czech Republic, and Italy.

Then, rather suddenly, Paul decided to leave Berry. The decision was shocking if you didn't know Paul. It's not typical for a physicist to give up a tenured faculty position to go to seminary, but that's what Paul did and, for him, it made sense. His decision, though, left me in some difficulty. First and foremost my friend and mentor was leaving, but his departure caused some practical problems too. Paul had been the only person to teach astronomy at Berry for many years. Should we hire an astronomer to replace Paul? It made more sense for the department to hire a physicist, but we needed *someone* to teach astronomy.

I made what was, in hindsight, a rash decision, but one that I have never regretted. I decided that I would teach the astronomy courses at Berry. Although my doctoral degree is in physics, my undergraduate degree was in both physics and astronomy. I was confident that I could do it, but I only wanted to do it if I got to teach The Copernican Revolution. I had fallen in love with Paul's course. I loved it because it told a *story*, a story of science. Human beings love stories, but most students are never exposed to the story of science. They are taught science from a dry textbook that reads more like an encyclopedia than a novel. Teaching the history of science brings science to life. It shows students that science is a human activity, driven by the human passion to understand. It makes science interesting not just for the scientific knowledge, but for the *struggle* to gain that knowledge.

Moreover, I thought Paul's course was exactly what nonscience majors needed. Students who plan to be scientists need training in the most current scientific knowledge, but students who don't plan on careers in science have different needs. They still need to know about science, but what they really need is an understanding of how science works. Students can gain that understanding by actually *doing* science, as the science majors would eventually do, but students majoring in other areas don't really have time to develop the expertise needed to conduct real scientific research. Paul's course led me to believe that the best way to teach those students about the nature of scientific inquiry was to teach them the historical development of some important piece of science; to pick some fundamental bit of scientific knowledge and teach them *how* we gained that knowledge.

It doesn't get much more fundamental than the idea that the Earth rotates on its axis and orbits the Sun. We learn these facts as young children and most of us never question them, but this is not knowledge that was easily obtained. It took thousands of years of inquiry into the workings of Nature before humans came to understand the basic functioning of our solar system. There were strong reasons to reject the idea of a moving Earth and it took a great deal of effort to overcome those objections. It is a fascinating and important story, but also an accessible one. Black holes and string theory are exciting, but a real understanding of those topics takes years of study and significant mathematical background. A deep understanding of the motions of the Sun or the planets can be gained much more quickly and with only a modest amount of mathematical knowledge. There is a great joy to be had in that kind of deep understanding, regardless of the topic. And let's face it: knowing how the Sun moves across the sky is likely to be of more practical value for most people than knowing the latest theory of quantum gravity.

I have taught The Copernican Revolution eight times since Paul left Berry and I have loved it every time. Although I developed my own curricular materials (more on that soon), I continued to use the textbook that Paul had assembled for his course. That book was a compilation of writings from various sources as well as a significant amount of Paul's own original writing. I added to it and modified it but it was never published. Paul had moved on to other things, and I was too busy teaching my courses to have time to revise and publish Paul's book.

Then I was awarded a sabbatical for the spring of 2017. Finally I had the opportunity to turn the book into something that could be published, but I decided I didn't want it to be a traditional textbook. I was convinced that the story of the Copernican Revolution should be told to the widest possible audience, so I decided, with Paul's blessing, to rewrite the book for a general audience of readers interested in astronomy and its history. I am incredibly grateful that Cambridge University Press was receptive to my idea, and the result is now in your hands.

The purpose of this book is, in the first place, to tell the scientific story of the Copernican Revolution with a sensitivity to the historical context in which that story took place. The focus of the book is on the evidence and ideas that led to our modern understanding of the solar system, but my aim is to present those ideas and that evidence with the necessary historical background so that readers can understand why some ideas that we now know are true were initially rejected, and why some ideas (and evidence) that we now know are false were initially accepted.

To really understand the story of the Copernican Revolution we must begin at the beginning, with naked eye observations of the skies. Chapter 2 gives a detailed account of what we can observe of the stars and the Sun, and how ancient astronomers devised a simple theory to account for all of these observations. Chapter 3 describes observations of the Moon and the mysterious wandering stars (what we would now call planets). Chapter 4 explains how ancient astronomers attempted to account for the strange behavior of these wanderers and how they tried to fit their theories into a bigger picture of how the universe as a whole functions. The efforts of these ancient astronomers and philosophers were incredibly successful and that success served as the background against which the Copernican Revolution played out. Without understanding the astronomy of the ancient Greeks, it is impossible to understand the Copernican Revolution.

In Chapter 5 we finally get to Copernicus himself. That chapter provides a detailed account of his revolutionary theory of the Earth's motions and how his ideas provided an entirely new perspective on what we see in the heavens. Chapter 6 examines the work of Tycho Brahe who, like so many others of his time, rejected the Copernican theory but whose meticulous observations of the heavens laid the groundwork for the eventual success of Copernicus' main ideas. It was Tycho's one-time assistant, Johannes Kepler, who would transform the insightful but flawed theories of Copernicus into a recognizably modern theory of the solar system. Kepler's work is detailed in Chapter 7.

Kepler's theory of *how* the planets move holds up well today, but his ideas about *why* the planets move that way have been discarded. A full understanding of the movements of the solar system required the development of a new physics. Galileo Galilei, whose work is discussed in Chapter 8, would provide the first steps toward that new physics as well as a host of telescopic discoveries that won many converts to the Copernican cause. It was the fully developed universal physics of Isaac Newton, detailed in Chapter 9, that would conclude the Copernican Revolution by providing an explanation of why the Earth *must* orbit the Sun. With Newton's physics the Copernican theory (as modified by Kepler) emerged triumphant, even though there was still no direct evidence for the motions of the Earth. Chapter 10 gives an account of the additional evidence, gathered after the Revolution was already complete, that confirmed the ideas of Copernicus and Newton.

Although my main aim was to tell a particular scientific story, I also wrote this book to help readers understand the nature of science in general. To that end, Chapter 1 serves not only as an introduction to the book, but also as a commentary on what scientific theories are supposed to do and how they are judged. In addition, each of the remaining chapters ends with a section titled "Reflections on science." These sections highlight important lessons about the nature of science that can be drawn from the story in each chapter. I hope this material will help readers come away from the book with a better understanding of why science is so difficult, but also so *interesting*.

I have tried to make this book as accessible as possible, while still providing an accurate and detailed account of the scientific story I want to tell. Unlike the great *De revolutionibus* of Copernicus, this book was not written only for experts in astronomy. Although astronomy is a mathematical subject, and readers who wish to understand the development of astronomy must be prepared to tackle some mathematical arguments, I have tried to present these mathematical arguments in a form that is easy for nonexperts to understand. There are no equations in the main text. Instead, equations and detailed mathematical calculations are relegated to the Appendices. (I do, however, encourage everyone to read these Appendices!) Likewise, this book is not intended only for historians of astronomy. Although I provide citations to my sources, those citations are given in the Notes at the back of the book where they will not distract the more casual reader. Explanatory footnotes, on the other hand, are given at the bottom of the appropriate page of the main text.

As mentioned above, this book arose from a course that Paul and I taught at Berry College. I fully intend to use this book as the textbook for that course when I teach it again in the future. Anyone who wishes to use this book as a textbook for a course is invited to use my course materials, including the many open-source computer simulations and classroom activities that I have created to help students work through the story of the Copernican Revolution. I have also designed a series of class projects that allow students to make observations of a fictitious solar system and develop their own (Ptolemaic and Copernican) models for that system. All of these curricular materials are available on my website.[i] Many of the activities make use of the open-source planetarium program *Stellarium*.[ii] The use of *Stellarium* to make simulated observations of the skies,

---

[i] http://sites.berry.edu/ttimberlake/teaching/copernican-revolution/
[ii] http://stellarium.org/

or of my simulations for visualization of astronomical theories, will benefit any reader whether they are using this book as a textbook for a course or just reading it for their own interests.

Readers who want to learn more about the story of the Copernican Revolution should consult the References section, but let me take this opportunity to make a few specific recommendations. Thomas Kuhn gave a classic account of the transition from Ptolemaic to Copernican astronomy in his *The Copernican Revolution*. I. B. Cohen provided an excellent overview of the transition to the new physics of Galileo and Newton in his *The Birth of a New Physics*. To some extent this book is intended to combine those two classic works and update them using recent scholarship in the history of science. Millevolte's *The Copernican Revolution* complements this book by telling much the same story, but with greater emphasis on cultural aspects and less on the technical science. Koestler's *The Sleepwalkers* provides an entertaining account of that same story, although I disagree with many aspects of Koestler's presentation. Two other useful overviews of this material are Toulmin and Goodfield's *The Fabric of the Heavens* and Crowe's *Theories of the World from Antiquity to the Copernican Revolution*. Owen Gingerich's *Eye of Heaven* is a wonderful compilation of some of his brilliant essays on these topics. Hirshfeld's *Parallax* provides an engaging account of some of the material in Chapter 10.

I also highly recommend reading the words of the great scientists who played pivotal roles in the Copernican Revolution. English translations of many important works are available and some of these are accessible to a nonexpert reader (and perhaps after reading this book you will have enough expertise to tackle even the more difficult works!). Book I of Ptolemy's *Almagest* (translated by Toomer) is highly readable, as is Book I of Copernicus' *De revolutionibus* (translated by Rosen). Fuller, but still brief, accounts of Copernicus' theories are given in his *Commentariolus* and in the *Narratio Prima* of Rheticus, English translations of which can be found in Rosen's *Three Copernican Treatises*. I particularly recommend the works of Galileo, including his *Sidereus Nuncius* and *Il Saggiatore* (English translations available in Drake's *Discoveries and Opinions of Galileo*), but especially Galileo's *Dialogo* (Drake's translation). A fairly accessible account of Newton's work in his own words is given in his *A Treatise of the System of the World*.

For those seeking a more detailed account of the lives of these scientists, several accessible biographies are available. I recommend Sobel on Copernicus, Ferguson and Love on Tycho and Kepler, Wootton on Galileo, and Gleick on Newton. Those seeking more scholarly biographies should seek out the classic accounts by Armitage on Copernicus, Thoren on Tycho, Caspar on Kepler (English translation by Hellman), Drake on Galileo, and Westfall on Newton.

I have many people to thank for making this book possible, with the usual caveat that none of them should be held responsible for the errors I have, inevitably, made. I certainly never would have written this book without my co-author Paul Wallace. Although I take primary responsibility for the book you now hold, what I wrote was based on Paul's course and his original textbook (including his excellent figures and diagrams) and he has certainly earned his place as a co-author. His work laid the foundations for this book, and his guidance, advice, mentorship, and friendship have helped to sustain my interest in this project over many years. I wish Paul had never left Berry, but I am trying to view this book as a silver lining to that cloud.

I would also like to thank the many historians of science who have helped to uncover and illuminate all the bits and pieces that I have tried to assemble into this story of the Copernican Revolution. I am not a historian myself and I could not have hoped to write this book without relying on their work. My indebtedness to them will be apparent from my citations and the References. I would also like to acknowledge those with whom I have interacted directly, mostly though my participation in the Biennial History of Astronomy Workshops at the University of Notre Dame or through the H-ASTRO listserv. Owen Gingerich, Chris Graney, and Mike Crowe have been particularly helpful, not just with their knowledge of the history of astronomy but with their encouragement of my own efforts. I owe special thanks to Matt Dowd for organizing the Notre Dame workshops and welcoming an outsider into the fold. I thank Owen Gingerich and Johan Kärnfelt for feedback on an early draft of the book.

I also wish to thank all of my Astronomy 120 students over the past several years who have helped me to refine my explanations of the material in this book, and in some cases to improve my own understanding of those topics. Particular thanks go to my former student Tricia Steele for helping me to better understand Aristotle. Likewise, I am grateful to my former professors at Vanderbilt University, especially David Weintraub and Richard Haglund, who helped build my love for astronomy and for the history and philosophy of science. My thanks go to Berry College for awarding me the sabbatical semester during which I wrote the first full draft of this book and to the editorial staff at Cambridge University Press (especially Vince Higgs, Lucy Edwards, Margaret Patterson, and Esther Miguéliz Obanos) who expertly guided me through the process of turning a manuscript into a published book. I thank Mike Bailey, Sandy Meek, and Kalen Maloney for permission to use their photographs. I owe tremendous thanks to JoAnn Palmeri and the History of Science Collections at the University of Oklahoma Libraries for their generous permission to use many images from their fantastic collections.

Of course, no project of this magnitude is possible without the support of those closest to the author. I thank my wonderful wife, Karen, who ably steered the ship of our family while her husband was busy playing with armillary spheres and reading 17th century astronomical treatises. I thank my sons, Max and Pete, whose interest in all things science gave me extra motivation when my energy began to flag. Finally, I thank my parents, Jack and Susan, who gave me the opportunity to pursue my dream of being a scientist. I only wish that my mom was still around to hold a copy of this book in her hands.

T. T.

1

Introduction: mysterious skies

1.1   Three mysteries

Throughout recorded history humans have watched the sky. They have marveled not only at the beauty of the Sun, the Moon, and the stars, but also at the motions of these objects across the sky. In tracking the motions of these heavenly objects they encountered three fundamental mysteries. This book is about how those mysteries were solved ... and then solved again (and again).

The first mystery is revealed by even casual observation of the heavens. The Sun moves westward across the sky throughout the day. The Moon displays a similar westward motion that may be visible during the day or the night. Likewise, the stars move westward throughout the night but they do not seem to move *relative to each other*. They maintain fixed patterns that we have come to associate with pictures known as constellations. Why do these lights in the sky move in this way? That's the first mystery.

Uncovering the second mystery requires much more than an occasional glance at the sky, but careful observations made over weeks or months show that the Sun moves relative to the fixed pattern of the stars. Even easier to spot is the motion of the Moon relative to the stars. Much harder to see, but still discernable to the careful observer, is the fact that five of the stars don't maintain their positions in the fixed pattern held by the thousands of other stars visible to the naked eye. Like the Sun and Moon, these five "wandering stars" move around relative to the "fixed stars." Why do these seven objects move around relative to the thousands of fixed stars that seem to form a static pattern? That is the second mystery.

Once these seven wanderers were identified, people began to track their motions against the background of the fixed stars. They found that the Sun and Moon move steadily eastward against this background. In fact, the Sun and Moon both move through the same set of constellations so they not only move in the same direction against the starry background but they also follow nearly the same path. The five wandering stars are usually seen to trudge along that same path, moving eastward against the starry background like the Sun and Moon. Occasionally, though, one of these wanderers will halt its eastward motion through the fixed stars, move westward for a while, stop again, and then resume its eastward motion. What could possibly make these wandering stars move in this bizarre way? That is the third, and perhaps the deepest, mystery. It was the key that unlocked the secret of the heavens.

The story begins with careful observations of the sky, like those made by ancient Babylonian astronomers. From the 2nd century BC to the 2nd century AD, ancient Greek astronomers built on this observational foundation and created sophisticated geometrical models to explain the mysterious motions. Their models assumed what seemed to be obvious: that the Earth sat stationary while the heavens moved around it. The mysteries, it seemed, were solved.

The solutions proposed by the ancient Greeks were so successful that they went largely unchallenged for 1400 years. Then, in 1543, Nicolaus Copernicus published a book that offered a new solution to these mysteries. The model proposed by Copernicus made sense of the strange forward-and-back motions of the wandering stars in a way that the ancient Greek models did not, but Copernicus' theory was not easy to accept. He proposed something that went against common sense: he suggested that the Earth moves. He claimed that the Earth spins around and also that the Earth and all of the wandering planets, but not the Moon, orbit around a stationary Sun that rests near the center of the "solar system."

At first astronomers and natural philosophers could not make sense of Copernicus' idea and his theory was not widely adopted. But a few astronomers found Copernicus' model compelling and they worked to refine and further explain his ideas until they became acceptable. By the end of the 17th century astronomers had reached a deep understanding not only of how the Earth and the wandering stars (or planets) move, but why they move that way. The ancient Greek theory was dead and the motions of the Earth became common knowledge.

The transition from the Earth-centered (or "geocentric") universe of the ancient Greeks to the Sun-centered (or "heliocentric") solar system that we accept today is known as the "Copernican Revolution." This book tells the story of the Copernican Revolution from ancient observations of the skies to the explanation of planetary motions in terms of a universal gravitational force in the 17th century and beyond.

The full story of the Copernican Revolution involves politics, religion, social and economic change, literary traditions, translation, intercultural exchange, patronage, personal rivalries, war, plague, and death. No single book could hope to address all aspects of this story. While this book will touch on these issues, it aims primarily to tell the *scientific* story of the Copernican Revolution, focusing on observations and experiments, mathematical models and scientific theories, instruments and measurement techniques, and the principal works of the scientists who sought to understand the operations of the heavens.

1.2   Why should you read this book?

Why, though, would you want to know the scientific story of the Copernican Revolution? After all, you already know the punch line: the Earth really does spin on its axis and orbit the Sun. Copernicus was right about that. You learned that in grade school. Case closed. But do you know *why* we believe that the Earth moves? If you think about it, the motions of the Earth are certainly not obvious.

In a way, this book is like a mystery novel in which you already know the answer to the puzzle. Even though you know the answer, the story of how that answer was discovered is fascinating. One thing that makes it particularly fascinating is that the answer we accept today was not the first answer to this puzzle. How did the ancient Greeks reach conclusions that differ so dramatically from our modern understanding? Once they had found their solution, how (and why) did we come to abandon that solution and accept a completely different one?

That story is a fascinating tale full of twists and turns, heroic effort, and brilliant insight. It is one of the great human stories. The geocentric theory of the universe developed by the ancient Greeks was one of humanity's great intellectual achievements. The heliocentric theory proposed by Copernicus and finalized by Isaac Newton is an even greater achievement. The change from a geocentric to a heliocentric perspective radically altered the way people thought about the universe and our place in it. That change affected far more than just astronomy: it had a tremendous impact on religion, philosophy, and other facets of society and it paved the way for modern science as we know it.

Let's briefly consider just one impact of the Copernican Revolution: the possibility of extra-terrestrial life. In the ancient Greek cosmos Earth was a unique place. The heavens were fundamentally different and could not serve as a home to "life as we know it." But in the heliocentric system proposed by Copernicus the Earth was just one planet among many. If there was life on Earth, then why not on other planets? Furthermore, if the Sun was just one star among many then why couldn't those other stars have planets, and life, of their own? The idea that there might be other life "out there," perhaps even creatures more intelligent than us, gave us a completely new perspective on ourselves. The Copernican Revolution didn't just change astronomy, it changed our views on what it means to be human. The story of such a profound change is one worth knowing.

It is also worth knowing the story of the Copernican Revolution because it is good to occasionally question the things you have been told. Most people learn about the motions of Earth from trusted authorities such as their parents or teachers. There is nothing wrong with believing what you are told by people you trust. In practice, we all must accept many things that we are told or we couldn't get on with life. But every now and then it is good to examine the evidence for yourself and see if you are convinced, and evaluate the arguments to see if they are valid. This exercise will improve your critical thinking skills and help you spot flawed arguments and invalid claims in other parts of your life. Inevitably, you will encounter such arguments and claims and it helps to be prepared for them.

You may have even heard invalid claims about the Copernican Revolution. In one version of the story the ancient Greek theories of the universe are silly and obviously wrong, and the only reason people did not immediately accept the Copernican theory was because of opposition by religious authorities. You may have heard that Copernicus was persecuted for proposing his heliocentric theory, or that people objected to the heliocentric theory because it "demoted" the Earth from its prime location at the center of the universe. None of these things is true. The real story is much more complicated, but also much more interesting.[1]

Perhaps the best reason to learn the *scientific* story of the Copernican Revolution, though, is that it will help you understand the nature of science. Science is a complex activity that cannot be reduced to a short list of rules and procedures, in spite of what your grade school teachers may have told you (see, sometimes you have to question authority!). That simplified version of the "scientific method" might be appropriate for young students first learning about science, but real science is much more complicated, messy, creative, and exciting. Learning the scientific story of the Copernican Revolution will help you better understand how science is really done, how scientific theories are proposed and evaluated, and how our scientific knowledge grows.

In fact, you can gain a better understanding of the nature of science by reading about the history of science than you can from reading a standard science textbook. Most science textbooks focus on the end products of science, the knowledge that is provided by our best current theories. An understanding of the end products of science is important if you want to *use* scientific knowledge, so science textbooks have good reason for focusing on current knowledge, but knowing how to use something is different from knowing how it was created. You may be able to drive a car, but do you know how that car was built? If you want to know how scientific knowledge is obtained, the best approach may be to do some scientific research yourself, but that option is not available to most people. The next best way to learn about the nature of science is to learn about the history of science.

The historical approach to science emphasizes *how* we learned what we know rather than just *what* we have learned. It shows that science is the creation of human beings, not something that fell from the sky. Science requires a lot of hard work and creativity. Science is difficult and the methods of science are far from infallible. Sometimes scientists get things wrong. By learning about the history of science you can glimpse this human side of scientific inquiry. The history of science shows how difficult it is to gain new scientific knowledge and how easy it is to go astray. It also shows how great effort and persistence can pay off as we gain not just new scientific knowledge, but also insight into how to determine when that knowledge is reliable and when it is not.

Another advantage of the historical approach to learning about science is that it automatically starts from relatively simple ideas and works up to more complicated and difficult theories. Building from the ground up makes it easier to understand the science at a deep level, and therefore it puts you in a better position to understand what the scientists were doing at each stage of the story. Importantly, a historical account of science *is* a story. We often learn best through stories, and the story of the Copernican Revolution provides an excellent opportunity to learn about the nature of science.

1.3   The nature of science

The scientific story of the Copernican Revolution describes the change from one scientific theory, the geocentric theory of the universe, to another, the heliocentric theory of the solar system. But what are scientific theories? What are they used for and how do we judge them? Scientific theories are complex things and this book cannot hope to provide a thorough discussion of every aspect of scientific theories. For the most part, we will simply examine particular theories in their historical context and see how they were used and evaluated by actual scientists. However, it may help to start by briefly considering in general what scientific theories are supposed to do.

First and foremost, scientific theories are supposed to fit with observed phenomena. That might mean that we expect our theories to reproduce previously observed phenomena in a qualitative way (e.g. the Sun has risen in the east every morning of your life). It might also mean that we expect them to provide qualitative predictions for as yet unobserved phenomena (e.g. the Sun will rise in the east tomorrow morning).

For some types of theories, particularly mathematical theories, we expect the theory to provide a quantitative fit to the data. That means our theories should be able to produce numerical values that agree with prior quantitative measurements, as well as successfully predict the numerical values of future measurements (e.g. the Sun will rise tomorrow at 6:43 AM and will be first visible at a point on the horizon that is $12°$ south of due east). Of course, any quantitative theory is likely to have uncertainties and errors due to flaws or approximations in the theory or because of inaccuracy of the numerical parameters that go into the theory. Note that, whether a theory is qualitative or quantitative, it is only meaningful if it is possible for observations to contradict the theory. A theory that can be made to fit with any conceivable observational result is not much of a theory at all.

We may also expect our theories to do more than just fit with our observations. We may expect them to *explain* what we observe. Ideally we would like for the observed phenomena to follow necessarily if the world is the way our theory says it is. In that case the theory doesn't just reproduce our observations, it purports to tell us what is really going on that results in the phenomena we observe. Of course, it is possible that more than one theory can successfully explain the same set of observations, so even if a theory seems to explain what we see that doesn't mean the world really works that way.

If we do expect our theories to tell us how the world really works, then we must demand that our successful theories don't contradict each other. If two theories give different predictions for some observable phenomenon, then they can't both be correct descriptions of how the world works. We would like for our best scientific theories to fit together to provide a coherent picture of the physical world. If our theories do contradict, then at least one of them must not be a correct description of the world, but the contradiction alone does not tell us which one is wrong and which one (if any) is right.

Finally, we may expect our theories to be beautiful. Beauty, of course, is subjective. However, we often expect our theories to be simple, to not involve too many "adjustable parameters," to not have many exceptions or caveats, to be, in a word, elegant. Many of our best theories, once we fully understand them, make us say "of course it must be that way!"

It can be helpful to think about scientific theories in terms of an analogy. Scientific theories are, in some ways, like maps. Like maps, scientific theories are created for a specific purpose (or set of purposes). Different purposes require different types of maps. If you are driving through a city you might want a street map, but if you are planning a hike in the wilderness a topographic map might be more useful. Maps are supposed to mimic some aspects of the physical world, but they also ignore many other aspects that aren't important for the map's intended purpose. For that reason maps, like scientific theories, are always approximate. They are never exact in every possible detail, and that's a good thing. Think about how useless would be a map of London that was exact in every detail. For one thing, it would have to be constantly updated in order to be accurate. Even worse, it would be so complicated that using the map would be no easier than simply walking around London itself!

This map analogy leads to some interesting questions. How might we expect different maps of the same area to relate to each other? What happens if we try to use a map for some purpose other than that for which it was intended? Could we create a map that, although it would not be exact in every detail, might provide concise and accurate information suitable for a wide variety of purposes? Even if we could, would we be able to claim that such a map was true? What does it even mean for a map to be true? We can ask similar questions about scientific theories.

We can also consider some helpful analogies for the process of doing science. In some ways science is like putting together a jigsaw puzzle. We want to fit all of the pieces together to form a coherent and sensible picture. Sometimes you can't tell if the pieces form a sensible picture until several more pieces are added to the puzzle.

If science is like solving a jigsaw puzzle, then that puzzle is an extremely challenging one. For one thing, the puzzle doesn't have well-defined boundaries. You certainly can't look at the box cover to see if you are "getting it right." There are lots of missing pieces … and there always will be. In fact, in real science the pieces don't come pre-cut. Scientists have to cut their own pieces by performing experiments and making observations. Each piece serves as a tiny window into the nature of the physical world, but the shape of those pieces and the picture they show depends very much on choices made by the scientist: what they choose to observe and how they make their observations. It is possible that these "slices of reality" could be cut along natural "seams," but we have no way to know in advance what those seams are.

When we find that some pieces don't fit together it may be because they really don't connect to each other in the puzzle, but it may also be that we have just cut the pieces the wrong way. We might even end up with false pieces that aren't part of the puzzle at all. How can we ever know if our puzzle is correct or complete? In practice, we probably can't know. However, we can still feel that we are making progress if we are able to fit more and more pieces into the puzzle in a way that seems to form a coherent and sensible picture.

Doing science is also a bit like cooking. Great chefs follow recipes, to be sure, but they also modify recipes and even invent entirely new ones. They use some standard cooking techniques (baking, grilling, frying, etc.) but the results they get will depend on how they blend different techniques and different ingredients to create something new. In a similar way, scientists use some standard methods to make observations, perform experiments, and build models and theories. But how it all comes out will depend on the creative ways in which they combine these methods and the ingredients (data, theories, assumptions) they use in their work. Just as the greatest chefs may invent new techniques for cooking, so too the greatest scientists sometimes invent new methods for scientific inquiry. Success in science, as in cooking, can be subjective, but often we can achieve widespread agreement about failure. Skilled scientists, like skilled chefs, obtain their skill by practicing constantly and overcoming repeated failures.

Finally, science is like art or literature. Artists create art because it pleases them to do so. Likewise, scientists do science because it brings them joy to make a novel observation or develop a successful new theory. Scientific work can be tedious and difficult, but at times it is thrilling.

Like a great work of art, great science can (and should) be appreciated by others. Just as knowledge of artistic and literary techniques can help someone appreciate a work of art or literature, so knowledge of the nature of science can help someone appreciate a great scientific accomplishment like the Copernican Revolution. It is our hope that this book will help you appreciate this great human achievement, just as you should appreciate the great works of art and literature.

1.4   Changing knowledge

This book describes theories that were developed to explain the observed motions of the heavens, as well as the process by which those theories were developed and tested. However, this book is primarily about the change from the geocentric picture of the world to the heliocentric picture of the solar system. To understand the process of theory change it may help to consider what happens when we abandon one theory in favor of another.

As mentioned earlier, we often expect our theories to *explain* observed phenomena. We want the theory to tell us what is really going on that led to our observations. A theory gives meaning to our observations. When we make observations we *see* a phenomenon, but a theory allows us to see that phenomenon *as* something meaningful. Theories let us see our observational data as the outcome of processes that are not directly visible to us.

Because more than one theory can explain the same set of data, it is possible to see a certain set of observations as being two (or more) different things. When

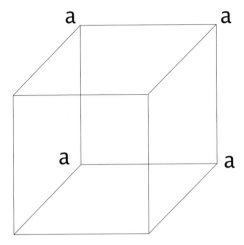

Figure 1.1 A Necker cube. The face whose corners are labeled with the letter a could be at the front or at the back of the cube.

we abandon one theory and adopt another we go from seeing our data as one thing to seeing them as something else.

Something analogous can happen in our visual perception. We can view a certain visual image and see it as a certain object, but then we may find that it is possible to view the same visual image as a different object. Changing from seeing the image as one thing to seeing it as a different thing is known as a "gestalt shift." One of the classic gestalt shifts can be generated by looking at the "Necker cube" shown in Figure 1.1. The image itself consists of several straight line segments, but our visual system tends to assign meaning to these line segments by seeing them as a projection of a three-dimensional cube. But there is some ambiguity in our interpretation of the image. One of the "faces" of the cube in Figure 1.1 has labeled corners. It is possible to see that labeled face as being at the front of the cube, but it is also possible to see the labeled face as being at the back of the cube.

In both cases the viewer is seeing the same visual image, but the visual image is interpreted differently. It is possible for a viewer to switch back and forth between the two different interpretations, first seeing the labeled face as the back of the cube, then seeing it as the front of the cube, and so on. Although this is a particularly simple example of a gestalt shift, it is analogous in some ways to the shift between the geocentric and heliocentric viewpoints in astronomy. Both geocentric and heliocentric astronomers saw the same lights in the sky moving in the same way, but they interpreted those motions very differently.

In an isolated image like the one in Figure 1.1 it may be impossible to decide which interpretation is "correct." However, if that image is put into a

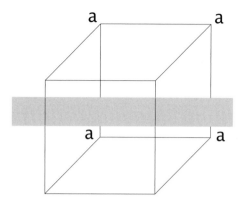

Figure 1.2 A gray bar passing through the Necker cube in Figure 1.1 provides clues to the cube's orientation. In this case the labeled face must be at the back of the cube.

relationship with other images then that relationship can provide clues about how best to interpret the original image. For example, Figure 1.2 shows the original Necker cube but this time with a gray bar passing through the image. Careful inspection of which lines are blocked by the gray bar, and which are not, suggests that in this Necker cube the labeled face is at the back of the cube so that we can consistently interpret the gray bar as passing through the middle of the cube.

In a similar way, astronomers had no good way to decide between the geocentric and heliocentric viewpoints when all they considered was the motions of lights in the heavens. However, astronomers did not judge their astronomical theories in isolation from everything else. They judged them in the context of other knowledge, particularly theories about how things move. Originally these theories of motion seemed to indicate that the geocentric viewpoint was correct. The heliocentric theory was inconsistent with ancient physics in the same way that the appearance of the gray bar in Figure 1.2 is inconsistent with the labeled face being at the font of the cube.

When we change from one theory to another, that change can lead to conflict with other knowledge, just as viewing the Necker cube as having the labeled face in front is inconsistent with the appearance of the gray bar in Figure 1.2, or it may help to resolve conflicts that already existed. When theory changes occur, scientists are left to sort out all of the conflicts with existing knowledge that may arise from the new theory. As we will see, an important part of the story of the Copernican Revolution deals with the way scientists resolved the conflict between the heliocentric model and ancient theories about motion. Viewing the heavenly motions from a new perspective ultimately led us to think differently about *all* motions.

The Necker cube is a simple example in that it presents two obvious interpretations and we just have to choose between those two. Real science deals with much more complicated situations, and it can be hard to formulate even one theory that can adequately explain the available data while remaining consistent with our other knowledge. Because of that difficulty, once we have found an interpretation that seems to work we may be very reluctant to let go of it. The gestalt shift from a geocentric universe to a heliocentric solar system was not an easy change to make, and Copernicus' proposal was not widely accepted until about two hundred years after he published it.

To understand why the Copernican Revolution took so long, and why it happened at all, we need first to understand the geocentric theory that was overthrown in that revolution. We need to know why the ancient Greeks believed in a geocentric universe. To understand their reasons for adopting a geocentric theory we must first carefully examine what they saw in the skies.

2

Two spheres: modeling the heavens and the Earth

2.1 Watching the stars

Let us begin by observing the stars in the night sky. We want to put ourselves in the shoes of an ancient observer watching the stars wheel overhead. For this purpose we will forgo the use of a telescope or binoculars and just look with our own eyes. In a way, we have put ourselves at a disadvantage relative to that sky-watcher of the distant past: it is much harder to see the stars today because of light pollution, at least if you are observing near an urban area. On the other hand, we will have an advantage over that ancient observer because we can talk about our observations using terminology and concepts that have been handed down to us by preceding generations of astronomers. Since our story will begin with civilizations that flourished near the Mediterranean Sea, let's assume that we have found a dark location at approximately the same latitude as the Mediterranean. From this dark spot, what can we see?

2.1.1 The circling skies

The night sky above us is speckled with stars, some bright, others dim. Looking closely we see that some of them have an orangish tinge, while others seem a bit blue. The stars are spread across the sky in no obvious pattern. There doesn't seem to be much structure there, although we do notice the dim and hazy band of white light known as the Milky Way that cuts across the night sky. Let us set aside the Milky Way for now and focus on the stars. If we watch them long enough we will see that they move.

Although the arrangement of the stars shows no obvious pattern, careful observation over an hour or more shows us that their motions do show a definite pattern. If we can figure out the cardinal directions (north, south, east, and west) we notice that the stars seem to move from east to west as a general rule. We can investigate the motion in more detail by looking in specific directions. Turning ourselves to the east (the general direction from which the Sun rises each day) we see that the stars in this part of the sky all move upward and to the right. New stars appear over the eastern horizon as we watch. Turning around and facing west we see the stars move downward and to the right. Here stars vanish from sight as they drop below the horizon.

What about the view to the south? In this direction the motion is quite different. In our southern sky we see stars move in clockwise circular arcs, beginning low on the horizon toward our left and arcing upward and to the right, then back down to vanish below the horizon toward our right. The view toward the north is even more striking. In that direction the stars appear to move in circular arcs as well, but these circles run counterclockwise and are entirely above the horizon. If we watch long enough we may see a star begin a bit to our left (toward the northwest), arc down toward the horizon, and then curve back upward as it moves a bit to our right (toward the northeast). In our imagination we can trace out the rest of its circular path, but the coming of daylight will spoil our chance to see it go all the way around this circle. If we carefully observe several stars in the north we notice that they all move in counterclockwise circular paths and, what is more, those circles all have the same center. Sitting near that center point is a moderately bright star. Of all the stars we see in the sky, only this one star seems to sit still throughout the night. This is the star we now know as Polaris, also known as the North Star because it always sits there above the horizon to the North.

These regular patterns of motion suggest that the stars might move as a collective whole, rather than as individual objects. To test this idea we can carefully observe a particular group of stars, noting (and perhaps carefully drawing) their arrangement on the sky. As the hours pass, the stars in this group will move, but do they all move together and keep the same arrangement, or do they move individually and break up the group? After some time we become convinced that they are moving collectively: the arrangement of the group remains fixed, although the orientation of the entire group may change by tilting one way or another.

The observation that the stars move collectively suggests the idea of mapping out the arrangement of all the stars, perhaps mentally forming them into smaller groups to make them easier to track. That idea forms the basis of what we now call **constellations**: groups of stars with a fixed arrangement. Ancient observers associated particular images with these star groups in order to make them easier to remember, and often the images associated with neighboring groups were related to each other by some story. The modern constellations used by Western astronomers are adapted from constellations defined by the ancient Babylonians and Greeks. The constellation of Orion the hunter is accompanied in the sky by the constellations of his hunting dogs, Canis Major (the big dog) and Canis Minor (the small dog), all easily visible in the southern skies of winter. In one mythological story Andromeda, daughter of King Cepheus and Queen Cassiopeia, is rescued from a sea monster by the hero Perseus. Thus, the constellations Cepheus, Cassiopeia, Andromeda, and Perseus all lie near each other in our northern skies.

However we define our constellations, if we carefully note their arrangements we will find that the same groups of stars appear again the next night. This observation tells us something very important about the stars: they seem to be permanent fixtures of the sky. The stars that disappear on our western horizon must come back somehow. We can test this theory. Perhaps on a winter night we notice the modern constellation of Lyra the harp, with its bright star Vega, drop below our northwestern horizon early in the night. But later that night, not long before the Sun rises, we see Lyra rise in the northeast. These stars that are appearing in the northeast must be the same ones that we saw vanish in the northwest: their arrangement is identical. So the stars are not destroyed when they drop out of sight. Although we can't see them, we suspect that they must still be there below our horizon. Why else would the same arrangement of stars appear later in the night? But we cannot be sure: after all, we cannot see the stars when they are below the horizon so we don't really know what they are doing then.

Our observations are beginning to suggest that some larger pattern is at work. Perhaps there is some way to tie together all of these pieces into a coherent picture that explains what all of the stars are doing. First, though, we need to make some more careful observations. Armed with our new knowledge of the constellations, and knowing that the same constellations appear each night, we decide to identify a few particular stars and carefully track their motions over the course of several nights. These observations give us some new pieces for our puzzle. We find that any individual star will always rise at the same point on our eastern horizon, and likewise it will always set at the same point on our western horizon. However, we may not be able to see both the rising and setting of a given star in a single night. That is because some stars stay above the horizon longer than others. Stars in our southern skies stay up for the shortest time, while those toward the north stay up longer. Some stars in the northern sky, the ones that seem to circle around that stationary star, are always above the horizon during the night. Perhaps they are also up during the day, but we can't be sure. We suspect that our stationary star, the star known as Polaris in the constellation Ursa Minor, is always up since it never seems to move from its appointed spot.

We also find that some stars reach higher points in the sky as they move from east to west. Generally the southern stars do not get very high, but stars that rise in the northeast may pass almost directly overhead before they descend toward the northwest.

The general pattern we see is that every star moves along part of a circular path, and each star moves along the same partial circular path every night. On any given night we cannot see any star move all the way through its circle. For those stars circling around Polaris the entire circle is above the horizon, but we can't see those stars go all the way around because daylight hides the stars from our view. Other stars seem to follow a circle that is cut in two by our horizon. We can watch the stars move along the part that is above the horizon, but what happens when they go below? We suspect that they continue to follow the same circle below the horizon. We can't know this for sure, but it seems like a reasonable idea.

We have also seen that the stars seem to move collectively, as though they were all part of a single object. So the stars seem to be part of a larger thing that spins around such that each individual star moves along a repeating circular path. If we know a little geometry (and the ancient Greeks were experts in geometry) we might recognize that the entire pattern of star motions can be explained if we assume that the stars are fixed to the inside of a giant sphere that is spinning around us (see Figure 2.1). This enormous sphere rotates about an axis that passes through our location and also through a point on the sphere very close to the star Polaris. The rotation of this sphere has little effect on Polaris since it sits so close to the axis of rotation, but the stars near Polaris will move in counterclockwise circles (as seen from the inside, where we are). Stars farther south will move on circles that have a small part cut off by our horizon, while stars even farther south will have most of their circles below the horizon with only a small part visible above. In other words, this sphere idea predicts exactly the motions we see.

The model shown in Figure 2.1 is our first example of a scientific theory. It does a good job of explaining the motions of the stars that we have observed, but it also makes definite predictions about some things we have not seen. As an example of a prediction made by this theory, notice the dotted-dashed circle near the lower right portion of the ball in Figure 2.1. This circle shows a star moving in a clockwise circle (as seen from inside), but the entire circle is below our horizon. So this star is entirely invisible to us! Does such a star actually exist? We know there are some stars that never set (the ones near Polaris), and some stars that do rise and set (all the others that we can see), so couldn't there be some stars that never rise at all? If our model is correct, then such stars may exist. This is a very definite prediction. Is it testable? Can we find out if such

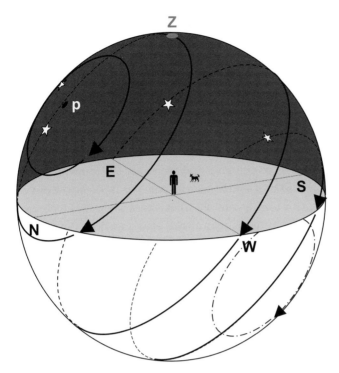

Figure 2.1 The spinning sphere of the stars. The point p indicates where the axis of rotation crosses the sphere. The diagram shows the star paths that result from the sphere's rotation. Only the portion of these paths above the horizon plane will be visible.

stars exist? At this point we are not sure, but perhaps we can find a way to test this prediction later.

There is another subtle point to be made about this model. The model encourages us to think of this sky-sphere almost as if it really is an actual *thing* to which the stars (whatever they are) are physically attached. Shall we make this ball-realness part of our theory? It's such an intuitive and simplifying idea that we might be tempted to go with it, until we have need to drop it. After all, if we refused to make a real ball part of our model, we would have to explain why the thousands of individual stars make independent circular treks *on their own*. If you stick all the stars to a single sphere and let it turn, then the circular motion of that one ball alone accounts perfectly for the thousands of stellar motions we see. This idea is simplifying and efficient, but it may be wrong. In fact, the whole model may be wrong; maybe the stars zigzag and do loops beneath the horizon. Maybe they are extinguished nightly and re-lit the next day. We have no way of knowing what is happening with the stars when we can't see them. But this star-sphere model is such a simple and efficient way to explain all of the star motions we do see that for now it seems reasonable to accept the theory in full until we have some reason to doubt it.

If we are going to stick to this theory for now we should give it a nice name: we will call our star-ball the **celestial sphere**.[1] We would like to know some more details about this celestial sphere. We know it spins about an axis that goes through our location and through a point near Polaris. We will call that point near Polaris, where the rotation axis crosses the celestial sphere, the **north celestial pole**. The stars appear to circle counterclockwise around the north celestial pole, and the stars close to the pole that never set are known as **circumpolar stars**. There should also be a south celestial pole but it lies below our horizon to the south. The rotation axis of the celestial sphere seems to be at an odd angle relative to our horizon. What is that angle? How fast does the celestial sphere spin? Does it spin around at a constant rate, or does it speed up and slow down? How big is this celestial sphere? Note that these are quantitative questions. To find answers we are going to need to put some numbers on the sky.

2.1.2   *Numbers on the sky*

Let's start by establishing a quantitative method for precisely describing the location of a star in our sky. We can't give a location in terms of distances because we don't yet know how far away the stars are. All we can really talk about is the direction in which we see a star. To give a quantitative description of direction we are going to need angle measurements. The system of angle measurements we will use is the one used by ancient Babylonian astronomers and later adopted by the ancient Greeks. In this system angles are given in **degrees** (°), where an angle of 360° takes us all the way around a circle. If we need greater precision we can use fractions of a degree. Since the Babylonians used a system of counting based on the number 60, they specified fractions of a degree in sixtieths. One **minute** (often called a "minute of arc" to distinguish it from a minute of time, and symbolized by ′) is one sixtieth of a degree, so $1° = 60′$. Likewise a **second** ("second of arc", ″) is one sixtieth of a minute: $1′ = 60″$. A detailed discussion of angular measure is provided in Appendix A.1.

Now, how are we going to assign numbers to different points on our sky? First we need to decide what shape the sky has. Our celestial sphere theory suggests that we can think of the sky as a hemisphere, since our sky is just that part of the celestial sphere that is above our horizon, and our horizon seems to cut that sphere in half. Actually, our real horizon doesn't cut the sphere exactly in half. Our real horizon is bumpy and irregular: there may be trees that stick up, or mountains in the distance, maybe even a nearby building that blocks part of our view of the sky. But if we think of an idealized, mathematical horizon it will be a circle that cuts the celestial sphere into two equal parts. So our sky

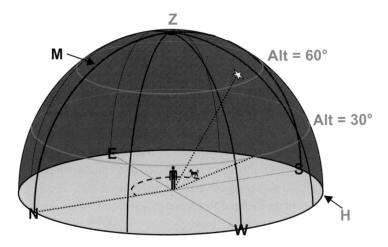

Figure 2.2 Numbering the dome of the sky. The altitude-azimuth (alt-az) coordinates allow us to specify particular points on the sky dome. Altitude is measured up from the horizon, so the horizon (H) has altitude 0° and the zenith (Z) has altitude 90°. Azimuth is measured eastward from north, so the azimuths of the cardinal directions are 0° (N), 90° (E), 180° (S), and 270° (W). The meridian line (M) runs along the line of 0° azimuth up to the zenith and then down the line of 180° azimuth. The figure shows an example star with approximate coordinates az = 155° and alt = 41°.

is a hemispherical dome. This might seem obvious even without our celestial sphere theory, but it really isn't so obvious. Many ancient cultures assigned other, nonspherical shapes to the sky.

Before we start trying to put numbers on this dome, let's identify a few special points. First of all, we can mark the four cardinal directions at the appropriate points on our horizon as shown in Figure 2.2. Remember that if we are facing that stationary star Polaris then we are facing north, so east will be to our right, west to our left, and south directly behind us. Next we define the **zenith**, which is the imaginary point Z on the sky directly overhead. Using our direction points and the zenith we can also define the **meridian**, which is an imaginary line M that runs from the southernmost point on the horizon S to the northernmost point on the horizon N through the zenith Z.

Now we are ready to start assigning numbers to the points on our sky, a process known as creating a **coordinate system**. Since we have decided that our sky has a spherical shape, we need to use a coordinate system that is appropriate for a sphere. One way to specify the location of points on a sphere is to use two angles, like the longitude and latitude angles we now use for the Earth's surface. We can apply this idea to the points on our sky dome by specifying the angle of a point *above* the horizon as well as the angle of the point *around* the horizon. The first angle is called the **altitude** and the second is called the **azimuth**; the altitude-azimuth (alt-az) grid is shown in Figure 2.2. Recall that the horizon makes a circle around us, and the angle all the way around that circle is 360°. By convention azimuth is measured eastward from north: so north is 0°, east is 90°, south is 180°, and west is 270°. If we keep going around the horizon the azimuth angle approaches 360° as we approach the north point. When we reach the north point the azimuth then resets to 0°. All points on the horizon are said to have an altitude of 0°; a star halfway between the horizon and the zenith has an altitude of 45°, and the zenith itself always has an altitude of 90°. Figure 2.2 shows how the alt-az system can be used to specify the location of a star at alt $= 41°$, az $= 155°$.

Now we can start to use our coordinate system to investigate the details of our celestial sphere model. Let's start with the stationary point in our northern sky: the north celestial pole (NCP). What is the azimuth of the NCP? Since the NCP is always due north, its azimuth is always 0°. In fact, the direction of the NCP *defines* the direction north. North is north because it is the direction of the NCP's azimuth. But what about the altitude of the NCP? This requires a measurement, but if we have a good way to measure angles we will find that for our Mediterranean observer the NCP has an altitude of about 35°. We might wonder why it has that particular value, but for now we will just accept it as a fact. Polaris, which is very close to the NCP, will always have an altitude of about 35° and an azimuth close to 0° (or 360° if it is slightly to the west).

What about the other stars? Our alt-az coordinate grid is fixed on our sky, so since the stars move across the sky their altitude and azimuth coordinates must continually change, but there are some regularities. A given star will always rise at the same azimuth (and its altitude must necessarily be 0° at rising since it is crossing the horizon). That azimuth will always be between 0° and 180°, on the eastern half of the horizon. That same star will always set on the western half of the horizon at an azimuth that is 360° minus its rising azimuth. In addition, a given star will always cross the meridian line at the same point. When a star crosses the meridian line it is called a **transit**. When a star transits its azimuth must be either 0° or 180° (since the meridian runs north–south). The star's altitude reaches its maximum value when the star transits, and a given star always transits at the same altitude.

As an example of these patterns, let's consider the star Mintaka, one of the stars in the belt of Orion. For our Mediterranean observer, Mintaka always rises due east (az $= 90°$), transits the south part of the meridian at alt $= 55°$, and sets due west (az $= 270°$). On the other hand, the star Sirius rises in the southeast at an azimuth of about 113°, transits the southern meridian at an altitude of about

38°, and sets in the southwest at an azimuth of about 247°. The star Vega rises in the northeast near az = 40°, transits the northern meridian near alt = 86°, and sets in the northwest near az = 320°.

We know that different stars stay above the horizon for different amounts of time. For example, using a modern clock we could find that Mintaka is above the horizon for 12 hours, but Sirius is only above the horizon for just over 10 hours, while Vega is up for about 16.5 hours. However, our celestial sphere model suggests that all stars should complete their full circular motions in the same amount of time. To check this prediction we can use our clock to measure the time between successive transits of a star. We start our timer when the star crosses the meridian and stop it when the star returns to the meridian (except in the case of circumpolar stars, which cross the meridian twice for each rotation of the celestial sphere). We find that the time between transits is the same for all stars: it is always 23 hours, 56 minutes. We will refer to this length of time as one **sidereal day** (the word sidereal means "with respect to the stars"), but it must also be the rotational period of our celestial sphere. Additional measurements could convince us that the celestial sphere rotates at a constant rate. For example, the time between rising and transit for a star is always equal to the time between transit and setting for that same star, which suggests that the celestial sphere is not speeding up or slowing down. Ancient astronomers did not have accurate clocks with which to make such measurements, but the motions of the stars seemed so regular that they assumed that the motion of the celestial sphere was uniform.[2]

The alt-az coordinates of a star change constantly because the coordinates are fixed to the sky but the stars are not. However, our celestial sphere theory says that the stars *are* fixed to the celestial sphere. So if we laid out a system of coordinates on that celestial sphere then the coordinates of a star would be fixed (at least according to our theory). Figure 2.3 illustrates just such a coordinate system. Note that the coordinate system is aligned with the rotation of the celestial sphere, so it is tilted relative to our alt-az coordinate grid. Figure 2.3 shows that the north celestial pole and the as-yet-unseen south celestial pole occupy special places in this coordinate system. Additionally, there is a special circle that is equidistant from the two poles. This great circle is known as the **celestial equator**.[i] Note that the celestial equator intersects the horizon at the due east and due west points, and it crosses the meridian at an altitude of 55° (= 90°−35°).[ii] Exactly half of the celestial equator lies above the horizon.

---

[i] A great circle is any circle on a sphere that divides the sphere into two equal hemispheres.
[ii] For our Mediterranean observer. As we shall see, the angle will be different for observers at other latitudes.

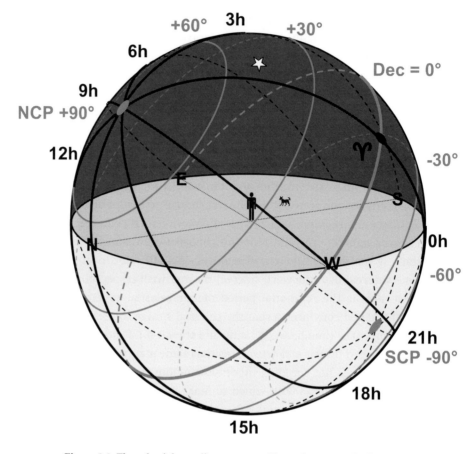

Figure 2.3 The celestial coordinate system. Lines of constant declination are parallel to the equator. Lines of constant right ascension (hour lines) run from the NCP to the SCP. The star shown is on the back side of the celestial sphere and has RA = 5^h, Dec = $15°$.

Now for the grid of our new coordinate system. We imagine a set of circles that are parallel to the celestial equator and are centered on the poles, much like circles of latitude on the Earth. Five of these (including the equator itself) are shown in Figure 2.3. Circles north of the celestial equator are labeled by positive angles and circles south of the celestial equator are labeled by negative angles; the celestial equator itself is at $0°$. The celestial poles are $90°$ away from the celestial equator, and are assigned values of $+90°$ (north) and $-90°$ (south). These labels are in keeping with those of the Earth's north and south poles; this coordinate on the celestial sphere is therefore fully analogous to terrestrial latitude. It is called **declination** (Dec).

Our second coordinate is similar to longitude on the Earth and is called **right ascension** (RA). Imagine that the celestial sphere is divided up into 24 pieces, like segments of an orange. Some of these segments are shown in Figure 2.3. They are labeled not by angle but by *hour*. We will see why later. There is a 0 hour (^{h}) line, a 1^h line, a 2^h line, all the way up to 23^h, after which we return to 0^h. For simplicity, Figure 2.3 shows these hour lines in 3 hour increments. Notice that these hours are marked by semicircular arcs, not full circles; for example, the 0^h arc runs from the NCP to the SCP, but it does not continue back around to the NCP. The continuation of the 0^h line would be the 12^h line. Because all of the lines of right ascension meet at the NCP and SCP, the RA coordinates of the NCP and SCP are undefined.[i]

The system of RA and Dec coordinates is known as the **celestial coordinate system**. Notice that all lines of RA meet all circles of declination at right angles, and that any point on the sphere is uniquely specified by a given pair of coordinates (RA, Dec). Also notice the dot next to the ♈ symbol at the point where the 0^h line meets the celestial equator. This is the origin of the celestial coordinate system: (RA, Dec) = $(0^h, 0°)$. It is a special point and is called the "vernal equinox" for reasons we will discuss later. It makes sense to use the celestial equator as our circle for Dec $= 0°$ because the celestial equator is a unique circle, but the hour line that passes through ♈ is chosen as the line of RA $= 0^h$ for historical reasons (just as the longitude line through Greenwich, UK, is chosen as the line of zero longitude for historical reasons). We will see later that the ♈ point does have a special significance, but our coordinate system would have worked just as well if we had chosen a different hour line for our origin.

Look at Figure 2.3 and try to imagine the celestial sphere spinning. The lines of constant declination just spin in place. They don't move around at all on the sky. But the lines of constant right ascension move across the sky from east to west. During each rotation of the celestial sphere there will be a time when a particular RA line will line up perfectly with our meridian. Let's say that at some particular moment the RA $= 3^h$ line lies along the meridian. That means every star that has RA $= 3^h$ is transiting at that moment. If we wait an hour we will find that the RA $= 4^h$ line is aligned with our meridian, so all of the RA $= 4^h$ stars are transiting.[ii] Now we see why it makes sense to use hours for RA rather than degrees. If we want to know the time between the transits of two stars all we have to do is find the difference in the stars' RA values. For example, the star

---

[i] In the alt-az coordinate system the azimuth of the zenith is undefined for the same reason.
[ii] Actually the 4^h line will not quite be lined up with our meridian because the celestial sphere rotates once per sidereal day (23 hours, 56 minutes) rather than once every 24 hours, but it will be close.

Sirius (with RA $= 6^h 45^m$) will transit about 1 hour and 13 minutes later than the star Mintaka (with an RA of about $5^h 32^m$). Note that we cannot generally find the time between risings or settings in this way because the RA lines never line up with our horizon, but this method will work to find differences in rising and setting times if the two stars have the same declination.

Our new celestial coordinate system can help us to understand why different stars may have different motions. Each star moves along a path that matches its declination line. From Figure 2.3 we see that the declination circles for Dec > $55°$ are entirely above the horizon, so stars with Dec > $55°$ will be circumpolar stars that never set. Stars with declinations between $0°$ and $55°$ (like Vega with a declination of about $39°$) will rise in the northeast, transit at a high altitude, and set in the northwest. We can see that these stars will be above the horizon for more than 12 hours because the portion of their declination circle that is above the horizon cuts across more than 12 hours of RA. Note that a star with a declination of $35°$ would pass right through our zenith, so the declination of our zenith is $35°$, equal to the altitude of the NCP.

A star that is on the celestial equator (with Dec $= 0°$, like Mintaka) will rise due east, transit at $55°$ altitude, and set due west as seen from the Mediterranean. Such a star will be up for 12 hours because the visible part of the celestial equator cuts across exactly 12 hours of RA. Finally, stars below the celestial equator (like Sirius, with Dec $= -17°$) will rise in the southeast, transit at low altitudes, and set in the southwest. These stars will be up for less than 12 hours since their visible Dec circle cuts across less than 12 hours of RA. Finally, stars with Dec < $-55°$ will always be below the horizon and thus invisible to us. If we could see these "south circumpolar stars" (assuming they even exist) they would appear to circle clockwise about the SCP.

We find that our model of a celestial sphere of stars, rotating with a period of 23 hours and 56 minutes about an axis oriented $35°$ above due north on our horizon, seems to explain everything we have observed about the motions of the stars as seen from the Mediterranean. That's pretty impressive: there are thousands of stars visible in our sky, and this one simple model explains how all of them move. The stars, though, are not particularly important for our day to day life. The Sun, on the other hand, is of great practical importance.

2.2   Tracking the Sun

Up to this point we have focused on the stars, but of course the most obvious thing in our sky is the Sun. The Sun regulates human life in many ways. The daily cycle of light and darkness sets the rhythm of our activity and our rest. The Sun dictates the annual cycle of seasons which governs our agriculture.

Understanding the motions of the Sun gave ancient humans the power to predict these cycles and prepare for the changes they would bring. Undoubtedly, humans would have closely observed the motions of the Sun long before they had the ability to write down what they saw. The ancient Greeks, armed with their geometry and their theory of the celestial sphere, developed a detailed understanding of the Sun's motions.[3] How did they fit the motions of the Sun onto their celestial sphere model, and what observations did they use as a guide?

2.2.1   Changing seasons

During a single day, from our Mediterranean location, we see the Sun rise somewhere on the eastern horizon, reach its highest point in the sky as it crosses the meridian to the south, and then set somewhere in the west. We will refer to the time when the Sun is highest in the sky, just as it transits the meridian, as **local noon**. From a single day of observation we might conclude that the Sun moves just a like a star, and therefore we can treat the Sun as though it is fixed to the celestial sphere. Long-term observation, however, reveals that this is not so.

For example, if we carefully watch the point on the horizon at which the Sun rises we will see that it changes over time.[4] There is a day in late March (on our modern calendar) on which the Sun rises due east. This is the day we call the **vernal equinox**, or spring equinox. As the days go on we will see the Sun rise at points that are progressively farther north. This northward trend continues until late June when, on the day we call the **summer solstice**, the Sun rises at its northernmost point. After that the point of sunrise moves southward, hitting due east again in late September on the **autumnal** (or fall) **equinox**. The southward motion continues until the sunrise reaches its southernmost extreme on the date of the **winter solstice** in late December. Then the sunrise point turns north again, reaching due east on the next vernal equinox approximately 365 days after the previous vernal equinox. The pattern then repeats.

Stars always rise at the same point on the horizon, but the Sun doesn't obey that rule. So it seems that the Sun must be moving relative to the stars. Close observation of the stars visible just before sunrise, or just after sunset, allows us to estimate the Sun's location on the celestial sphere at any particular time. By carefully tracking that location over time we find that the Sun seems to move roughly eastward relative to the stars. Tracking the Sun in this way over many days reveals that the Sun moves generally eastward along the celestial sphere in a great circle path that repeats about every 365 days. This roughly 365-day period of the Sun's motion relative to the stars is known as the **sidereal year**.

The Sun moves slowly eastward on the celestial sphere while the celestial sphere rotates westward, so the Sun lags behind the stars as they move westward across the sky. Therefore, it must take the Sun a little bit longer than the stars to go all the way around our sky. It takes the celestial sphere 23 hours and 56 minutes to spin around once, but during that time the Sun has moved almost one degree eastward on the sphere (since it moves 360° in 365 days). The celestial sphere must rotate about one extra degree in order to get the Sun back to its starting point. In fact, using a modern clock to measure the time between consecutive local noons (transits of the Sun) we find that it takes 24 hours for the Sun to go all the way around the sky, on average. The time from one local noon to the next is known as a **solar day** to distinguish it from the sidereal day, but if the word "day" appears without a modifier then it refers to the solar day. The length of the solar day varies a little bit over the course of a year, but our clocks are designed such that the average length of a day corresponds to exactly 24 hours. In other words, our clocks are set to the Sun, not the stars.

Although the length of a full solar day (from one local noon to the next local noon) varies only a tiny bit, the amount of daylight within a single day varies considerably over the course of a year. When the Sun rises in the northeast it stays up longer and gets higher in the sky. But when the Sun rises in the southeast it doesn't stay up long or get very high on the sky. This explains the seasonal variations in temperature that we experience. In summer when the Sun is up longer and higher, temperatures will be warm. In winter the Sun is up shorter and lower, so temperatures are colder. Note that in the summer the Sun behaves like a star that is north of the celestial equator (Dec > 0°): it rises northeast, transits at high altitude, and sets northwest more than 12 hours after rising. The winter Sun behaves like a star that is south of the celestial equator (Dec < 0°): it rises southeast, transits at lower altitude, and sets southeast less than 12 hours after rising. Figure 2.4 summarizes these observations. This behavior suggests that the Sun is sometimes north and sometimes south of the celestial equator. The Sun's path on the celestial sphere must be tilted relative to the equator. But by how much?

2.2.2   *Sticks and shadows*

To track the motions of the Sun in greater detail we can use shadows.[5] If we place a straight rod or stick so that it sticks up vertically from the ground, the shadow cast by this stick can tell us the azimuth and altitude of the Sun. Our stick is known as a **gnomon**. It is easy to determine the Sun's azimuth from the gnomon's shadow. Whatever direction the shadow points, the Sun is in the opposite direction. So if the shadow is pointing 20° north of west (az = 290°) then

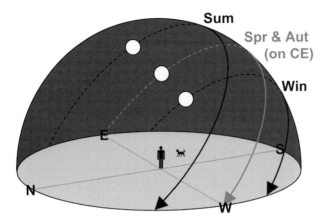

Figure 2.4 The seasonal paths of the Sun. In summer, the Sun is north of the celestial equator and it moves like a star at positive declination. In winter, the Sun is south of the equator and it moves like a star with negative declination. On the equinoxes the Sun lies on the celestial equator.

the Sun must be 20° south of east (az = 110°). Over the course of a day we will see the shadow of our gnomon change. The shadow becomes visible shortly after sunrise. At that time the shadow will be long and will point roughly westward. Over the course of the day the shadow will grow shorter and swing around to point northward, then grow long again as it swings to the east.[i]

The changing azimuth of the gnomon's shadow provides the fundamental mechanism for a sundial. The shadow rotates from west to north to east as the day progresses and the direction of the shadow at any time tells us how much of the daytime has passed. The shadow will be shortest when it points due north, just as the Sun is crossing the meridian in the south, and the time is exactly (local) noon. The ancient Babylonians, and later the ancient Greeks, divided the daytime into 12 equal hours and likewise divided the night into 12 equal hours. Because the amount of daylight and darkness varies over the course of a year, the length of one of these **seasonal hours** was not fixed. Instead, the seasonal hours in summer were long during the day but short during the night. The opposite was true in winter. Only on the equinoxes would the seasonal hour correspond to the hours used in modern clocks, which are sometimes referred to as **equinoctial hours**.[6] Although ancient Greek astronomers adopted the use of equinoctial hours, non-astronomers kept time according to seasonal hours, and most public sundials were designed to display the seasonal, rather than equinoctial, hour of the day.[7]

---

[i] Note that we are still assuming an observer at the latitude of the Mediterranean.

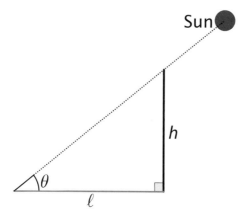

Figure 2.5 The right triangle formed by a vertical gnomon with height h, the gnomon shadow with length ℓ, and the ray of sunlight that just grazes the top of the gnomon and hits the end of the shadow. The Sun's altitude θ can be determined using trigonometry if h and ℓ are known.

So much for the Sun's azimuth. To extract the Sun's altitude from the gnomon's shadow we need to use trigonometry (see Appendix A.3). Basic trigonometry deals with right triangles. Every triangle has three sides and three interior angles that add up to 180°. In a right triangle, one of the interior angles is a right (or 90°) angle. That leaves us with two unknown angles and three sides of unknown length. Trigonometry allows us to determine the remaining unknown quantities if we are given two pieces of information: either the lengths of two sides, or the length of one side and one of the unknown angles. Figure 2.5 shows how the shadow of our gnomon, the gnomon itself, and a ray of sunlight form a right triangle. Trigonometry allows us to determine the altitude of the Sun (the angle θ in Figure 2.5) from the height of the gnomon and the length of its shadow. Details can be found in Appendix A.4.

The variation of the Sun's altitude over a single day is straightforward. The altitude is small just after sunrise. It then increases steadily until reaching its maximum value at local noon when the shadow points due north. After that the Sun's altitude decreases until the shadow vanishes as the Sun sets. None of this is particularly surprising, but what is more interesting is that the Sun's altitude at local noon changes from day to day.

Let's take a closer look at how the Sun's local noon altitude varies over the course of a year. Figure 2.6 shows a plot of the Sun's altitude at local noon versus the day of the year for our Mediterranean observer. On the vernal equinox (about March 20) the Sun has the same transit altitude as the star Mintaka (55°). This suggests that, like Mintaka, the Sun must lie on the celestial equator on the day of the vernal equinox. On the summer solstice (about June 21) the Sun reaches its

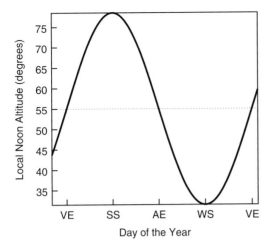

Figure 2.6 Plot of the Sun's local noon altitude throughout the year for an observer at 35° N latitude. The altitude is a maximum on the summer solstice, a minimum on the winter solstice, and obtains its median value of 55° on the equinoxes.

greatest local noon altitude, about 78°.5. At the summer solstice the Sun must be north of the celestial equator with a declination of 23°.5 (= 78°.5 − 55°). On the autumnal equinox, the local noon altitude returns to the median value of 55°, suggesting that once more the Sun lies on the celestial equator (Dec = 0°). Around December 21, on the winter solstice, the Sun's local noon altitude is only 31°.5, indicating a declination of −23°.5 (= 31°.5 − 55°). After the winter solstice the Sun's local noon altitude increases again until the next vernal equinox and the pattern repeats.

Like the motions of the stars, the motions of the Sun seem to follow a set pattern. But that pattern is not quite the same as the pattern for a star. How do we make sense of the Sun's motions using our celestial sphere model?

2.2.3   The ecliptic circle

The Sun's changing declination as revealed by its local noon altitude fits perfectly with the annual changes in the point of the sunrise and the length of daylight. In fact, all of our observations can be explained if the Sun moves along the celestial sphere in a great circle path that is tilted relative to the celestial equator by 23°.5.[8] This path is known as the **ecliptic** and the 23°.5 tilt is known as the **obliquity** of the ecliptic. The Sun moves eastward along the ecliptic, crossing the celestial equator at two different points. The Sun will reach one of these points on the date of the vernal equinox, and that point is denoted by the symbol ♈. It is this vernal equinox point that we used as the zero point for our right

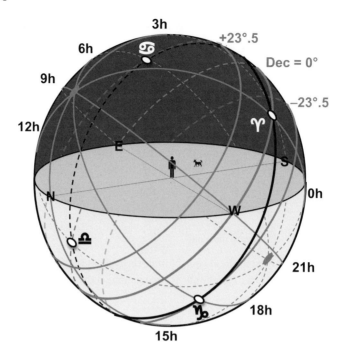

Figure 2.7 The Sun follows a path on the celestial sphere known as the ecliptic, shown in black in the diagram. The ecliptic is tilted $23°.5$ relative to the celestial equator. The diagram shows hour lines of RA in 3^h increments as well as the celestial equator and lines of declination at $\pm 23°.5$, all in gray. Also shown are the locations of the Sun on the vernal equinox (Υ, RA $= 0^h$, Dec $= 0°$), the summer solstice ($\odot$, RA $= 6^h$, Dec $= 23°.5$), the autumnal equinox ($\Leftinox$, RA $= 12^h$, Dec $= 0°$), and the winter solstice ($\triangle$, RA $= 18^h$, Dec $= -23°.5$).

ascension coordinate. So Υ lies at RA $= 0^h$, Dec $= 0°$. The other, autumnal, equinox point is symbolized by $\Leftinox$ and it lies in the opposite part of the sky from Υ at RA $= 12^h$, Dec $= 0°$. The ecliptic reaches its northernmost point at RA $= 6^h$, Dec $= 23°.5$. We will use the symbol $\odot$ for this northern extreme, which the Sun reaches on the summer solstice. On the winter solstice the Sun reaches the southern extreme of the ecliptic at RA $= 18^h$, Dec $= -23°.5$. The winter solstice point is indicated by $\triangle$. Figure 2.7 illustrates the path of the ecliptic on the celestial sphere.

Careful observations made over several years reveal that it takes the Sun about 365.25 days to travel from Υ back to Υ. This time period is known as a **tropical year**. Observations made over a human lifetime seem to indicate that the equinox and solstice points remained fixed relative to the stars, so for now there seems to be no difference between this tropical year and the sidereal year we defined earlier. We will examine how the ecliptic relates to the stars in more detail, but first let's look at some practical uses for our knowledge of the Sun's cycles.

The variations of the Sun's motion over the course of a tropical year give rise to the seasons and are thus critically important to agriculture. Ancient people devised methods to track the variations in the Sun's path in order to determine where they fell within the annual pattern, so that they could plant or harvest their crops at the right times. We have already seen that the local noon shadow of a gnomon can be used to mark out the solstices and equinoxes, and early Greek sundials were used mostly for this purpose rather than for measuring time during the day.[9] The ancient Greeks also observed the first morning rising of stars in order to keep track of the yearly cycle of seasons. As the Sun moves eastward along the ecliptic, it rises later and later relative to the stars. New stars become visible on the horizon just before sunrise blots them out. The ancient Greeks developed lists, known as *parapegmata*, of stars whose first morning risings occurred at various points throughout the year. Armed with one of these parapegma, an observer could determine the approximate date by taking a quick look to the east before sunrise.[10]

Governmental authorities developed more formal calendars to track the annual cycle of the Sun. The ideal calendar would have the solstices and equinoxes fall on the same date every year. The problem is that the tropical year is not equal to a whole number of days. The extra quarter of a day in the 365.25-day period causes trouble. The simplest solution is to round to the nearest day, and that is exactly what the ancient Egyptians did. They used a 365-day calendar, but as the years went on they found that the dates of the solstices and equinoxes (as well as the approximate dates of seasonal events like the flooding of the Nile) occurred later and later in their calendar. The city-states of ancient Greece did not develop a unified calendar. Instead, each city kept its own local calendar, and each of these calendars suffered from problems as great as those of the Egyptian calendar.[11]

The next big step forward in calendar keeping came by proclamation from Julius Caesar. The *Julian calendar* began, like the Egyptian calendar, with a year of 365 days. However, any year that was divisible by four was considered a "leap year" and included an extra day. Thus, the average length of a year in the Julian calendar is 365.25 days, closely matching the observed tropical year.[12] Unfortunately, the tropical year is not *exactly* 365.25 days. It is actually 365.2422 days. So even the Julian calendar was not perfect, and in the 16th century a new calendar was proposed by Pope Gregory XIII that better aligned the calendar to the tropical year. This Gregorian calendar, which we still use today, eliminated the leap years in years divisible by 100, except those divisible by 400. The average Gregorian year is thus 365.2425 days, very close to the true tropical year. We will encounter this Gregorian calendar reform again, later in our story.[13]

The tropical year that forms the basis for our calendars is the time for the Sun to complete one circuit around the ecliptic, from ♈ to ♈. We might reasonably assume that the Sun moves along the ecliptic at a uniform rate, but is that actually so? We can test this hypothesis by examining how long it takes for the Sun to go between the solstice and equinox points. Each arc from one of these points to the next one represents a 90° segment of the ecliptic. If the Sun moves uniformly along the ecliptic then we would expect the time to cover each 90° segment to be the same. A close look, however, shows that this is not the case.

The time from the vernal equinox to the summer solstice, which we will refer to as spring, is currently 92 and 3/4 days. Summer, from summer solstice to autumnal equinox, is 93 and 2/3 days. Fall, from autumnal equinox to winter solstice, is 89 and 5/6 days. Finally, the winter season from winter solstice to vernal equinox is 89 days. The Sun does not travel uniformly along the ecliptic. It must travel faster along the ecliptic (relative to the stars) during winter and fall so that it covers these two 90° segments in less time. Likewise, the Sun must travel slower along the ecliptic during spring and summer to explain why these seasons are longer.

We mentioned earlier that the length of the solar day varies a little bit throughout the year. One reason for this variation is the changing speed of the Sun along the ecliptic. In winter, when the Sun is moving faster along the ecliptic toward the east, the Sun travels farther along the celestial sphere in the course of a single day. As a result, the difference between the sidereal day and the solar day will be greater. Since the celestial sphere rotates uniformly, the nonuniform motion of the Sun will tend to make the solar day slightly longer in winter. Likewise, the solar day will be slightly shorter in summer.

However, there is another effect that causes changes in the length of the solar day: the obliquity of the ecliptic. Refer to Figure 2.7. Near the summer and winter solstices the Sun is far away from the celestial equator. At these larger declinations, the hour lines of right ascension are closer together than they are near the equator. As the Sun moves along the ecliptic, it will move through the hour lines more rapidly around the solstices than it does around the equinoxes. The difference between the sidereal and solar day is determined by how much the Sun's RA changes over the course of one day, so from this analysis we see that the obliquity of the ecliptic causes the solar day to be longer near the solstices and shorter near the equinoxes. These two effects (nonuniform motion and obliquity) combine to produce the observed variations in the length of the solar day throughout the year.[14]

2.2.4   *Constellations of the zodiac*

We have already mentioned a few of the constellations used by astronomers in the Western tradition. Now it is time to take a closer look at the constellations that lie along the ecliptic. These are the constellations of the **zodiac**. As the Sun slides along the ecliptic it moves through the following constellations, in order: Aries, Taurus, Gemini, Cancer, Leo, Virgo, Libra, Scorpius, (Ophiuchus), Sagittarius, Capricornus, Aquarius, and Pisces. The pattern then repeats. Ophiuchus is listed in parentheses because it is not an official constellation of the zodiac. According to tradition there are only twelve constellations in the zodiac.

The issue of Ophiuchus highlights the fact that the zodiac is not really so much a list of constellations as it is a set of landmarks for tracking the Sun's motion along the ecliptic. For this purpose, ancient astronomers defined twelve **signs** of the zodiac.[15] Each sign is a $30°$ segment of the ecliptic, so together they mark out the full $360°$ path of the ecliptic. When the system was originally devised, each sign was associated with the constellation of the same name (with Scorpius laying claim to the portion of the ecliptic in Ophiuchus). We will see later that the correspondence between signs and constellations became disrupted. For now, what is most important is that the equinox and solstice points have fixed locations within the signs. For example, the vernal equinox is at the beginning of the sign of Aries. In fact, the symbol that we use for the vernal equinox (Υ) is also the symbol for the sign of Aries. Likewise, the summer solstice lies at the beginning of the sign of Cancer ($\mathfrak{S}$), the autumnal equinox at the beginning of Libra ($\libra$), and the winter solstice at the beginning of Capricornus ($\capricorn$).

In a well-regulated solar calendar, the Sun always passes through a given sign on the same dates. For example, the Sun is in the sign of Sagittarius ($\sagittarius$) from about November 22 until the winter solstice. It is in the sign of Gemini ($\gemini$) from about May 21 to the summer solstice. At this point you may notice the connection to astrological horoscopes: your horoscope sign is set by the sign of the zodiac occupied by the Sun at the moment of your birth. We will discuss astrology, and its historical connection to astronomy, later in our story. For now, what matters is that we can use the signs of the zodiac to track the Sun's progress around the ecliptic.

To help us track the Sun's motion through the signs we can introduce a third coordinate system. Recall that our local alt-az coordinate system was fixed on our sky, and the coordinates were defined with reference to our horizon circle. Our celestial RA-Dec coordinate system was fixed to the rotating celestial sphere, with the coordinates defined relative to the celestial equator. Now we can define

ecliptic coordinates that are fixed to the celestial sphere but defined with respect to the ecliptic.[16]

Ecliptic latitude just measures the angle between a point on the celestial sphere and the closest point on the ecliptic (i.e. measured perpendicular to the ecliptic). Points north of the ecliptic have positive latitudes, those south of the ecliptic have negative latitudes. The Sun always lies directly on the ecliptic, so its ecliptic latitude is $0°$.

Ecliptic longitude indicates the location of a point measured along the ecliptic. To specify the ecliptic longitude of a point we make use of the signs of the zodiac. The ecliptic longitude of a point gives the sign that the point is in, as well as the number of degrees between the point and the western edge of the sign. For example, on the autumnal equinox the Sun lies at the beginning (western edge) of the sign of Libra, so its ecliptic longitude is ♎ $0°$. Ten days later the Sun will have moved about $10°$ east along the ecliptic, so its ecliptic longitude will be ♎ $10°$. After an additional 24 days the Sun will have moved into the sign of Scorpius (♏) and its ecliptic longitude will be ♏ $4°$.

Figure 2.8 illustrates this ecliptic coordinate system. As we have seen, only the ecliptic longitude is really useful for the Sun. However, both longitude and latitude will be helpful when we come to study the Moon and planets. For now, though, we have exhausted what we can accomplish by watching the skies from a single location. It's time to move around.

2.3   Around the Earth

From our assumed location in the Mediterranean region we have observed many patterns in the motions of the stars and the Sun, and we have constructed a theory to account for those motions. But what happens if we move to a new location on Earth? Will the patterns exhibited by the stars and Sun change, or will they be the same? If they do change, how can we modify our theory of the celestial sphere (with the ecliptic) to account for these changes? We will begin our investigation of these questions by observing what happens if we move to the north or south of our original location.

2.3.1   North and south

As we move northward from our starting point we notice that the pattern of stellar motions does change. In particular, the NCP (and nearby Polaris) gets higher in the sky. If we journey all the way to the north pole of Earth (latitude $90°$ N) we would find the NCP at our zenith and the stars circling counterclockwise around our horizon. Similarly, if we were to move southward from our starting point we would see the NCP and Polaris dip down toward the

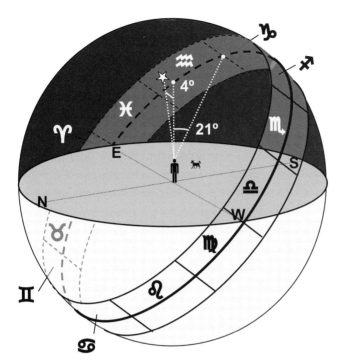

Figure 2.8 The zodiac and the ecliptic coordinate system. Each sign of the zodiac represents a 30° segment along the ecliptic. The ecliptic longitude of a point gives the sign occupied by the point as well as the angle between the point and that sign's western edge. The ecliptic latitude of a point measures the angle from the point to the ecliptic along an arc perpendicular to the ecliptic. The star shown has ecliptic coordinates $(\lambda, \beta) = (\approx 21°, 4°)$. The vernal equinox point ♈ lies on the eastern horizon at the time shown.

horizon. New stars would become visible on our southern horizon. If we traveled far enough to the south we would be able to see the south celestial pole, with stars circling clockwise around it.

The changes we observe as we move around on Earth illustrate something important about the two coordinate systems we defined earlier. The alt-az system is *local*. The altitude and azimuth of a particular star at a particular time depend on the location of the observer. For example, the altitude of Polaris seems to remain constant for any observer in the northern hemisphere of Earth, but the altitude will be different for observers at different latitudes. On the other hand, the celestial coordinates are *not* local. Polaris lies close to the NCP (Dec = 90°) for *all* observers, no matter their location on Earth.

The stars aren't the only things that change as we move around. The pattern of the Sun's motion also changes as we move north or south. Moving northward from our starting point we notice that the local noon altitude of the Sun decreases. When we reached a latitude of 66°.5 N (known as the **Arctic Circle**) we would find that the local noon altitude on the winter solstice has dropped so low that the Sun never rises above the horizon on that day. At that same latitude we would find that on the summer solstice the Sun never quite sets. Moving farther north we would find periods of darkness for several days around the winter solstice and periods of daylight for several days near the summer solstice. At the Earth's north pole this pattern goes to its extreme: 6 months of continual daylight followed by 6 months of night.

If, instead, we were to head southward from our starting point we would find that the local noon altitude of the Sun increases as we travel. At latitude 23°.5 N we would find that the local noon altitude of the Sun on the summer solstice reaches 90°: the Sun passes through our zenith at local noon. This latitude is known as the **Tropic of Cancer** because the Sun passes directly overhead on the summer solstice, when the Sun is at the beginning of the sign of Cancer (♋). Going farther south we would reach the equator of Earth, where the Sun would pass overhead on the equinoxes. At latitude 23°.5 S the Sun passes overhead on the (northern) winter solstice and this latitude is known as the **Tropic of Capricorn**, since the Sun is at the beginning of the sign of Capricorn (♑) on that date. Between the two tropics we would find that the local noon Sun is sometimes to the south (in winter) and sometimes to the north (in summer).

Traveling southward from the Tropic of Capricorn we would find our seasonal patterns of sunlight reversed. In June, on the northern summer solstice, the Sun will be up for less than 12 hours and it won't reach very high in the sky. That date would be the winter solstice for the southern hemisphere. In December, near the northern winter solstice, the Sun stays up for more than 12 hours and it gets high in the sky. That date is the southern summer solstice. Note that in this book we will generally refer to the solstices and seasons according to the convention for the northern hemisphere, so the summer solstice refers to the June solstice and the winter solstice refers to the December solstice (unless otherwise noted).[i]

At latitude 66°.5 S (the **Antarctic Circle**) we would find that the Sun never rises on the southern winter solstice (northern summer solstice) and it never sets on the southern summer solstice (northern winter solstice). At lower latitudes the periods of darkness around the southern winter solstice and daylight around the

---

[i] We apologize to our southern hemisphere readers, but astronomy has a historical bias in favor of the northern hemisphere and it would be excessively cumbersome to mention the conventions for both hemispheres every time we mention a solstice or season.

southern summer solstice become longer, until we reach the south pole of Earth (latitude 90° S) where we have 6 months of darkness during the southern winter followed by 6 months of daylight during the southern summer.

2.3.2 The Earth's curve

How can we explain the changes in the motions of the stars and Sun as we move north or south on the Earth? All of the changes can be accounted for if we simply tilt the celestial sphere by an angle that corresponds to our latitude. Our original location was at latitude 35° N, and we saw that the NCP was 35° above our horizon. As we move to new latitudes on the Earth, the entire celestial sphere seems to tilt such that the NCP moves along our meridian line. If we move to 55° N latitude we will find the NCP at an altitude of 55°. If we move to the Earth's equator (latitude 0°) the NCP will lie on our horizon to the north (and the SCP will lie on our horizon to the south). As we move into the southern hemisphere the SCP becomes visible due south, always at an altitude equal to our southern latitude angle. This tilting of the celestial sphere according to our latitude on Earth explains all the changes in the motion of the stars and the Sun as we move north or south.

But why should the celestial sphere tilt in this way? The ancient Greeks figured out that the tilting of the celestial sphere is really a consequence of the shape of Earth.[17] If the Earth is curved in the north–south direction like the outer surface of a circle, then as we move around on Earth our horizon plane changes its orientation with respect to the celestial sphere. Figure 2.9 illustrates this effect by showing the horizon plane for an observer at two different latitudes on Earth. The horizon plane is a flat plane that just grazes the spot on Earth where the observer is standing. Except for that one point, the entire Earth is below the horizon plane while the visible sky is above it. As the observer moves from latitude 35° N to the north pole, along the Earth's curved surface, the observer's orientation changes. Up, for an observer at the north pole, points in a different direction than up for the observer at 35° N. The changing orientation of the observer results in a tilting of the horizon plane.

As the horizon plane tilts, different portions of the celestial sphere will be visible to the observer. For an observer at 35° N the NCP will be visible 35° above the horizon, but for an observer at 35° S the NCP will be 35° *below* the horizon, and thus invisible. Since the Sun is always on the ecliptic, and the ecliptic is fixed to the celestial sphere, motion along the Earth's curved surface will also affect the apparent path of the Sun. For example, an observer at 35° N will see part of the Dec $= -23°.5$ circle (which traces the Sun's path on the winter solstice) above the horizon, but for an observer at the north pole that entire circle is below the horizon and the Sun will not be visible at any time on the

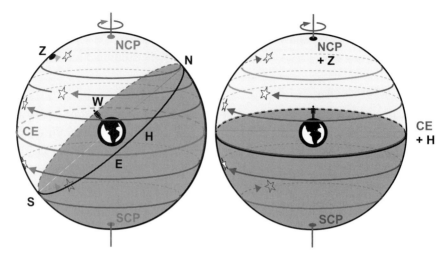

Figure 2.9 Our horizon plane changes as we change latitude. The figure on the left shows an observer at 35° N latitude. The horizon circle is indicated by H and the portion of the celestial sphere below the horizon is darkened. The figure on the right shows the same thing for an observer at 90° N (the north pole of Earth). Each observer sees a different portion of the celestial sphere. Note that the scale is not correct: the Earth should be a point in comparison to the celestial sphere.

winter solstice. All of the observations we have discussed are consistent with this model.

The apparent tilting of the celestial sphere as we move north and south shows that the Earth is curved along our meridian. But is the surface of Earth also curved east to west? In fact there is evidence that the surface of Earth is curved in every direction. Ships sailing away on the sea will gradually fall below the horizon, with the hull (the bottom) of the ship disappearing first and the sail last, no matter what direction they are sailing. This observation makes sense if the Earth is curved and convex (like the outer surface of a circle). If Earth was flat we would just see the ships get smaller and smaller as they got farther away, but they would not drop below the horizon. If the Earth was curved but concave (like the inner surface of a circle) then we would see the Earth's surface rise up as we looked out to greater distances.

It seems, then, that the Earth is curved and convex in all directions. The simplest object that is curved and convex along all directions is a sphere. This simple geometry appealed to the ancient Greeks and likely led them to the idea that the Earth was a sphere. There was, though, more direct evidence of the Earth's spherical shape. In a lunar eclipse, the shadow of the Earth falls on the Moon. The shape of that shadow can tell us something about the shape of the Earth. The ancient Greeks knew that the shadow cast on the Moon by the Earth during a lunar eclipse always had a circular shape. The only object that always casts a circular shadow, no matter what direction the light is coming from, is a sphere. So there were good reasons for believing the Earth was spherical. In the 4th century BC Aristotle provided clear evidence for a spherical Earth in his treatise *On the Heavens*.[18] From that point on the spherical shape of the Earth was widely accepted by the ancient Greeks and their intellectual descendants.

In fact, the ancient Greeks were not able to observe the stars and the Sun from above the Arctic Circle, nor from the southern hemisphere, but their limited observations led them to believe that the Earth was spherical. Armed with the idea of a spherical Earth sitting at the center of the celestial sphere, they were able to *predict* all of the phenomena that we have described. Ancient geographers even used these predictions about the Sun's behavior at different latitudes to divide the Earth into five (or more) "climate zones," such as: the Arctic (above 66°.5 N), the Northern Temperate (23°.5–66°.5 N), the Tropics (23°.5 S to 23°.5 N), the Southern Temperate (23°.5–66°.5 S), and the Antarctic (below 66°.5 S). Note how all of these climate zones are related to the obliquity of the ecliptic (23°.5) and its complement (90°–23°.5 = 66°.5). [19]

Ancient geographers didn't stop at mapping out climate zones. They also devised methods for determining the size of the spherical Earth. In *On the Heavens* Aristotle said that mathematicians had estimated the circumference of the Earth to be 400,000 stades (a stade is a unit of distance), but he didn't explain the methods used.[20] In the 3rd century BC, Eratosthenes of Cyrene devised a way of using the Sun's local noon altitude at two different latitudes to determine the size of Earth.[21] Eratosthenes lived in Alexandria, Egypt, where, at some point, he served as the chief librarian of the Great Library. He believed that the city of Syene (now Aswan), south of Alexandria along the Nile River, was located on the Tropic of Cancer. Therefore, on the summer solstice the Sun would pass through the zenith of an observer in Syene. He also knew that the local noon altitude of the Sun on the summer solstice in Alexandria was less than 90° by about 1/50 of a circle (7°.2). Eratosthenes recognized that this must mean that Alexandria lay at a latitude 7°.2 north of Syene.

He believed that Alexandria was due north of Syene, which meant the two cities lay along the same longitude line on Earth. Eratosthenes also knew that the distance from Syene to Alexandria was estimated to be 5000 stades. His shadow measurements indicated that a line running due north from Syene to Alexandria would constitute 1/50 of a great circle, because longitude lines are all great circles. Therefore the circumference of this great circle was $50 \times 5000 = 250,000$ stades. Unfortunately, we don't know exactly how long a stade was in modern units, but it may have been about 1/10 of a mile. In that case, Eratosthenes found the circumference of Earth to be 25,000 miles. Since the circumference of a circle is π times its diameter, that means the Earth radius must be $25,000/\pi \approx$ 8000 miles. That result is remarkably close to the accepted modern value of about 7,900 miles. In fact, there are some errors in Eratosthenes' data (Syene and Alexandria are not on the same longitude line – Syene is farther east – and Syene was not truly on the Tropic of Cancer) but they tended to cancel each other and in any case his general method was sound.

Later geographers gave different estimates for Earth's diameter. Claudius Ptolemy, whose astronomical work we will discuss in detail later, estimated the Earth's diameter to be about 57,000 stades (or 5700 miles if our conversion factor is correct). This low result, along with an overestimate of the combined width of Europe and Asia, would later be used by Christopher Columbus to justify his proposal to sail from the western coast of Europe to the eastern coast of Asia.[22] The inaccurate numbers led him to greatly underestimate the westward distance from Europe to Asia and he was quite fortunate that there was another continent in the way!

2.3.3   A matter of time

So far we have focused mostly on the effects of Earth's curvature along a north–south meridian line. But does the Earth's east–west curvature have astronomical effects? Indeed it does, and with modern clocks we can easily show that the Earth's east–west curvature affects the timing of local astronomical events. For example, if we measure the time of local noon at two different longitudes using synchronized clocks (which read the same time regardless of where we are on Earth) we will find that local noon occurs later for the location that is farther west. Likewise, star transits occur later the farther west we go. Careful measurements would show that these local astronomical events happen one hour later for every $15°$ of westward longitude.

The ancient Greeks, however, had no way to compare times with synchronized clocks. Sundials could only tell local time: a sundial will always read noon at your local noon, no matter where you are on Earth. There was a way, however, to compare the timing of events at different longitudes. Instead of using global clocks to measure local astronomical events, they used local clocks to measure a global astronomical event. Lunar eclipses, when they are visible, are visible to all observers on Earth simultaneously. So observers can determine the time of a lunar eclipse relative to their own local noon and compare with the same measurement made by observers in other locations.

The ancient Greeks found that an eclipse that began shortly after sunset as seen in Sicily would have started more than an hour after sunset as seen from a more easterly location like Asia Minor.[23] As measured by a global clock the eclipse began at the same time for all observers, but according to that same global clock sunrise comes later for the observer in the west and thus the eclipse will happen earlier in that observer's night. This is exactly what one would expect for a spherical Earth.

This discussion of local versus global time illustrates a conundrum that we face in defining time on Earth. The ancients told time using instruments such as sundials that gave only local time. As we have seen, they measured time in seasonal hours whose length depended on the time of year and the location on Earth. They were aware that the local times of two different locations would be out of sync, but the difference in local times was not really noticeable over the distances that they could travel in a single day. In modern terms, their "jet lag" was negligible.[24]

The advent of long-distance travel (via railroad) and communication (via telegraph) in the 19th century created a need for time to be standardized from place to place and throughout the year. Part of the solution was to move from seasonal hours to equinoctial hours so that one hour would amount to the same amount of time regardless of location or date, but there was still the problem of using local time, which depends on the longitude of the observer. If each city kept its own local time the train timetables would be very confusing! However, defining a global time for the entire Earth wouldn't make much sense either: for some people noon on the global clock would occur in the middle of the night![i]

The solution we have devised is the use of **time zones**. There are 24 time zones on Earth and in principle each time zone is a span of 15° in longitude. Within a zone all clocks are synchronized to the same time, but as we move westward from one time zone to the next the clock time decreases by one hour. That way local noon will occur close to noon on the clock for observers in all time zones. It is true that observers near the western edge of the time zone will see their local noon occur later on the clock than observers near the eastern edge, but that difference should not exceed one hour.

Of course, the pattern of adding an hour to the clock every time we move one zone to the west can't continue indefinitely. If it did, then in going all the way around the Earth and back to our starting point we would have to set our clocks back by 24 hours, but that makes no sense. At some point we have to disrupt the pattern. We do this by establishing an International Date Line (IDL). The simplest version of this line would run down the middle of a time zone. As we move westward across this line we subtract 24 hours (one day) from our

---

[i] Astronomers, however, do use a global clock so that they can easily compare the times of observations made by different observers. The global time they use is known as Coordinated Universal Time (UTC), which is closely related to Greenwich Mean Time (GMT), the local (mean) solar time as measured at the Royal Observatory in Greenwich, UK.

clocks. That way if we go all the way around the globe our clocks will get back in sync with the clocks that stayed at our starting point.

For political and geographical (but not astronomical) reasons the actual time zones and IDL that we use are much more complicated than the simple versions described above. The IDL and the boundaries of the time zones are irregular: they don't lie along longitude lines. Some times zones have appendages that extend completely across adjacent time zones. For example, if you travel from Russia to the USA across the Bering Strait you will cross the IDL but your clock time will change by only 21 hours rather than 24. Although the details of our system for telling time are dictated by modern politics, the basic principles of that system are based on things that were known to the ancient Greeks: the apparent motions of the Sun and the spherical shape of Earth.[i]

2.4   Precession: a third sphere

In the 4th century BC Aristotle laid out the general structure of the cosmos in his *On the Heavens*. We will explore Aristotle's cosmos in greater detail in Chapter 4, but in general outline it consisted of two spheres: a celestial sphere at the outer limit, and a much smaller spherical Earth at the center. The Earth was completely stationary, while the celestial sphere rotated once every 23 hours, 56 minutes about an axis that passed through the celestial poles. The stars were fixed to the celestial sphere but the Sun moved around on that sphere, following its ecliptic path that was tilted 23°.5 relative to the celestial equator. The Sun completed its annual motion around the ecliptic in about 365.25 days, all the while participating in the daily rotation of the celestial sphere. This simple model was incredibly powerful: it could explain all of the observations we have discussed so far.

A new discovery forced astronomers to add another sphere to this cosmos. In the 2nd century BC the astronomer Hipparchus investigated the possibility that stars were not completely fixed to the celestial sphere. He may have been motivated by the appearance of a new star in the heavens, but whatever his reasons Hipparchus carefully measured the positions of many stars on the celestial sphere and compared his measurements to those of his predecessors Timocharis and Aristyllos in the 3rd century BC. He found that stars near the ecliptic, such as Spica in the constellation Libra, had shifted about 2° eastward relative to the

[i] One other modification that people in some areas make to their clocks is a one hour adjustment to "daylight saving time" during part of the year. Daylight saving time serves to increase the hours of daylight in the evening during summer months, but that is simply a convenience and not astronomically necessary.

equinox points. Since the signs of the zodiac are defined relative to the equinox and solstice points, that meant that the stars had moved a little bit eastward *through the signs*. The motion of the stars relative to the equinoxes/signs is now known as the *precession of the equinoxes*, or just precession. Hipparchus estimated that the rate of precession was about 1° per century.[25]

Hipparchus found evidence that this very gradual motion of the stars with respect to the signs was shared by all stars collectively, but if precession was really a slow rotation of the entire sphere of stars, about what poles did it turn? There were two obvious choices: the celestial poles or the ecliptic poles. Hipparchus suspected that precession was a very slow rotation of the sphere of stars about the ecliptic poles. This hypothesis was verified about 265 years later by Ptolemy, who compared his own measurements with the catalog of star positions made by Hipparchus. Ptolemy found that during that 265-year period the ecliptic latitude of Spica had not changed, but that Spica's declination had changed. This result showed that precession could be viewed as a slow rotation about the ecliptic poles. Furthermore, Ptolemy confirmed Hipparchus' estimate: the ecliptic longitudes of all stars seemed to increase by about 1° per century.[26]

We now know that the Hipparchus/Ptolemy precession rate is too low. The real rate of precession is about 1°.4 per century.[i] The fixed stars seem to rotate about the ecliptic poles once every 26,000 years. However, the daily rotation of the entire celestial sphere, including the Sun and the stars, still takes place around the celestial poles. Ancient astronomers still believed that the stars were fixed onto a sphere, but after the discovery of precession that "sphere of fixed stars" had to be separated from the celestial sphere with its daily rotation.

One consequence of the separation of the sphere of fixed stars from the sphere of daily motion is that the celestial poles move around among the stars. We know that the angle between the north celestial pole and the north ecliptic pole (NEP) must be equal to the obliquity of the ecliptic, which we have found to be 23°.5. Hipparchus and Ptolemy believed that the obliquity was constant (though more on that later). So the NCP will always be 23°.5 away from the NEP, but as the sphere of fixed stars spins around, eventually *any* point that is 23°.5 from the NEP will be lined up with the NCP at some time. Currently the NCP sits very close to the star Polaris, but over long periods of time the

---

[i] Some recent authors have claimed that Ptolemy fabricated the measurements he used to derive his precession rate by simply copying Hipparchus' positions and adjusting them for precession at a rate of 1° per century. Others have pointed out that Ptolemy's erroneous precession rate can be explained as a result of honest errors (for example, errors in his determination of the equinoxes). We will not delve into this controversy here.[27]

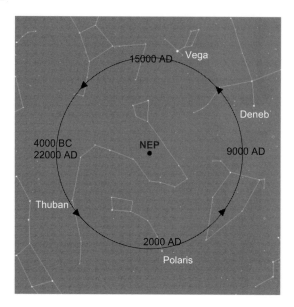

Figure 2.10 Motion of the NCP due to precession. Over a period of 26,000 years the NCP traces out a counterclockwise circle of points located 23°.5 from the NEP. In different eras, different stars may serve as "pole stars."

NCP will move counterclockwise around the NEP as shown in Figure 2.10. In about 12,500 years the NCP will be far from Polaris but close to the very bright star Vega.

Another consequence of precession is that the constellations move relative to the equinoxes and the signs of the zodiac. The vernal equinox point (♈) lies, by definition, at the beginning of the sign of Aries. In Hipparchus' day that point was also at the western edge of the constellation of Aries, but that is no longer the case. Now the vernal equinox point (♈) and most of the sign of Aries lies within the constellation of Pisces because, since the time of Hipparchus, the stars in Pisces have moved about 30° eastward, passing the vernal equinox point in the process. Likewise, the stars of the other zodiac constellations have shifted by about 30°, or one full sign. When you were born, the Sun was probably *not* in the constellation that corresponds to your astrological sign.

Precession has one other consequence that we should mention. Recall that the sidereal year is the time for the Sun to complete one circuit around the sphere of the fixed stars, while the tropical year is the time for the Sun to complete its motion from the vernal equinox point (♈) back to ♈. Now we know that the stars move relative to ♈. As a result, the sidereal year and the tropical year are not the same. Since the Sun moves eastward along the ecliptic, and the stars move eastward relative to the equinoxes, it takes a little bit longer for the Sun to lap the stars than it does for the Sun to lap the equinoxes. Think of the stars, Sun, and the vernal equinox as people on a running track. The equinox is just sitting in one place on the track. The stars are walking VERY slowly around the track, while the Sun is running around the track in the same direction. If all three start at the same spot, then the Sun will get back to the vernal equinox a little bit before it laps the stars. The tropical year is, therefore, a bit shorter than the sidereal year. Modern measurements indicate that the tropical year (by which we set our calendars) is 365.2422 mean solar days, while the sidereal year is 365.2564 days, for a difference of about 20 minutes.

Not many ancient Greek astronomers seem to have paid much attention to Hipparchus' discovery of precession. Ptolemy, however, not only acknowledged Hipparchus' discovery but helped to confirm it and clarify it.[28] As we will see, Ptolemy would become the primary source of astronomical information for astronomers in the Middle Ages. As a result, precession would play an important role in medieval astronomy.

2.5   Reflections on science

In this chapter we have encountered our first scientific model, the celestial sphere. We have also seen how additions and modification to that model, such as the ecliptic, the spherical Earth, and a sphere to account for precession, can allow that model to explain a greater variety of phenomena. This is a good time to look back at what we expect from a scientific theory or model and see how our celestial sphere model measures up.

What do we want a scientific model or theory to do? Of course we want our theory to fit with our observations. In astronomy this is sometimes called "saving the appearances." We have seen that our two-sphere model with ecliptic accounts for a tremendous variety of phenomena: the motions of the stars and Sun as seen from any location on Earth over short time periods. With the addition of a sphere for precession the model also matches long-term observations. That's pretty impressive. We would also like for our model to *explain* the natural phenomena that we observe. If we believe that these heavenly spheres are really up there, spinning around, then our model does provide an explanation for the motions of the stars and Sun. If those spheres are really up there doing their thing, then we should see what we do, in fact, see.

Our celestial sphere model is very impressive, but no model can explain everything. Our celestial sphere model doesn't tell us anything about why the clouds move across our sky, or why it rained yesterday. Every scientific theory has limited *scope*: it only explains a limited range of natural phenomena. In fact, our scientific models are usually built to explain a specific set of observations or measurements. In building a scientific model we have to decide what things to pay attention to and what things to ignore. Our celestial sphere model does a great job of explaining the positions of the stars and Sun on our sky, and how those positions change, but we have completely ignored the color and brightness of the stars, or why they have the particular arrangement we see. Our theory is silent on those issues.

It is inevitable that we ignore some things in building our scientific theories: it is impossible to try to account for *everything* right from the start. Often, though, our selection of what to pay attention to and what to ignore is closely connected to the physical and intellectual tools we have available. For example, the ancient Greeks developed a strong interest and expertise in geometry, which led them to develop geometrical models such as the celestial sphere. This preference for geometry also led them to devise impressive geometrical instruments such as the armillary sphere, which is essentially a physical realization of Figure 2.7 with circles representing the celestial equator, the tropics, the Arctic and Antarctic circles, and the ecliptic. This device could be used as a physical model to aid in understanding the celestial sphere theory, but it could also be used as a measurement instrument.[29] In contrast, the Babylonians made extensive observations of the stars and planets and developed sophisticated arithmetical theories to predict their motions, but they did not create geometrical models.[30]

In addition to explaining observed phenomena, we would like our scientific theories to predict things that we have not yet observed. We have seen that the celestial sphere model was used to make such predictions. The ancient Greeks never traveled south of the equator, but their theory told them that if they could they would see stars circling clockwise around the south celestial pole. They could not test that prediction, but the celestial sphere model was so successful in explaining the things that they could see that ancient Greek astronomers were confident that its untested predictions were accurate. We now know that their confidence was justified.

It is important to distinguish two different aspects of a scientific theory like the celestial sphere. First, there is the basic structure of the theory. This basic structure will determine the general *qualitative* behavior of the theory. For example, the idea of a rotating celestial sphere dictates that points on the sphere will move in circular paths with the exception of two special "poles" that won't move at all. These qualitative behaviors are direct consequences of the model's overall structure.

On the other hand, there are the specific numerical values that we put into the theory so that it can be used to make *quantitative* predictions. These values are known as the **parameters** of the theory. For example, the period of the celestial sphere's rotation and the obliquity of the ecliptic are numerical parameters that are important for defining our celestial sphere model. Even the locations of the celestial poles and the vernal equinox point (♈) among the stars can be considered numerical parameters, since we will need to use angle measurements to precisely indicate where these points lie. If we want our theory to match our observations it is important that we use the correct values for the parameters. A theory with the right basic structure but the wrong parameter values may be able to reproduce some qualitative features of our observations, but it won't be able to reproduce precise, quantitative observations.

In order to accurately determine the parameters' values we will need to make measurements. We usually try to define our parameters in terms of the procedures used to measure them. For example, the sidereal day (which is also the rotational period of the celestial sphere) is defined as the time between consecutive transits of a given star. The definition tells us how to measure the quantity. In any case, it is necessary to clearly establish a procedure for measuring our parameters or calculating them from other known quantities. For example, we define the obliquity as the angle between the ecliptic and the celestial equator. We can't measure that angle directly, but we have seen that measurements of local noon shadows throughout the year can allow us to calculate the obliquity.

Once we have determined the parameter values for our theory, we can use it to make quantitative predictions. For example, I could use the celestial sphere model to predict the number of hours of daylight on New Year's Day at latitude 50° S on Earth. I have never been to latitude 50° S, and I may never go to that latitude. It is very likely that I will never be there on New Year's Day. Even so, I can be confident that my prediction for the number of hours of daylight will be correct. The ability to confidently predict things that you have not observed, and perhaps cannot observe, can have practical benefits, but even if the predictions are not practically important, they provide a sense of power and control over Nature. Nature's secrets are no longer hidden. The joy of uncovering Nature's secrets motivates scientists to continue with their difficult, and sometimes tedious, work.

A good theory explains a lot, but also raises questions for further study. Our celestial sphere model raises some interesting questions. How *big* is the celestial sphere? Why is the ecliptic tilted relative to the celestial equator? Is the obliquity of the ecliptic really constant? Are the stars truly fixed on the celestial sphere? Why doesn't the Sun move along the ecliptic at a uniform rate? Is there anything *outside* of the celestial sphere? What about the Moon and the planets? How do they fit into our celestial sphere model? A good scientific model seems to reveal the inner workings of Nature, but it also points us toward further mysteries to be explored. It is to some of these mysteries that we now turn.

3

Wanderers: the Moon and the planets

Our word **planet** is derived from a Greek word meaning "wanderer." In the context of astronomy a wanderer is any heavenly body that moves across the celestial sphere, in constrast to the fixed stars that remain stationary on that sphere. We have already seen that the Sun moves along its ecliptic path, so the Sun is a wanderer. There are six other lights in the sky, visible to the naked eye, that wander amongst the fixed stars. The most obvious of these is the Moon. In addition, there are five objects that look very much like stars but that move around on the celestial sphere. These are what we would now call the five naked-eye planets: Mercury, Venus, Mars, Jupiter, and Saturn. We can refer to these five objects as "wandering stars." In this chapter we will examine the behaviors of the Moon and the wandering stars.

Note that our modern use of the word "planet" is different from that of ancient Greek astronomers. To the ancient Greeks, there were seven planets: the Sun, the Moon, and the five wandering stars. The Earth was not considered a planet but, as we will see, was thought to be something fundamentally different from the heavenly bodies. It is important to remember the difference between the ancient and modern use of the term "planet" because, in a sense, this entire book is about how the change in that term's definition came about.

3.1 The ever-changing Moon

3.1.1 *The Moon's motion against the stars*

The Moon is the second brightest object in our sky after the Sun. It can be seen during the day, but dominates the night sky when it is above the horizon. The Moon's light, though, is not bright enough to blot out the stars so it is easy to track the Moon's motion across the celestial sphere. Observations show that the Moon moves roughly eastward along the ecliptic, like the Sun, so it is useful to track its motion with ecliptic coordinates. However, the Moon does not exactly follow the ecliptic path and it does not repeat the exact same path each time around the celestial sphere. The Moon can be found on either side of the ecliptic, but its ecliptic latitude is always between about $-5°$ and $+5°$.

On average, it takes the Moon 27.32 days to go all the way around the ecliptic and return to its starting point among the fixed stars. This time is called a **sidereal month**.[i] The word "month" derives from the word "Moon." In a single day the Moon moves about 13° relative to the stars, but like the Sun it does not move across the celestial sphere at a uniform rate but instead exhibits small variations in speed. Keep in mind that, as the Moon moves eastward relative to the stars, the entire celestial sphere continues in its daily westward rotation. Therefore, the Moon lags behind the stars in its motion across our sky.

We will discuss the appearance of the Moon in greater detail soon, but for now we note that the Moon has an apparent diameter (the angle between a point on the edge of the Moon and the opposite point, see Appendix A.2) of about $0°.5$. The Moon's apparent diameter varies slightly, from about $29'.3$ to $34'.1$ (recall that $1° = 60'$). These values are strikingly similar to the apparent diameter of the Sun, although the Sun's apparent diameter doesn't vary much (from $31'.5$ to $32'.6$). Ancient astronomers were aware of variations in the apparent size of the Moon, but not the Sun.[1] These changes in apparent size are correlated with the Moon's changing speed relative to the stars: the Moon moves faster relative to the stars when it appears larger. The variations in size and speed form a repeating pattern with a period of 27.55 days, or one **anomalistic month**.

3.1.2 Lunar phases

The Moon moves about 13° per day along the ecliptic, while the Sun moves only about 1° per day. Therefore, the Moon will periodically lap the Sun in their race around the celestial sphere. If we measure the time from when the Moon passes the Sun to the next time it passes the Sun, a period known as a **synodic month**, we find that it is 29.53 days, on average. Relative to the Sun, the Moon moves about 12° eastward on the celestial sphere during one day. As a result, the Moon rises about 50 minutes later each day.

As the Moon moves around relative to the Sun its appearance changes. Sometimes the entire face of the Moon is brightly lit, but at other times only a portion

---

[i] Because of precession the sidereal month differs slightly from the **tropical month**, which is the time for the Moon to go from ♈ back to ♈. However, this difference is negligible for our purposes.

of that face is bright while the remainder is much darker (but still dimly visible). In the 5th century BC the Greek philosopher Anaxagoras explained that the Moon shines by reflecting sunlight, and it is the geometry of the Earth, Moon, and Sun that determines the Moon's appearance as seen from Earth.[2] The top of Figure 3.1 shows a simplified version of this geometry, as seen from above the Earth's north pole. The Sun is assumed to remain stationary off to the right of the diagram, while the Moon moves around the Earth. The side of the Moon facing toward the Sun is always brightly lit while the opposite side is dark. However, the side of the Moon that is facing toward Earth may include the brightly lit half, the dark half, or some portion of both. The different appearances of the Moon are known as its **phases**. The Moon phases are illustrated in the bottom of Figure 3.1.

When the Moon is at position 1 in Figure 3.1 the dark half of the Moon is facing toward Earth. In this case the Moon will be in the same part of the sky as the Sun. It will rise and set with the Sun, but it won't be visible. This phase of the Moon is known as a **new Moon**. When the Moon moves to position 2 an observer on Earth will see the Moon about 45° to the east of the Sun on the sky. A portion of the Moon's western edge will be lit while the rest is dim. When it is in this **waxing crescent** phase, the Moon will rise around 9 AM and set around 9 PM, although it will be easiest to see after sunset when the skies are dark. At position 3 the Moon will appear 90° east of the Sun. The Moon's western half will be lit while its eastern half is dim. This **first quarter** Moon will rise around midnight and set around noon. This general pattern continues as the Moon moves to positions 4 (**waxing gibbous**, 3 PM–3 AM), 5 (**full**, 6 PM–6 AM), 6 (**waning gibbous**, 9 PM–9 AM), 7 (**third quarter**, midnight–noon), and 8 (**waning crescent**, 3 AM–3 PM). After a full synodic month the Moon returns to position 1 and the pattern repeats. (Note that the rising and setting times given here are only very rough estimates.)

Although the brightness of the Moon's face changes with the phases, it is interesting to note that the features visible on the Moon's face do not change. When the surface of the Moon is lit by sunlight we can see that some portions of that surface are darker while others are brighter, but the pattern of these dark and light patches is always the same. This observation suggests that the same portion of the Moon's surface always faces toward Earth. The Moon's unchanging features probably helped inspire the idea that celestial bodies are fixed to giant transparent spheres that spin around the Earth. Imagine picking up this book with the front cover facing you. If you spin around the book will move in a circle around you, but the front cover will always be facing you. In much the same way, the Moon seems to move around Earth while always showing us the same face, just as though it were in the grasp of a rotating transparent sphere.

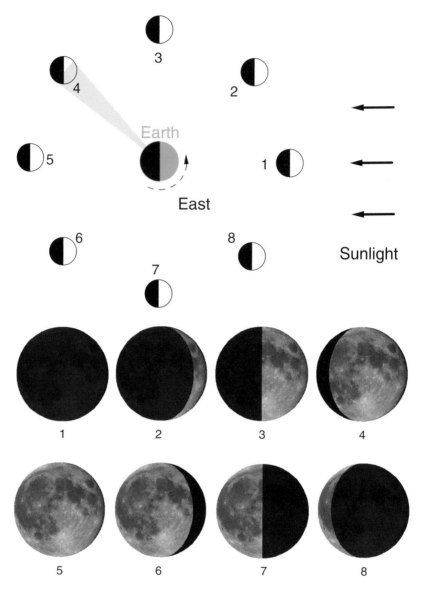

Figure 3.1 Phases of the Moon. The top portion of the figure shows the (simplified) geometry of the Earth–Moon–Sun system, with the Sun located far off the right side of the figure. The light gray triangle indicates the portion of the Moon visible from Earth when the Moon is in position 4. The bottom portion of the figure shows the corresponding appearance, or phase, of the Moon when it is in each of the positions labeled in the top portion. The diagram is not to scale.

3.1.3   Lunar calendars

The changing phases of the Moon provided a convenient way for ancient observers to keep track of the passage of time. If you understand the cycle of phases, a quick look at the Moon can give you a good estimate of where you are in that cycle. Since the cycle repeats every 29.53 days, it is possible to construct a calendar based on the synodic month. Typically such a calendar will begin a month with the first sighting of a crescent Moon after the new Moon. The month will then consist of a period of 29 or 30 days, usually alternating but with an occasional extra 30-day month in order to keep the average number of days in a month equal to one synodic month. In this way, a given phase of the Moon (say, third quarter) will always occur on the same day of the month (give or take a day).[3]

This lunar calendar is convenient on short time scales, but it doesn't line up well with the annual cycle of the seasons. A tropical year consists of 12.4 synodic months, so you can't fit a whole number of months into a year. One solution devised by ancient cultures was to use years of twelve months, alternating between full (30-day) and hollow (29-day) months. In this case the calendar year fell short of a tropical year and the seasons would gradually move later and later in the calendar. Before the seasons got too far out of sync with the calendar, an extra month was inserted (or *intercalated*) to bring things back into reasonable alignment. A calendar that attempts to align with both lunar and solar cycles is known as a lunisolar calendar.[4]

One particularly good way to design a lunisolar calendar is to make use of the **Metonic cycle**. The discovery of this cycle is attributed to Meton of Athens in the 5th century BC, although it was known earlier to Babylonian astronomers.[5] The cycle is a period of 19 tropical years (6939.6 days), which is very nearly equal to 235 synodic months (6939.7 days). A calendar based on the Metonic cycle consists of 12 years of 12 months and 7 years of 13 months, for a total of 235 months. Of these months, 125 are full (30 days) and 110 are hollow (29 days) so that the total number of days in the cycle is 6940, in close agreement with the actual time for 19 tropical years and 235 synodic months.

Our modern civil calendar is the Gregorian calendar discussed in Chapter 2, which aligns well with the seasons but not with the phases of the Moon. The average number of days per month in the Gregorian calendar is 30.44, so the Gregorian calendar gets steadily out of sync with the 29.53-day synodic month. However, the calendar of holy days in Islam is a purely lunar calendar, the holy days of Judaism follow a lunisolar calendar, and many eastern cultures still keep traditional holidays according to lunar or lunisolar calendars.

3.2   Eclipses: hiding the Sun and Moon

The phases of the Moon were regular and predictable, so much so that ancient cultures could use them to keep track of time. Much less predictable were eclipses. Eclipses are among the most dramatic events that can occur in our skies. In a **solar eclipse** the Sun becomes darkened and obscured. A total solar eclipse can entirely block out the Sun and make the daytime go dark for a few minutes. A partial solar eclipse (as shown in Figure 3.2) only obscures a portion of the Sun's face, but during a solar eclipse the Sun may be partially obscured for a few hours. In one type of solar eclipse, known as an annular eclipse, the central portion of the Sun's face is obscured but the outer edge of the Sun is still visible as a thin ring (or annulus) for a short time. Total and annular eclipses, when they occur, are visible from only a limited portion of the Earth, while partial eclipses are visible over a wider region. Solar eclipses only occur when the Moon is in its new phase.

During a **lunar eclipse** an otherwise full Moon goes dark. As with solar eclipses, lunar eclipses can be total (in which the entire Moon is darkened) or partial (in which only part of the Moon's face is darkened, as in Figure 3.2). In some cases the darkened portion of the Moon's face may take on a reddish color. A total lunar eclipse can last more than an hour, while the Moon may be partially obscured for several hours. Lunar eclipses, when they occur, are visible to anyone who can see the Moon.

The basic causes of eclipses were known to Anaxagoras in the 5th century BC.[6] He understood that solar eclipses occur when the Moon, which is closer to Earth than is the Sun, passes through the line between the Earth and Sun and

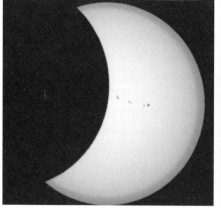 

Figure 3.2 Left: photograph of a partial solar eclipse (photo credit: Sandra Meek). Right: photograph of a partial lunar eclipse (photo credit: NASA/Bill Ingalls).

blocks the Sun's light from reaching Earth. That is why solar eclipses only occur during a new Moon: it is only in that phase that the Moon can be between the Earth and Sun (see Figure 3.1). We have seen that the apparent diameter of the Moon varies and that sometimes it is greater than the Sun's apparent diameter and sometimes less. If the Moon's apparent diameter is greater than the Sun's a total eclipse is possible, but if the Moon's apparent diameter is smaller then instead of a total eclipse we would see an annular eclipse since the Moon isn't large enough to obscure the entire face of the Sun. In a total solar eclipse the shadow of the Moon forms a spot on the Earth that is only about 100–200 km across and that spot moves across Earth's surface. A total solar eclipse is only visible from within that spot, but observers in a much larger area outside of the spot will see a partial eclipse.

A lunar eclipse occurs when the Earth falls on the line between the Moon and Sun. The Earth then blocks sunlight from reaching the Moon. Since the Moon shines by reflecting sunlight, if there is no sunlight reaching the Moon then the Moon will be dark. Lunar eclipses can only occur when the Moon is full because in that phase the Moon will be on the opposite side of Earth from the Sun (see Figure 3.1). The shadow of the Earth at the distance of the Moon is larger than the Moon itself, so although the Moon moves through the shadow during an eclipse it can take some time for it to move all the way across the shadow.

A look back at Figure 3.1 makes it seem as though a total (or perhaps annular) solar eclipse should occur *every* time the Moon reaches new phase, and likewise that a total lunar eclipse should occur each time the Moon reaches full phase. However, the diagram in Figure 3.1 provides only a simplified two-dimensional picture of the Earth–Sun–Moon geometry. We know that the Sun is always seen on the ecliptic, which forms a great circle on the celestial sphere. Imagine a flat plane that slices through the celestial sphere, intersecting the sphere exactly along the ecliptic circle. We will call this plane the **ecliptic plane**. If the Moon was always in the ecliptic plane, as it seems to be in Figure 3.1, then we would get a solar eclipse every new Moon and a lunar eclipses every full Moon. However, we know that the Moon is not always in that plane, because sometimes the Moon is seen north of the ecliptic and sometimes south.

The Moon's motion above and below the ecliptic plane explains why eclipses don't occur every month. If the Moon is in full phase and *also* in the ecliptic plane, then we will see a lunar eclipse. However, if the Moon is full but slightly above or below the ecliptic plane we will only see a partial eclipse. If the full Moon is even farther from the ecliptic plane, as it usually is, then we won't see any eclipse at all. Similarly, during a new Moon we may see a total or annular solar eclipse if the Moon lies in the ecliptic plane, but if the Moon is outside of that plane we may get a partial eclipse or, more commonly, no eclipse at all. The fact that lunar eclipses are rare, and solar eclipses even more rare since they cannot be seen by everyone when they do occur, indicates that the Moon is usually well outside the ecliptic plane when it is in full or new phase.

How often does the Moon cross the ecliptic plane? A point where the Moon (or another celestial body) crosses the ecliptic is known as a **node**. If the Moon crosses from south (below) to north (above) then the point is called an **ascending node**, and if it crosses from north to south it is called a **descending node**. Careful observation of the Moon's motion reveals that the time between consecutive nodes of the same type averages 27.21 days. This period is known as the **draconic month**. The draconic month is only slightly different from the sidereal and anomalistic months, but it differs significantly from the synodic month that regulates the Moon's phases. So if a lunar eclipse occurs during a particular full Moon, we can be sure that we won't have an eclipse during the next full Moon because the Moon will have already passed through its node two days prior.

Knowing the draconic month and synodic month we can predict when eclipses are likely to occur. For example, 12 synodic months (354.4 days) is very close to 13 draconic months (353.7 days). So if we see a lunar eclipse we might expect to see another one in 354 days. Over longer periods we can predict not just individual eclipses, but whole cycles of eclipses. Ancient Babylonian astronomers were aware of what is now called the **saros** cycle. They realized that a period of 6586 and 1/3 days was nearly equal to 223 synodic months (6585.32 days) and 242 draconic months (6585.36 days).[7] Patterns of solar and lunar eclipses will seem to repeat after this cycle of just over 18 years. In fact, the saros cycle is nearly equal to 239 anomalistic months (6585.54 days), so the apparent sizes of the Sun and Moon won't change much from one eclipse to the next within a given saros cycle. That means that if an eclipse is annular, then the next eclipse in that saros will probably be annular as well.

3.3   Solar and lunar distances

3.3.1   *Aristarchus on the distances and sizes*

The Babylonian's ability to predict eclipses was impressive, but the geometrical expertise of the Greeks allowed them to use eclipses and other observations as tools for measuring the distances to the Moon and Sun. In the 3rd century BC astronomer and mathematician Aristarchus of Samos wrote a work entitled *On the Sizes and Distances of the Sun and Moon*.[8] Aristarchus began by using geometry to find the relative distances of the Sun and Moon. He realized that when the Moon is exactly in quarter phase the angle between the Earth–Moon line and the Moon–Sun line must be 90°, thus forming a right triangle as shown in Figure 3.3. If the angle θ could be determined, then trigonometry would give

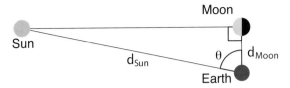

Figure 3.3 Geometry of Aristarchus' method for finding the relative distances of the Sun and Moon. The drawing is not to scale.

the ratio d_{Sun}/d_{Moon} where d_{Sun} is the distance of the Sun from Earth and d_{Moon} is the distance of the Moon from Earth. The details of this method are given in Appendix A.5.

Aristarchus estimated that when the Moon was in quarter phase the angle between the Moon and the Sun was less than a right angle by one-thirtieth of a right angle. In other words, $\theta = 90° - 3° = 87°$. From trigonometry we then find that $d_{Sun}/d_{Moon} \approx 19.1$, so according to Aristarchus the Sun is about 19 times as distant from Earth as the Moon is. Furthermore, if the Sun is 19 times as distant as the Moon but has the same angular size as the Moon, then the diameter of the Sun must be 19 times as large as the Moon's diameter.

We now know that Aristarchus' estimate for the angle θ was inaccurate. The correct value is $\theta \approx 89° \, 51'$. Although this angle is only slightly different from Aristarchus' estimate, that small difference leads to a large change in the result: the Sun is really about 380 times as far away as the Moon. Although Aristarchus' method is sound, in practice it is very difficult to determine the exact time when the Moon reaches quarter phase. It is also challenging to accurately measure the angle between the centers of the Sun and Moon. These practical challenges would prevent astronomers from obtaining a better estimate for d_{Sun}/d_{Moon} for nearly two thousand years.

Aristarchus also developed a method to determine the absolute sizes and distances of the Sun and Moon. We will provide only a brief outline of his method here. His method relied on his observation that during a lunar eclipse the apparent diameter of the Earth's shadow (at the distance of the Moon) seemed to be about two times the angular diameter of the Moon. Combining that observation with his estimate of the relative distances described above, and the fact that the Moon and Sun have nearly the same apparent diameter (which Aristarchus took to be 2°), Aristarchus used a geometrical argument to determine the distances to the Sun and Moon.

Aristarchus found that the Moon's distance was 20 times the radius of the Earth, while the Sun's distance was 382 Earth radii, but his results were based on some bad data. The apparent diameters of the Sun and Moon are really about 0°.5 and the Earth's shadow is really about 2.6 times the size of the Moon. Using these improved values, Aristarchus' method gives a lunar distance of 67 Earth radii and a solar distance of 1273 Earth radii. This value for the Moon's distance is close to the modern value, but because of the problems in determining the relative distance the value for the solar distance is far too small.

3.3.2 Parallax: watching from two places at once

Another way to determine the distance to a celestial object is to use the phenomenon of **parallax**. Parallax is easy to demonstrate. Close your left eye, then hold your thumb upright at arms length and look at it through your right eye. Note the position of your thumb against the background. Now close your right eye and open your left eye. Look at your thumb again: it has shifted to the right relative to the background. This demonstration perfectly illustrates the parallax phenomenon. When an object is viewed from two different locations it will appear in different locations relative to the background.

Now move your thumb closer and repeat the experiment. You should find that your thumb seems to shift by a greater amount, relative to the background, when it is closer to your eyes. This result illustrates the inverse relationship between the parallax effect and the distance to the object: more distant objects will show less parallax. We can use this relationship to determine the distance to an object from a measurement of its parallax shift.

In the 2nd century BC Hipparchus used parallax to estimate the distance to the Moon.[9] The geometry of his method is illustrated in Figure 3.4. Two observers view the Moon from different locations on Earth. The observer at O_1 sees that Moon at position M_1 on the celestial sphere, while the observer at O_2 sees the Moon at M_2. The angle between the Moon's two apparent locations is 2θ. The distance between the observers, known as the **baseline**, is denoted b while the distance to the Moon is denoted d.[i]

Note that the angle θ, half of the apparent shift 2θ, appears as an angle in a right triangle in Figure 3.4. The sides of that triangle are $b/2$ and d. If we know the values for θ and b, we can use trigonometry to find d. The details are given in Appendix A.6, but Figure 3.4 suggests that θ will increase if d decreases or b increases.

Hipparchus knew of a solar eclipse in which the Sun was totally eclipsed as seen from the Hellespont (now known as the Dardanelles in modern Turkey). For observers in Alexandria (Egypt) the Sun was at most four-fifths obscured during the same eclipse. Hipparchus estimated that the parallactic shift (2θ) of the Moon

---

[i] Technically d is the distance from the center of the baseline to the center of the Moon, but this will be approximately equal to the distance between the centers of the Earth and Moon.

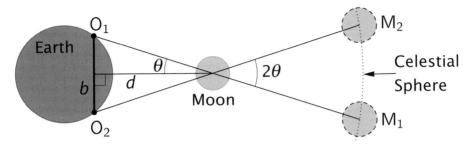

Figure 3.4 The geometry of parallax. Observers at locations O_1 and O_2 on Earth see the Moon at apparent locations M_1 and M_2 on the celestial sphere. The distance of the Moon (d) can be determined from the parallax angle (θ) and the baseline distance between the observers (b). The diagram is not to scale. Adapted from Timberlake (2013) with the permission of the American Association of Physics Teachers.

for these two locations was one-fifth of the Sun's apparent diameter, or about $6'$. Using an estimate of the distance between the locations for his baseline (b), and assuming that the parallax of the Sun was so small as to be undetectable (because the Sun is much more distant than the Moon), he found that the distance to the Moon was about 77 Earth radii.

In the 2nd century AD, the astronomer Claudius Ptolemy estimated the Moon's parallax by comparing the Moon's observed position as seen from Alexandria to his own theoretical prediction of where the Moon should be as seen from Earth's center.[10] His result indicated that the Moon lies about 59 Earth radii away, or about 240,000 miles if we use Eratosthenes' estimate for Earth's diameter (see Section 2.3.2). Using Aristarchus' value for the ratio of Sun and Moon distances, Ptolemy went on to estimate the Sun's distance at about 1210 Earth radii (nearly 5 million miles). Since both Sun and Moon have apparent diameters of about $0°.5$, these distance estimates indicated that the Sun's physical diameter was 5.5 times that of Earth (44,000 miles), while the Moon's physical diameter was less than one-third that of Earth (2300 miles). These estimates for the sizes and distances of the Sun and Moon, particularly those of Ptolemy, were generally accepted by astronomers in the Western tradition for the next 1500 years.

3.4   The wandering stars

3.4.1   Strange motions

In addition to the Sun and Moon, Babylonian astronomers were aware of five other lights in the sky that looked like stars but did not behave like fixed stars. These wandering stars came to be associated with divine beings, and tracking their motions became an important task within the Babylonian religious tradition. As a result, the Babylonians kept extensive records of their observations and developed sophisticated arithmetical rules for predicting the positions of these wandering stars on the celestial sphere. They recorded their observations and predictions on clay tablets, some of which survive today. Much of this Babylonian astronomical knowledge was inherited by the Greeks, who assigned the names of their own gods to these mysterious wanderers.[11] The Greeks passed on their knowledge to the Romans and we now know these objects by the names of Roman gods: Mercury, Venus, Mars, Jupiter, and Saturn. For now we will use the term "planet" to refer to one of these five wandering stars, until we find that we need a new definition for that term.

We will explore the observable behavior of the planets in some detail, because it is those behaviors that later astronomers would seek to predict and explain. These observations provide important clues that helped astronomers develop theories to account for the motions of the planets. Eventually those clues would help lead us to our modern view of the solar system.

The planets are not fixed to the celestial sphere, nor are they fixed relative to each other. Instead, each of the planets seems to move independently of the rest. They generally move eastward along the ecliptic (while participating in the daily rotation of the celestial sphere, just like everything else in the heavens). Like the Moon, the planets do not move exactly along the ecliptic: sometimes they are above the ecliptic and sometimes below. In what follows we will examine the motion of the planets on the celestial sphere, ignoring the daily rotation. Since the planets are always close to the ecliptic we can track their motions using ecliptic coordinates.

The most striking thing about the planets is that occasionally they will halt in their eastward motion relative to the fixed stars, move westward on the celestial sphere for a while, and then turn to move back toward the east. This westward motion is unlike anything we have seen so far: the fixed stars don't move at all on the celestial sphere, while the Sun and Moon move only eastward. This unusual westward motion on the celestial sphere is called **retrograde** motion, to distinguish it from the more typical **prograde** motion toward the east. Prograde (eastward) motion is said to be *in the order of the signs* because the ecliptic longitude of the planet increases steadily during prograde motion, moving from one sign to the next just like the Sun and Moon. Retrograde motion, on the other hand, is *contrary to the order of the signs*, because the planet's ecliptic longitude decreases during retrograde motion.

Figure 3.5 shows the main features of Mars' motion along the celestial sphere. Throughout most of its journey Mars is in prograde motion, moving eastward in the order of the signs (as in position 1 in the figure). When it reaches

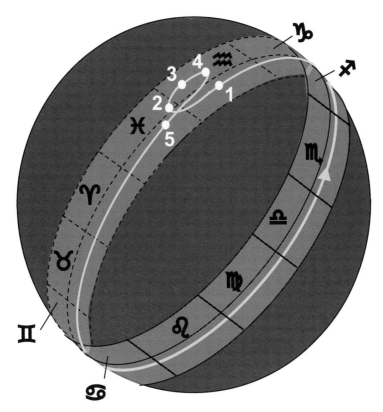

Figure 3.5 The motion of Mars. The light gray curve illustrates Mars' path through the zodiac. At position 1 Mars moves eastward in prograde motion. Mars reaches a station at position 2, is at the center of its westward retrograde motion at position 3, and reaches another station at position 4. At position 5 Mars has returned to its eastward prograde motion.

position 2, Mars seems momentarily to halt its eastward motion before beginning its westward retrograde motion. Position 2, at which Mars' motion switches from prograde to retrograde, is known as a **station**. At position 3 Mars is at the center of its retrograde motion, moving westward contrary to the order of the signs. At position 4 Mars reaches another station and its motion switches from retrograde back to prograde. From there Mars continues its usual motion in the order of the signs (as at position 5). The difference in ecliptic longitude between the two stations (positions 2 and 4) is known as the **retrograde arc**. In Figure 3.5 Mars has a retrograde arc of about 18° (a bit more than half of a sign).

The other planets show these same general features: mostly prograde motion around the ecliptic but with occasional periods of retrograde motion. The path of a planet does not truly repeat: each circuit around the celestial sphere will be slightly different. However, the general pattern of prograde and retrograde

Table 3.1 *Observational data for the planets, showing the tropical period T_t, the synodic period T_s, the maximum elongation ϵ_{max}, and the time from opposition to quadrature T_q. Also shown are the symbols used to denote each planet.*

Planet	Symbol	T_t (days)	T_s (days)	ϵ_{max}	T_q (days)
Mercury	☿	365	116	18°–28°	–
Venus	♀	365	584	45°–47°	–
Mars	♂	687	780	–	106
Jupiter	♃	4,333	399	–	87.5
Saturn	♄	10,759	378	–	86.9

motion does repeat periodically. We can characterize the motion of a planet by specifying two different periods for its motion: the **tropical period** and the **synodic period**.

The tropical period of a planet is the average time it takes for the planet to complete a loop around the celestial sphere. In other words, it is the average amount of time that it takes for the planet to travel all the way around the signs of the zodiac. This time can vary significantly from one trip around the celestial sphere to the next, so in order to find the true tropical period we must take an average over many such trips. The tropical periods of the five planets are given in Table 3.1. Note that the tropical periods of Mercury and Venus are both equal to one year. We will explore this coincidence shortly.

The synodic period of a planet is the time between one retrograde and the next, measured from the midpoint of the retrograde motion (position 3 in Figure 3.5). Measuring the synodic period in this way can be difficult, because it is hard to determine when the planet is at the exact midpoint of its retrograde. We will see shortly that there is a better way to measure the synodic period. Although the synodic period does not change very much from one cycle to the next, it is still necessary to average over several cycles to get the true synodic period. The synodic periods of the five planets are given in Table 3.1.

3.4.2   Inferior planets

All five of the "wandering stars" display the general characteristics discussed in Section 3.4.1. However, the five planets separate into two distinct groups when we consider other characteristics of their motions. The first group we will consider are the **inferior planets**: Mercury and Venus. We will discuss why these planets are called "inferior" later, but for now let's take a look at what distinguishes these two planets from the other three.

The most striking thing about these two planets is that they are always seen near the Sun in the sky. To quantify their nearness to the Sun we can define a measurement called **elongation**. The elongation of a planet is simply the difference between the planet's ecliptic longitude and the Sun's. In other words, elongation is the angle between the Sun and the planet, measured along the ecliptic. We will also define special terms to refer to specific elongations. A planet is said to be in **conjunction** when it has the same ecliptic longitude as the Sun, while it is in **opposition** when it is 180° away from the Sun along the ecliptic. A planet is in (eastern or western) **quadrature** when it is 90° (east or west) of the Sun.[i]

The interesting thing about Mercury and Venus is that their elongations are always small. Consider the motion of Venus. Suppose Venus begins to the west of the Sun, moving eastward (prograde) along the ecliptic faster than the Sun. Eventually Venus will catch up to the Sun. At this point Venus is in conjunction, at the same longitude as the Sun. Venus will then pass the Sun, and its eastward elongation will increase. After a while Venus' motion along the ecliptic begins to slow down and eventually it slows enough that it is moving at the same speed as the Sun. At that moment Venus is at its maximum eastern elongation: it is as far east of the Sun as it is going to get. As its speed continues to drop, the Sun will begin to catch back up with Venus and the elongation of Venus will decrease.

Venus' speed along the ecliptic continues to drop until it finally reaches a point where its motion comes to a momentary halt relative to the fixed stars: Venus is now at its eastern station. After that Venus will move westward along the ecliptic, in retrograde, while the Sun continues to move eastward. Soon the Sun catches back up to Venus and Venus is once again in conjunction. Venus continues its retrograde motion, falling behind (to the west of) the Sun until it reaches its western station. At that point Venus resumes its prograde motion but continues to fall farther behind the Sun because the Sun is moving faster to the east. Venus' speed along the ecliptic increases until it once more matches the Sun's speed: now Venus is at its maximum westward elongation. From there Venus continues to speed up. It closes the gap with the Sun and the whole cycle repeats.

Mercury exhibits the same general pattern of motion. The major difference between Venus and Mercury is that they have different values for maximum elongation. The motions of these planets do not strictly repeat, so the maximum elongation of the planet changes somewhat from one cycle to the next. However,

---

[i] Note that elongation can be useful in discussing the phases of the Moon. The Moon is full when it is in opposition, new when it is in conjunction, and in quarter phase when it is in quadrature.

Mercury's maximum elongation varies from about 18° to 28° while Venus' varies from 45° to 47°. Both planets stay near the Sun at all times, but Venus can stray farther from the Sun than Mercury can.

It's not hard to see that the motions of these inferior planets are in some way controlled by the Sun. They move almost like two excited dogs on leashes held by the Sun. The Sun strolls along the ecliptic at a fairly steady rate, while the inferior planets alternately run ahead and drop behind. On average, both planets move along the ecliptic at the same pace as the Sun, which explains why both Mercury and Venus have tropical periods of one year (see Table 3.1). The main difference between the two is that Mercury is on a shorter leash (it has a smaller maximum elongation) and it runs back and forth more frequently (it has a shorter synodic period).

Note that the inferior planets have two different types of conjunctions. In one case the planet is moving eastward and passes the Sun. This type of conjunction is called a **superior conjunction**. In the other type, known as an **inferior conjunction**, the planet is in retrograde and it falls behind the eastward moving Sun. Since the inferior conjunction happens right at the middle of the planet's retrograde arc, we can measure the synodic period of an inferior planet by measuring the average time between consecutive inferior conjunctions. In fact, the average time between consecutive superior conjunctions is also equal to the synodic period.

Because the inferior planets are always near the Sun they are difficult to observe. They can only be seen shortly after sunset in the west, or shortly before sunrise in the east. That is why the bright planet Venus was originally thought to be two different objects, known as Hesperos (evening star) and Phosphoros (light bringer) to the Greeks, although before the 4th century BC Greek astronomers recognized that the "morning star" and "evening star" were really the same object.[12] Mercury is particularly difficult to observe because it is much dimmer than Venus and also doesn't get as far from the Sun. The difficulty of observing Mercury would plague astronomers until the modern era.

In addition to their motion along the ecliptic, the inferior planets can also move a little bit above and below the ecliptic. Mercury's ecliptic latitude varies from about −5° to +4°. Venus strays farther from the ecliptic, reaching ±9°. These planets show their greatest (and least) ecliptic latitudes when they are in retrograde and in a particular sign of the zodiac. The retrograde motions of an inferior planet look a little bit different each time. Sometimes the planet executes a loop up above the ecliptic, sometimes it loops below the ecliptic, and sometimes it follows a zigzag path. However, if two retrogrades occur in the same part of the ecliptic (i.e. in the same sign) they will be similar.

One final point about the inferior planets: their appearance changes slightly as they move around the celestial sphere, but these changes are very hard to notice. For one thing, Venus and Mercury always appear to the naked eye as tiny circular disks and the size and brightness of these disks does not change in any easily noticeable way. Using modern measuring devices we know that Venus, at its brightest, is about three times as bright as when it is at its dimmest. Mercury's brightness changes more significantly. However, these planets have very small apparent diameters and their changes in brightness are perceived by the human eye as very slight changes in these apparent diameters. Given the proximity of these inferior planets to the Sun on the sky it was nearly impossible for ancient observers to be certain that the brightness/size of the inferior planets changed at all.

3.4.3   Superior planets

While Venus and Mercury are classified as inferior planets, the remaining three visible planets (Mars, Jupiter, and Saturn) are categorized as **superior planets**. The superior planets are not leashed to the Sun like the inferior planets are. The superior planets can be found at any elongation including quadrature (90°) and opposition (180°). However, the patterns of their motions are still linked to the Sun in a strange way.

Let's examine the motion of Mars on the celestial sphere. Mars' motion was illustrated earlier in Figure 3.5, but now we want to see how the motion of Mars relates to its elongation, or position relative to the Sun. Suppose Mars begins at eastern quadrature, 90° to the east of the Sun. Mars will be moving eastward (in the order of the signs) but slower than the Sun, relative to the fixed stars. Eventually the Sun will catch up to Mars and Mars will be in conjunction. The Sun will pass Mars and Mars will fall behind to the west. The prograde motion of Mars slows down as it passes through western quadrature and eventually it will momentarily halt as Mars reaches its first station. Mars will then move westward, contrary to the order of the signs. When it reaches the middle of its retrograde arc, Mars will be in opposition, 180° away from the Sun on the ecliptic. Mars will then complete its retrograde motion, reach its second station, and resume its prograde motion toward the east. Mars will pick up speed as it moves in the order of the signs, but it is still moving slower than the Sun, which will now approach Mars from the west. Soon Mars will be back to eastern quadrature, having been lapped by the Sun, and the cycle will repeat.

Jupiter and Saturn display a similar pattern of motion. As with the inferior planets, the motion of the superior planets does not strictly repeat. Each retrograde looks a bit different, looping upward when it occurs in some signs of the zodiac, looping downward in other signs, and performing a zigzag in others. Table 3.1 shows that of the three superior planets Mars travels the fastest around the ecliptic (since it has the shortest tropical period) and Saturn is the slowest. On the other hand, Mars retrogrades less often and Saturn most often. These two facts are related because superior planets always retrograde when they are in opposition. A superior planet will retrograde once for each time it is lapped by the Sun. Since Mars travels around the ecliptic slower than the Sun but faster than Saturn, it makes sense that it takes longer for the Sun to lap Mars than to lap Saturn.

Because superior planets are in opposition during the middle of their retrograde motion, we can measure their synodic periods by measuring the times between consecutive oppositions. We know a planet is in opposition when it rises in the east just as the Sun is setting in the west. We could also measure the synodic period by measuring the times between consecutive conjunctions, as this turns out to be the same (on average) as the time between oppositions.

In addition to their different periods, the superior planets also show differences in the size of their retrograde arcs. Although each retrograde arc is different, Mars has retrograde arcs that vary from about 12° to 18°. The retrograde arcs of Jupiter are always around 10°, while those of Saturn are about 7°. Notice the pattern: Mars has the shortest tropical period, the longest synodic period, and the largest retrograde arc. Saturn has the longest tropical period, the shortest synodic period, and the smallest retrograde arc. Jupiter is in the middle for all three measurements.

Another important quantity that we can measure for the superior planets is the time it takes them to go from opposition to eastern quadrature (or from western quadrature to opposition). Although this quantity is not of obvious importance now, we will see that it plays an important role in both Ptolemaic and Copernican theories of planetary motion. Table 3.1 gives the times from opposition to quadrature for the three superior planets. Note that the pattern is similar to that for synodic period: Mars has the largest value while Saturn has the smallest.

Like the inferior planets, the superior planets move above and below the ecliptic. Mars can be as far as 6° from the ecliptic, Jupiter only about 2°, and Saturn 3°. As with the inferior planets, the superior planets reach their largest ecliptic latitudes when they are in retrograde and in certain signs of the zodiac.

Because they can be found far from the Sun, the superior planets are easier to observe than the inferior planets. Like the inferior planets, the superior planets always appear as tiny circular disks. However, the brightness of these disks changes as the planets move around relative to the Sun. Using modern instruments we know that, at its brightest, Mars is nearly 70 times brighter than at its dimmest. Furthermore, Mars is always brighter when it is in opposition (and therefore also in retrograde motion). Jupiter and Saturn are also somewhat brighter at opposition/retrograde, but their variations are not as dramatic as those of Mars.

As noted earlier, the human eye perceives these changes in brightness as very slight changes in size. Although ancient observers could not have measured the changes in the brightness of Mars with much quantitative precision, they must have been aware that Mars varied in its appearance. When Mars rises at sunset it appears equal in size to Jupiter, while in other situations Mars appears much smaller than Jupiter. It is not likely, though, that they would have noticed the subtle changes in the apparent size of Jupiter and Saturn.

3.5   Reflections on science

3.5.1   Categories and classification

In this chapter we have focused on observations of the Moon and planets. But science is not just about making observations. It is also about organizing our observations and trying to use them to understand what is really going on in the world. One important way of organizing observations about objects is to classify objects into different categories. We try to define our categories such that objects within a given category share a set of observable, qualitative characteristics. The hope is that the categories we define will somehow line up with "natural" categories that exist in the world.

Of course, there are many possible ways to define the categories we use to classify objects. Ancient astronomers distinguished the fixed stars from the planets because of the way they move: the fixed stars all behave as though they are stuck to the rotating celestial sphere, while the planets (wandering stars) seem to move around on that sphere. This distinction between fixed and wandering stars seems like a natural one, but it is not the only possible one. For example, we could have defined categories according to color, in which case we might group Mars and the star Antares (both of which have a reddish tint) into one group, while Saturn and Canopus (both of which are yellowish-white) might form another group. There is no way to know in advance which classification scheme is "better," but some schemes will prove more useful than others.

As we gather more data we may find the need to create subcategories in our classification scheme. In this chapter we have seen that the planets can be divided into inferior and superior classes. That division seems to be a natural one. The inferior planets are always near the Sun and retrograde in conjunction, while the superior planets can stray far from the Sun and always retrograde (and appear brightest) in opposition. Once again we cannot be certain that these are "natural" categories, but if we are trying to predict and understand the apparent motions of the planets then it does seem like we will have to account for the difference between these two types of planets.

Most of the time we want to define our categories so that all objects within a given class share the same *general* characteristics. We don't want to be so specific that we end up with only one object in each category. Using such a classification system would be no better than just referring to each object by its proper name. In creating our categories we must ignore some differences between objects, such as the differences in color between fixed stars. On the other hand, we may find that *some* objects are sufficiently unique that they merit their own special class. Ancient astronomers considered the Sun and Moon to be planets (because they moved on the celestial sphere) but they did not classify them as inferior or superior, nor did they put them into a group together. Rather, the Sun and Moon were unique planets that defied further categorization.

3.5.2   Correlation and causation

Organizing observable phenomena into categories can help us to identify connections between different phenomena. When different observable phenomena seem to be linked in some way we say they are **correlated**. For example, we have seen that the appearance, or phase, of the Moon is correlated to the Moon's elongation from the Sun. When the Moon's elongation is 180° it will be full, when its elongation is 90° it will be in quarter phase, and so on. Likewise, we have seen that eclipses are correlated with both lunar phase and the lunar nodes.

Recognizing these correlations can help us to predict future phenomena. A careful measurement of the synodic month will allow us to predict the future phase of the Moon. Combine that with a measurement of the draconic month and we can predict future eclipses. However, we want our science to do more than just predict phenomena. We also want our scientific theories to explain phenomena. In particular, we would like for our theories to explain any correlations we observe. One way our theories can explain a correlation is to propose a causal explanation. A causal explanation suggests that a certain set of circumstances causes a particular observable phenomenon to occur.

We have already seen a causal explanation for the phases of the Moon. If we assume that the Moon is closer than the Sun to Earth and that it shines by reflected sunlight, then the geometry shown in Figure 3.1 explains *why* the phases of the Moon are correlated to its elongation from the Sun. Given our assumptions, it necessarily follows that the Moon's appearance must correspond to its elongation in exactly the way we observe. Similarly, if we assume that solar eclipses are caused by the Moon blocking sunlight from reaching Earth, then it must be the case that such an eclipse can only occur during new Moon when the Moon is at a node.

Now consider another set of correlations discussed in this chapter. Inferior planets always retrograde in conjunction and their tropical periods are equal to that of the Sun. In a similar fashion, superior planets always retrograde when they are in opposition. Why is the motion of each planet linked to the motion of the Sun? These correlations would become one of the central mysteries of astronomy for more than a thousand years, and the eventual discovery of a causal explanation for these correlations would change our picture of the universe.

3.5.3   The power of mathematics

Although we have tried to keep most of the technical mathematics confined to the Appendices, mathematical concepts and calculation techniques play a crucial role in astronomy. We have seen that trigonometry can help us determine the altitude of the Sun from a gnomon's shadow, which in turn allowed Eratosthenes to measure the Earth's diameter. In this chapter we have seen that trigonometry allows us to measure distances to celestial bodies such as the Moon and Sun. Mathematics is an incredibly powerful tool for doing science. When we can determine the mathematical relationships between measurable quantities, we can use those relationships to predict measurements that we have not yet done and perhaps cannot do! For example, parallax measurements helped ancient Greek astronomers determine an accurate distance to the Moon even though it would be more than 2000 years before humans could actually travel there.

The use of mathematics is not without its dangers, though. For one thing, the results of our mathematical calculations are, at best, as good as the numerical values we put into them. Aristarchus' used what we now know was an inaccurate estimate for the angle between the Sun and Moon during quarter Moon phase. As a result, he vastly underestimated the relative distance to the Sun. There was nothing wrong with his mathematical method, but his bad data led him to an incorrect conclusion. We can't really fault Aristarchus for his error: given the tools available to him he did the best he could. Errors like this tend to persist until new measurement tools are devised that can make more accurate measurements, or else new methods are found for calculating the same result from different data.

In addition to relying on good measurements, our mathematical methods also rely on certain assumptions. For example, the trigonometric methods we have discussed all rely on the idea that light travels in straight lines. While this seems to be true here on Earth, ancient astronomers could not be certain that the same held true in the heavens. Likewise, Aristarchus' method for determining the relative distances of Sun and Moon breaks down if the Moon doesn't actually shine by reflecting sunlight. As long as assumptions like these seem to work, we can continue using them, but as we extend our science we may find that some of our assumptions are false. If that happens we must re-evaluate any methods that were based on those assumptions.

3.5.4   *Astronomical vocabulary*

You may be concerned about the bewildering variety of new terms that have been presented in the first few chapters of this book. That concern is reasonable. It is harder to understand what you are reading when the text is filled with unfamiliar words. However, these unfamiliar terms are not meant to confuse but rather to clarify.

In everyday language the use of the word "month" is pretty straightforward. When we speak of a month we are usually referring to a specific month of the calendar ("What month is it?" "It is February.") or we may be referring to a period of roughly 30 days ("My birthday is a month from today."). The context makes it clear what we mean, and we are not using the word month in a way that requires great precision.

In a science like astronomy, we *do* want to be precise. We must define our scientific terms such that it is clear *exactly* what we mean by them. In the context of astronomy, the term "month" is vague. It indicates a period of time associated with the Moon, but there are several such periods and each has its own precisely measured value. We must distinguish between, say, the draconic month and the synodic month because the phases of the Moon vary according to the synodic month but not the draconic (or sidereal or anomalistic) month.

In other cases we introduce new terms to make it easier to talk about our subject. We could repeatedly say "the angle between the Sun and a planet as measured along the ecliptic," but it is much easier to say "elongation." Similarly, it is easier to say "retrograde motion" than "unusual westward motion relative to the stars." The introduction of a new term as a shorthand for a longer phrase is common to all areas of life. Astronomy is no exception.

The good news for you, the reader, is that we have already defined most of the terms we will need in this book. Although later chapters will introduce some new terms, the pace of new vocabulary will slow down. Furthermore, we will continue to use the terms we have already defined so that you are reminded of their meaning. After a while, the unfamiliar will become familiar and you will be fluent in the language of astronomy.

4

An Earth-centered cosmos: astronomy and cosmology from Eudoxus to Regiomontanus

4.1 Spheres within spheres

We saw in Chapter 2 that the two-sphere model, consisting of a celestial sphere centered on a spherical Earth, does an excellent job of accounting for observations of the stars and Sun from any location on Earth. This two-sphere model may have been first developed by the Greek astronomer Eudoxus in the 4th century BC.[1] Perhaps inspired by the success of the two-sphere model, Eudoxus went on to develop the first geometrical model to account for the apparent motions of the planets.

4.1.1 The homocentric spheres of Eudoxus

Unfortunately none of Eudoxus' works have survived to the present day. We only know about his model for the motions of the Sun, Moon, and five wandering stars because of comments made by other authors. Aristotle, who was a contemporary of Eudoxus, gave a brief description of the Eudoxan system in his *Metaphysics*.[2] The only other account we have of the Eudoxan system comes from Simplicius, writing in the 6th century AD, about 900 years after Eudoxus lived. Simplicius' account is more detailed, but it is less trustworthy because it was written so long after the fact.[3]

Fortunately, the basic structure of the Eudoxan model is well understood.[4] Eudoxus built his models using sets of nested spheres. The spheres are all **homocentric**, meaning they all have the same center. In this case all of the spheres are centered on the Earth. The outermost sphere rotates westward once per sidereal day about an axis through the celestial poles. Inside that sphere is another sphere which also rotates, but in the opposite sense (eastward). This second sphere rotates about a different axis, one that passes through the poles

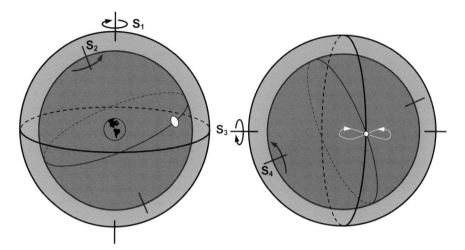

Figure 4.1 The Eudoxan model. The image on the left shows the spheres that account for the daily motion and the motion along the zodiac: the outer sphere rotates westward once per sidereal day about the celestial poles, while the inner sphere rotates eastward about the ecliptic poles. The image on the right shows the spheres used to produce the "hippopede" for the planets.

of the ecliptic. The axis of the second sphere is attached to the inner surface of the first sphere, so points on the second sphere participate in two simultaneous motions: a daily westward rotation about the celestial axis and a slower eastward motion about the ecliptic axis. The basic structure of the Eudoxan model is illustrated on the left in Figure 4.1. The model can be extended by including additional spheres, either inside the second sphere or between the first and second spheres.

It is not hard to see how this model could work for the Sun. If the inner sphere rotates once every sidereal year, then we can account for the mean solar motion.[i] Apparently Eudoxus thought that the Sun was sometimes found above or below the ecliptic, because he added a third sphere that would have produced very slight variations in the Sun's ecliptic latitude.[5] In fact, that sphere is unnecessary because the Sun is always exactly on the ecliptic.

The Moon, on the other hand, is sometimes found above and sometimes below the ecliptic. The Moon's nodes are the points where the Moon crosses the ecliptic. We have seen that the nodes must move relative to the stars because

---

[i] Eudoxus lived before Hipparchus discovered the precession of the equinoxes and accounted for the irregular motion of the Sun. So Eudoxus would not have distinguished between the sidereal year and the tropical year, nor would he have known about irregularities in the Sun's speed along the ecliptic.

the sidereal month (the time for the Moon to go around the celestial sphere) is not quite the same as the draconic month (the time for the Moon to go from a node back to the same node). To account for these facts Eudoxus must have had the Moon lie on an innermost sphere that rotates eastward once every sidereal month. Between that sphere and the sphere of daily rotation he could have placed another sphere that rotates westward once every 18.6 years, because it takes 18.6 years for the Moon's nodes to drift westward all the way around the ecliptic. This middle sphere would rotate about the ecliptic poles, but the innermost sphere would rotate about an axis tilted by about 5° relative to the ecliptic axis, in order to carry the Moon above and below the ecliptic plane each month.[6]

4.1.2   *The Eudoxan planets*

Based on the brief descriptions by Aristotle and Simplicius we cannot be certain exactly how Eudoxus constructed his model for the planets. We know that he started with the basic structure shown on the left of Figure 4.1 with the inner sphere rotating once per sidereal/tropical period of the planet. We also know that Eudoxus added two more spheres inside this structure, but the brief descriptions we have don't make it clear how these two spheres worked or even what their purpose was.

Assuming that the purpose of the additional spheres was to reproduce the retrograde motion of the planets,[7] the Italian astronomer Giovanni Schiaparelli proposed a reconstruction of the Eudoxan planetary theory in 1877.[8] We will present his reconstruction here, although other solutions are possible.[9] In Schiaparelli's version, the third and fourth spheres rotate about axes that are somewhat tilted relative to each other. Each sphere rotates once per synodic period of the planet, but the spheres rotate in opposite senses (one clockwise, the other counterclockwise). If the planet is placed on the equator of the innermost (fourth) sphere, then it will trace out a figure-eight path called the "hippopede" (horse fetter), as shown on the right in Figure 4.1.

Of course, the planets don't trace out figure-eights on the celestial sphere. However, if the axis of the third sphere is attached to the second sphere at points on the ecliptic, then as the planet executes its figure-eight it will also be carried eastward along the ecliptic. The combination of the hippopede motion and the motion along the ecliptic can cause the planet to move generally eastward but with occasional episodes of westward (retrograde) motion. The planet will also move above and below the ecliptic, just like the real planets do. Meanwhile the whole structure executes a daily motion on the sky as the outermost sphere rotates.

4.1.3   Refinements and flaws

According to Aristotle, Eudoxus used a total of 26 spheres to account for the motions of the Sun (3), Moon (3), and five visible planets (4 each). His student Callippus added additional spheres for Mercury, Venus, and Mars, as well as two spheres each for the Sun and Moon.[10] The additional spheres added by Callippus may have been intended to account for irregularities in the motions of the planets, such as the irregular motion of the Sun that leads to unequal seasons.[i] In any case, Callippus seems to have used a total of 33 spheres.

Eudoxus and Callippus seem to have dealt with each object separately. They had a model for the Sun, another model for the Moon, and another for each planet. Although these models shared many features in common, there was no attempt to fit these different models together into a single whole. Aristotle, however, wanted to produce a single mechanism that could account for all of the motions. To that end he placed the spheres for one planet inside the spheres of another. However, he did not want the motions of the outer planet to interfere with the motions of the inner planet. He added extra spheres between the planets that would "unroll" some of the motions of the outer spheres, rotating about the same axes but in the opposite direction. Aristotle wrote that the resulting system would include 55 spheres,[12] but in fact only 49 are needed. No unrolling spheres are needed for the innermost object, and once the daily rotation is accounted for by the outermost sphere there is no need to include a daily rotation sphere (or the corresponding unrolling sphere) for any other object.

These homocentric sphere models seemed to provide a plausible mechanism to account for the observed motions of the planets, at least qualitatively. However, the models cannot account for the motions of the planets in detail. In fact, if we put in the observed tropical and synodic periods for Venus and Mars we find that Eudoxus' model cannot even produce retrograde motion.[13] Although the model can produce retrogrades for the other planets, it cannot provide accurate predictions of planetary positions. However, Greek astronomers of the 4th century BC were not particularly concerned with detailed mathematical predictions. Eudoxus' contemporaries may have viewed his theory as a success because it managed to reproduce the general characteristics of planetary motion.[14]

Another problem for the Eudoxan models is that the planets are always at the same distance from Earth, because they lie on an Earth-centered sphere. That makes it hard to account for the changing brightness (or apparent size) of the planets.[15] We have seen that Mars appears larger/brighter when it is in

---

[i] Callippus may have been the first Greek astronomer to realize that the seasons were not all the same length.[11]

retrograde than at other times. That observation might seem to indicate that Mars is closer to Earth when it is in retrograde, but that would not be possible in the Eudoxan model. However, there are no records to indicate that the Greeks were concerned about, or even noticed, changes in the brightness or size of Mars and they were likely unaware of such changes in the other planets.[16]

Another observation that might have caused trouble for the Eudoxan model is the change in the apparent size of the Moon. We have seen that the Moon's apparent size can vary by roughly 15%. This change in size is particularly notice-able during solar eclipses. In a total solar eclipse the Moon is large enough to completely cover the face of the Sun, but in an annular eclipse the Moon is too small to completely obscure the Sun and the edges of the Sun remain visible. The changing angular size of the Moon seems to indicate that the Moon does not maintain a constant distance from Earth, but without any documentary evi-dence we can't be sure that this problem led astronomers to question Eudoxus' homocentric spheres. As far as we know, the failure of the Eudoxan model to account for apparent changes in size was not pointed out until long after the model was abandoned.

4.2   Ancient Greek cosmology: Plato and Aristotle

The homocentric models of Eudoxus and Callippus are examples of mathematical astronomy. These models were constructed in the hopes of matching the apparent motions of the planets and, perhaps, of revealing the actual motions of these objects through the heavens. As we have seen, each planet was treated individually. Eudoxus and Callippus did not attempt to merge these mathematical models into a unified description of the entire universe.

Aristotle, on the other hand, was not as concerned with the detailed math-ematical models for each planet. He simply adopted the mathematical models of Callippus without modification. However, Aristotle did seek to merge those models into a single whole. In that respect, Aristotle was involved in **cosmology** rather than astronomy. Cosmology treats the entire universe, the collections of all things that exist, as a single entity. The word derives from the ancient Greek term *cosmos*, which refers to the entire universe as a single orderly system. Cosmologists seek to understand the general structure and order of the universe on its largest scales without worrying about the small-scale details. They may also investigate the origin and eventual fate of the universe as a whole.

4.2.1   *Plato's cosmology*

Before we take a closer look at the cosmology of Aristotle we must first examine the cosmological stories of his (and Eudoxus') teacher Plato. In his

Republic, Plato relates a cosmological myth in which the overall structure of the universe is said to resemble a spindle: a rod used to wind thread, often surrounded by a wheel known as a whorl.[17] Plato described the whorl of the universe as a hemisphere divided into eight separate sub-hemispheres, which he referred to as circles, nested inside each other. In the story, all of these circles rotate about the axis of the spindle, which represents the celestial axis. This rotation, which is just the daily rotation of the celestial sphere, is assisted by Clotho, one of the three Fates, who occasionally pushes the outer circles with her right hand to keep them spinning. The inner circles, though, also rotate more slowly in the contrary direction. These circles are pushed along by Atropos with her left hand. The contrary motion of the inner circles represents the motion of the seven planets (including the Sun and Moon) along the zodiac, each at its own speed. Finally, Lachesis nudges each of these inner circles, alternately with her left and right hands. The alternating motion produced by the two hands of Lachesis presumably accounts for the irregularities in the motion of the planets along the ecliptic (and perhaps specifically the retrograde motion of the planets). Each of the circles is accompanied by a siren who sings a particular note and their combined voices produce the harmonious "music of the spheres."[i]

One thing that Plato made very clear was the ordering of the planets. He placed the stars on the outermost circle. Working inward from the stars the circles of the planets were in the following order: Saturn, Jupiter, Mars, Mercury, Venus, the Sun, and the Moon. The ordering was based on the average speed of each planet along the zodiac. Saturn, the slowest, was placed closest to the stars because its motion deviates the least from that of the celestial sphere. The Moon moves fastest along the zodiac so it was placed farthest from the sphere of the fixed stars. The only problem with this ordering principle was that the Sun, Venus, and Mercury all share the same tropical period so they all move along the zodiac at the same average rate. Plato acknowledged this fact, but did not explain how the ordering of those three circles was determined.

In Plato's *Timaeus* the universe is presented with a similar structure.[19] In this story the universe has a spherical shape and the tilt between the celestial equator and the ecliptic is described explicitly. In the *Timaeus*, Plato described the universe as a living being with its own soul and he discussed the creation of the universe by the demiurge, or divine craftsman. He stated that the demiurge created the Sun, Moon, and planets in order to measure and record the passage of time, so the heavens played the role of an enormous clock. Plato also claimed that the universe was composed of four different **elements**: earth, water, air, and

---

[i] The idea that each heavenly body emits a musical note, with all of the notes forming a celestial harmony, may go back to Pythagoras in the 6th century BC.[18]

fire. He thought that celestial bodies were made mostly of fire, with the other three elements predominant in the regions below the Moon.

For each element there was a smallest particle of that element, what we might now call an atom, and the particles of each element had a particular shape.[20] The shape of each particle was one of the **regular solids** (also called Platonic solids). A regular solid is built out of regular polygons, closed two-dimensional shapes composed of lines of identical length and with all interior angles equal (e.g. a square or an equilateral triangle). To build a regular solid, identical regular polygons are connected together to build a closed three-dimensional shape. There are only five regular solids that can be constructed in this way and Plato associated four of those with the elements: the cube (earth), the icosahedron (water), the octahedron (air), and the tetrahedron (fire). There was only one regular solid left: the dodecahedron, which Plato took to represent the universe as a whole since it approximates a sphere. Plato claimed that each of the regular solids could be broken down into triangles and those triangles could be reformed into new regular solids. This idea explained the process of generation and decay: elements could be broken down and reformed into new elements.

4.2.2   The Aristotelian cosmos

Plato's cosmology was highly speculative and it is not clear if he was really proposing these stories to serve as a scientific account of the universe. In constrast, Plato's student Aristotle, in his book *De Caelo* (On the Heavens), laid out a very specific and detailed cosmology that was tightly integrated with the astronomical science of his day. We have already seen that the general structure of Aristotle's cosmology was based on the two-sphere model of Eudoxus and incorporated Callippus' homocentric spheres for the Sun, Moon, and planets.

In ordering the planets Aristotle followed the same general rule as Plato: planets that move slowest along the zodiac are farthest out.[21] That put Saturn closest to the sphere of fixed stars and the Moon closest to Earth, just as in Plato's ordering. Aristotle further supported the placement of the Moon by noting that the Moon had been observed to **occult** the planets, meaning that the Moon sometimes blocked our view of the planets. For example, the Moon occults the Sun during a solar eclipse but Aristotle also mentioned the Moon passing "beneath" Mars (meaning between Mars and Earth).[22] However, Aristotle left open the question of the order of the Sun, Mercury, and Venus.[i]

---

[i] In the Middle Ages many scholars thought that Aristotle had agreed with Plato's ordering (Moon, Sun, Venus, Mercury, Mars, etc.). That was because this ordering is given in *De Mundo*, an ancient Greek treatise written by an unknown author who called himself Aristotle (sometimes referred to as Pseudo-Aristotle).

Aristotle believed that the universe was eternal: it was not created at some finite time in the past.[23] Although the temporal duration of the universe was infinite, its spatial size was finite. For Aristotle the sphere of the fixed stars was the ultimate boundary of the universe. Not only was there no *thing* outside the celestial sphere, there was not even any empty space outside that sphere. The outer boundary of the sphere of fixed stars was simply where the universe ended.[24] Aristotle believed that all motion in the universe was ultimately derived from the daily rotation of the celestial sphere, and the spinning of that sphere was caused by a Prime Mover. As we will see, Aristotle thought that all motions were caused by other motions, but the Prime Mover was the one exception: it was a source of motion, but it was itself unmoved.[25]

What we have described so far of Aristotle's cosmology may not seem to offer much that goes beyond Plato and Eudoxus, but the most powerful aspect of the Aristotelian cosmos was how it was tightly integrated with Aristotle's physics. We will turn now to that physics, which dominated Western thinking on the subject for more than 1500 years.

4.2.3   Aristotle's physics

Aristotle's physics was based on the ideas of purpose and place. Aristotle believed that all objects had a natural tendency to be in their correct place in the universe. By occupying its correct place an object fulfilled its fundamental purpose. Objects could be moved out of their correct place but they would seek to return to their proper place when they could. For Aristotle, the whole cosmos was a living organism and each part of the cosmos, even something that we would today consider an inanimate object, was imbued with a sense of purpose.[26]

Not all objects behave in the same way, though, so different objects must have a different purpose and a different proper place within the cosmos. These differences were closely tied to Aristotle's views on the elements. Aristotle adopted Plato's four elements (earth, water, air, and fire), but he rejected Plato's notion of atoms in the shape of regular solids. He also rejected the idea that celestial objects were composed mostly of the element of fire. Instead, Aristotle believed that celestial bodies were composed of a fifth element, often referred to as the **ether** (or sometimes as *quintessence*, which just means fifth element). Everything in the heavens was composed of this one, single element. Aristotle believed that the heavens were immutable, experiencing neither growth nor decay. That unchanging nature of the heavens made sense if they were composed of just one element that could not mix or combine with other elements.

In contrast, the region below the Moon (the **sublunary** sphere) was composed of earth, water, air, and fire. Each element had its own proper place within the sublunary region. Earth would naturally occupy a sphere that surrounded the center of the universe. Water would tend to occupy the space just outside the sphere of earth. Air would naturally form a sphere around the region of water. Finally, the sphere of fire stretched upward from the sphere of water to the inner surface of the Moon's sphere.

This general structure for the sublunary region mimicked the homocentric spheres of the heavens. The sphere of fire lay inside the Moon's sphere, the sphere of air was inside the sphere of fire, water inside air, and earth at the very center. It is important to understand, though, that there was a fundamental distinction between the heavens composed of a single pure ether and the sublunary region. The sublunary elements were allowed to mix with each other and it was the mixing and separation of these elements that produced all growth and decay. Likewise, the mixing of the elements prevented them from perfectly separating into their natural spheres, so sometimes water might be found below a stone or fire might be produced below the air. The elements would strive to move into their proper place (the stone would sink through the water or the fire would rise up in the air), but the constant change in the sublunary region prevented everything from settling down into a fixed state.

The distinction between the sublunary region and the heavens becomes even clearer when we look at how Aristotle explained motion. First of all, Aristotle identified two distinct kinds of motion: natural and violent. We will consider **natural motions** first. Natural motions were simply the fulfillment of an object's purpose. Every simple object (composed of only a single element) could have only one natural motion. For example, the natural motion of the ether was to circle the center of the universe at a constant speed. This uniform circular motion allowed celestial bodies to move, but to do so in an unchanging, repetitive pattern that would last for eternity.

The natural motion of the elements in the sublunary region was different. The sublunary elements sought their proper place in the universe with reference to the center of the universe, which served as the "bottom" of the entire structure. Heavy elements had a natural tendency to fall toward the bottom (center), while light elements had a tendency to rise upward. The heaviest element was earth, so earth would naturally tend to fall toward the bottom and accumulate in a sphere around the center of the universe. Water, another heavy element, also had a tendency to move downward to the bottom, but its tendency was not as strong as that of earth so it would tend to occupy the region around the sphere of earth. On the other hand, air and fire were considered light elements with a natural tendency to move upward, away from the center of the universe (with fire having the stronger tendency to move away). The outer surface of the sphere of fire served as the top of the sublunary region and it coincided with the inner surface of the Moon's sphere, which served as the bottom of the heavens.

Once an element reached its proper place it would naturally come to rest, staying exactly where it was supposed to be unless something else came along to move it out of place. However, the mixing of the sublunary elements prevented everything from settling into its assigned space, so the sublunary region did not have the perfect, spherical structure that the heavens had. Within the sublunary region individual objects might be composed of a mixture of different elements, and these objects would tend to fall or rise depending on their proportion of heavy or light elements.

The natural motions of the sublunary elements not only explained what was seen in daily life (rocks fall, flames rise), it also explained why, according to Aristotle, the spherical Earth lay stationary at the center of the celestial sphere. The Earth was simply a collection of the heavy elements: mostly it was a big chunk of the element earth, covered by a layer of water. Those heavy elements naturally collected at the bottom of the universe, so of course that was where the Earth had to be located. The motion toward the center would naturally tend to distribute the heavy elements evenly around that center, thus giving the Earth its spherical shape. Once those elements had collected near the center of the universe they would have reached their proper place and would no longer move at all. The immobility of the Earth was also important as a contrast to the eternal circular motions of the heavens.

Note that Aristotle's notion of weight was very different from our modern understanding. He thought of weight as a natural tendency of an object to seek the center of the universe. The more heavy elements there were in the object, the stronger that tendency would be. Therefore, Aristotle believed that heavier objects would fall faster, with the speed of fall proportional to the object's weight. Aristotle did acknowledge that heavy bodies seemed to speed up as they fell, which implies that the weight of the object increased as it got closer to the center of the universe.[27] In essence, the object would strive harder to reach its goal as it got closer to that goal.

So much for natural motions. In the sublunary regions, where the mixing of the elements led to constant change, objects could also be subject to what Aristotle considered unnatural motions, also known as **violent motions**.[28] Violent motions could only be caused by a mover, a force from some already moving object. In addition, violent motions were subject to resistance. The resistance depended on the weight of the object being moved, but also on the "thickness" of the medium, or substance, through which it moved.[i] Mathematically, Aristotle claimed that the speed of a violent motion was proportional to the force causing

---

[i] The notion of "thickness" is similar to our modern concept of density, or mass per unit volume. However, in some ways this concept is more like the modern notion of viscosity.

the motion, and inversely proportional to both the weight of the moving object and the thickness of the medium.[29]

There are several important consequences that arise from Aristotle's theory of violent motions. According to Aristotle, nothing could move (except in its natural motion) without something else pushing on it. Therefore, the Earth did not move because there was nothing pushing on it to make it move. If something did cause the Earth to move we would know it right away because we would feel the Earth move beneath our feet. Whatever was pushing the Earth would not also push us, and so the Earth would move out from beneath us. In particular, if the Earth rotated eastward so as to produce the apparent daily westward motion of the stars then we would all fall over toward the west, there would be a constant hurricane-force wind toward the west, and objects thrown straight up into the air would fall far to the west of their launch point. Since none of these things happened, the Earth must not be rotating or moving in any other way. In Aristotle's cosmos the Earth was *supposed* to be at rest at the center of the universe, but observations also seemed to show that it *was* in fact at rest there.

Another important implication of Aristotle's theory of violent motion was that there could be no empty space in the universe. Recall that the speed of a violent motion was inversely proportional to the thickness of the medium through which the object moves. If an object were in a vacuum, with zero thickness, then even the slightest nudge would cause the object to move infinitely fast. An infinite speed makes no sense in a finite universe, so Aristotle felt the need to exclude this possibility. Thus, he declared that no vacuum could exist. If matter was removed from some region it would immediately be replaced by matter from the surrounding regions. This notion is sometimes stated as "nature abhors a vacuum."

Aristotle's cosmological system was not without competitors in ancient Greece.[30] In particular, some Greek thinkers proposed systems that put the Earth into motion. Before Aristotle, the Pythagoreans of the 6th century BC suggested that Earth orbited around a "central fire." This central fire was not the Sun but was a body invisible to the inhabitants of Greece because their side of Earth always faced away from the fire. After Aristotle, in the 3rd century BC, Aristarchus suggested that Earth might rotate about the celestial axis and orbit around the Sun. Aristarchus' idea foreshadows the later theory of Copernicus, but as far as we know Aristarchus never developed his idea into a fully detailed astronomical system and the idea was not adopted by his contemporaries.

Aristotle's cosmology came to dominate ancient and medieval thought, in large part because it was so comprehensive and yet also coherent. Aristotle could provide at least qualitative explanations for a wide variety of observed phenomena, and all of the parts of his cosmological system fit together and supported each other. That made it hard for someone to question any piece of the Aristotelian cosmos, because altering one piece might mess up the whole thing. Even so, we will see that some of Aristotle's ideas were subject to criticism.

4.3   Heavenly circles and predictive astronomy

4.3.1   *Epicycles and eccentrics*

Although Aristotle's cosmology was in some sense built around the homocentric spheres of Eudoxus and Callippus, we have already seen that the model of homocentric spheres could not accurately account for the detailed motions of the planets. Later astronomers attempted to find other ways to explain the observed motions. Near the beginning of the 2nd century BC, Apollonius of Perga developed a theory based on combining two uniform circular motions.[31] The structure of his theory is shown in Figure 4.2. The planet moves uniformly along a small circle called an **epicycle**. At the same time, the center of the epicycle moves uniformly along a larger circle called a **deferent**. We aren't certain why ancient Greek astronomers felt that these circular motions had to be uniform, but that idea certainly fit with Aristotle's view of unchanging celestial motions.

Suppose the deferent and epicycle both lie in the plane of the ecliptic with the Earth at the center of the deferent. If the epicycle-center moves counter-

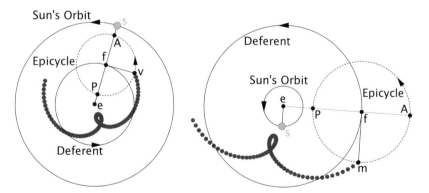

Figure 4.2 The deferent–epicycle model of Apollonius. The planet (v or m) moves uniformly along the smaller (dashed) epicycle circle, while the center f of the epicycle moves uniformly along the larger deferent circle which is centered on Earth e. The combination of the two motions can produce a general eastward motion along the zodiac with occasional westward retrogrades. The image on the left illustrates the geometry for an inferior planet (Venus), while the image on the right shows the arrangement for a superior planet (Mars).

clockwise as seen from above the north ecliptic pole, then observers on Earth will see the planet move generally eastward through the zodiac. If the planet also moves counterclockwise around the epicycle, then when the planet is at the innermost part of the epicycle observers on Earth will see it move westward, as illustrated in Figure 4.2.[i] Thus, the deferent–epicycle model can reproduce the general eastward, and occasional retrograde, motion of the planets.

One major difference between the deferent–epicycle model and the model of homocentric spheres is that in the deferent–epicycle model the planet's distance from Earth changes. When the planet is at point A on the epicycle in Figure 4.2 it is at its maximum distance from Earth. For that reason this point is called the **apogee**, meaning the highest point above Earth. Likewise, at P the planet is at its minimum distance and that point is called the **perigee**, the lowest point. The planet will be in the middle of its retrograde arc when it reaches perigee. Note that this model achieves a natural link between brightness/size and retrograde. The planet will be closest to Earth, and thus appear bigger/brighter, when it is in retrograde at perigee. This is just what we observe in the case of Mars, but we don't know if Apollonius or his immediate successors paid much attention to this fact.[32]

This model can also account for the connections between the motions of the planets and the motion of the Sun. We start by assuming that the Sun moves uniformly along a circle that is centered on the Earth. We know that inferior planets always appear near the Sun as seen from Earth. The deferent–epicycle model can reproduce this fact if the center of an inferior planet's epicycle always lies along the line from the Earth to the Sun (as in Figure 4.2, left). Likewise, we can guarantee that a superior planet will always retrograde when it is in opposition to the Sun if we make sure the line from the epicycle-center to the planet is always parallel to the Earth–Sun line (as in Figure 4.2, right). That way the planet will be at the perigee of the epicycle (and in the middle of its retrograde arc) exactly when the planet is on the opposite side of Earth from the Sun.

Apollonius likely thought of the deferent and epicycle circles as representing the equators of invisible spheres, much like the homocentric spheres of Eudoxus except that the epicycle's sphere was not centered on Earth. The motions of the planet were then carried out by the revolutions of these spheres. The deferent

---

[i] The westward retrograde motion also can be generated if the planet moves clockwise along the epicycle, but that model contradicts some details of planetary motion. In particular, the time for a planet to go from its mean speed relative to the stars to its minimum speed is always less than the time for it to go from mean speed to maximum speed. That is exactly what happens if the epicycle motion is in the same direction as the deferent motion, but it is the opposite of what happens if the epicycle moves the other way.

sphere carried the epicycle around the zodiac in the order of the signs, so the deferent sphere had to revolve once in the planet's tropical period. Since the epicycle-centers for the inferior planets were always lined up with the Sun, it made sense that the tropical periods of the Sun, Venus, and Mercury were all the same (see Table 3.1).

There was a subtlety in measuring the period of the epicycle's revolutions because while the epicycle sphere was revolving it was also carried around by the revolutions of the deferent sphere. In describing the period of the epicycle revolutions the ancient Greeks ignored the motion of the deferent, so the period of revolution for the epicycle is the time for the planet to go from one perigee to the next perigee (or one apogee to the next apogee). Because the planet was in the middle of retrograde at perigee, the period of the epicycle's revolution had to be equal to the synodic period of the planet. However, the orientation of the line from the epicycle-center to the planet was altered by the rotation of *both* spheres. On the other hand, if the period of the epicycle's rotation was measured relative to the fixed stars then the epicycle period for all superior planets would be one year.[i]

Apollonius' deferent–epicycle model can reproduce some of the qualitative features of planetary motions, and unlike the homocentric sphere model it still produces retrograde motion if the correct periods for Venus and Mars are used, but even this model cannot account for the planetary motions in an accurate, quantitative way.[33] That may not have been the goal of Apollonius at the start of the 2nd century BC, but later in the century Hipparchus sought to construct quantitative models that could accurately predict the positions of the planets within the zodiac.[34] Hipparchus' interest in predictive astronomy may have been inspired by an influx of astronomical data and calculation techniques from Babylonian astronomers.[35] We have already seen that Hipparchus compared his own measurements of the positions of the fixed stars to those of the Babylonians and, in the process, discovered the precession of the equinoxes. The Babylonians had developed sophisticated arithmetical procedures for accurately predicting the ecliptic longitudes of planets. Hipparchus tried to achieve the same accuracy using the geometrical models of the Greeks.

One area in which Hipparchus had particular success was in developing a model for the Sun's motion.[36] As discussed in Section 2.2.3, the Sun does not appear to move uniformly along the zodiac. That's why the seasons have different lengths. Hipparchus figured out a way to achieve the apparent irregularity

---

[i] There is a simple relationship between the synodic and tropical periods of a superior planet and the length of the tropical year: the reciprocal of the synodic period plus the reciprocal of the tropical period is equal to the reciprocal of the tropical year. You can check this relation yourself using the data in Table 3.1.

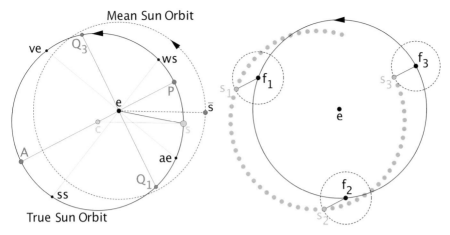

Figure 4.3 Eccentric solar orbits. The image on the left shows Hipparchus'
eccentric orbit for the Sun (with exaggerated eccentricity). The apogee A, perigee
P, solstice points (ss and ws), and equinox points (ve and ae), as determined by
Hipparchus, are shown. The image on the right shows how an equivalent orbit can
be constructed using a deferent and epicycle.

in the Sun's motion while still having the Sun move uniformly along a circle.
The trick was to locate the center of the Sun's circular motion somewhere other
than the Earth. This kind of circular orbit is known as an **eccentric** (off-center
circle) and it is shown on the left in Figure 4.3.

Just like the epicycle, the eccentric has an apogee (marked A in Figure 4.3) and
a perigee (marked P). The Sun moves along the circle at constant speed, so the
line from c to s rotates around at a uniform rate. That line indicates the direction
of the **mean Sun** ($\bar{s}$). In other words, it shows where the Sun would appear on the
celestial sphere if we could observe it from the center of its orbit. However, the
line from the Earth to the Sun (which shows the location of the **true Sun**) does
not rotate around at a uniform rate. As viewed from Earth it takes longer for the
Sun to move from the apogee A to the point Q_1 (known as the "first quadrant")
than it takes for the Sun to move from Q_1 to the perigee P, even though both of
these motions cover 90° as seen from Earth. In general, the true Sun will appear
to move slower when it is farther from Earth. So it seems slowest at apogee A
and fastest at perigee P.

We know that spring and summer are longer seasons and fall and winter
are shorter seasons. That's exactly what should happen in Hipparchus' model
if the true Sun is at summer solstice when it is near the apogee, and at winter
solstice near the perigee. Today summer is a bit longer than spring, and fall a bit
longer than winter, but in Hipparchus' day it was the other way around (spring
longer than summer, winter longer than fall).[37] Hipparchus determined that

the apogee of the Sun's eccentric lay a little bit west (clockwise) of the summer solstice. Hipparchus used the different lengths of the seasons to determine that the apogee lay about 25° west of the solstice. He found that the distance from the Earth to the center of the solar orbit, called the **eccentricity**, was about one-twenty-fourth of the radius of the orbit.[38]

Hipparchus knew from the work of Apollonius that the same type of motion could be produced using a deferent and epicycle.[39] The construction is shown on the right in Figure 4.3. The Sun moves uniformly along the epicycle, while the epicycle-center moves uniformly around a deferent centered on Earth. In this case the motion of the Sun along the epicycle is clockwise while the motion of the epicycle around the deferent is counterclockwise. The motions are synchronized such that the line from the epicycle-center to the Sun always points in the same direction. The end result is that the Sun's actual path is a circle whose center is shifted away from the Earth (the dotted circle in Figure 4.3, right). The Sun moves uniformly along this circle in perfect imitation of the eccentric model discussed above.

If these two models are exactly equivalent, which one should we use? Hipparchus preferred the epicycle model because he thought it important that the circles (and their associated transparent spheres) should share the same center as the celestial sphere.[40] In the epicycle model, at least one of the spheres is centered in that way. Hipparchus' successor, Ptolemy, preferred the eccentric model. He considered the eccentric model simpler because it involved only one motion, and he felt that this simplicity was an indicator of truth.

Hipparchus' solar model worked perfectly. It could be used to accurately predict the motion of the Sun along the ecliptic. Any errors in the model were smaller than the errors in observations of the Sun's position, so the observations were as likely to be at fault as the model. However, the planetary models still had problems. Armed with his wealth of Babylonian data and his own observations Hipparchus was aware that the simple deferent–epicycle model could not accurately account for the planetary motions, but he was unable to devise a better model. That goal was left for his famous successor, Claudius Ptolemy.

4.3.2   *The greatest: Ptolemy's Almagest*

The models of Apollonius and Hipparchus were synthesized and improved in one of the greatest astronomy books ever written. The book was written in the 2nd century AD by the astronomer Claudius Ptolemy. Its title can be translated as "mathematical treatise," but when it was later translated into Arabic it was called something like "the greatest treatise" and it has come to be known as the *Almagest*, a name derived from the Arabic phrase for "the greatest."

The *Almagest* provides a comprehensive account of how to model the apparent motions of the heavenly bodies, including how to determine the parameters of the models from observations. Ptolemy started from the general picture of the two-sphere model and provided evidence that Earth lay stationary at the center of the celestial sphere.[41] He embraced Hipparchus' eccentric circle model for the Sun's motion.[42] He also developed a more complicated theory of the Moon's motion.[43] We will focus our attention, though, on Ptolemy's theories for the planets.[44]

Ptolemy's starting point was the deferent–epicycle model of Apollonius,[45] but Ptolemy was seeking much more than a good qualitative explanation for retrograde motion. He wanted to construct models that would give accurate positions for the planets far into the future. In order to accomplish that goal he had to make several modification to the basic Apollonian model.

For one thing, Ptolemy needed to determine the exact parameters of the deferent–epicycle model for each planet. We have already seen that the periods of the two motions can be drawn directly from observations: the period of the deferent motion is just the tropical period of the planet, while the period of the epicycle (from apogee to perigee and back to apogee) is the planet's synodic period. In order to generate accurate predictions Ptolemy also had to get the sizes of the circles right.

It turns out that what really matters is the size of the epicycle relative to the deferent. If you double the size of both the deferent and epicycle it won't change any of the angles, as seen from Earth, so it has no practical effect on the theory. However, if you change the size of the epicycle without changing the deferent (or vice versa) then the predictions of the theory will change. What matters is the ratio: the radius of the epicycle divided by the radius of the deferent.

So how could Ptolemy find this ratio for each planet? The procedure is different for inferior and superior planets. For inferior planets the method hinges on finding the maximum elongation of the planet.[46] When the planet is at its greatest angle from the Sun, as seen from Earth, the geometry must be like that shown in Figure 4.4. In that figure there is a right triangle in which one side is the radius of the epicycle (r_E), while the hypotenuse (the longest side) is the radius of the deferent (r_D). If we know the maximum elongation angle (ϵ_{max} in Figure 4.4) we can use trigonometry to determine the ratio r_E/r_D. The details are shown in Appendix A.7.

The procedure is more complicated for superior planets. It involves observing the planet at opposition and again at quadrature (90° from the Sun), as shown in the two diagrams of Figure 4.5. In the diagram on the right side of Figure 4.5 we see that there is a right triangle with r_E as one side and r_D as the hypotenuse. Unfortunately, it is not possible to directly observe the angle θ in that diagram

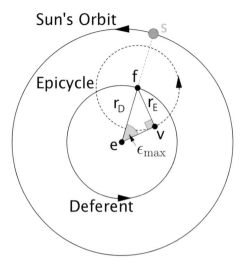

Figure 4.4 Finding the relative sizes of the epicycle and deferent for an inferior planet (Venus). The diagram shows the planet v at maximum elongation ϵ_{max} from the Sun s. The angle at the planet is a right angle. The epicycle radius is r_E and the deferent radius is r_D.

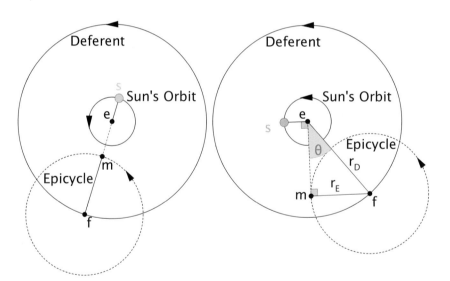

Figure 4.5 Finding the relative sizes of the epicycle and deferent for a superior planet (Mars). The diagram on the left shows the planet m at opposition. The diagram on the right shows the planet at quadrature (so the angle between the Sun s and planet m is 90° as seen from Earth). The line segment $\bar{m}f$ must be parallel to the line segment $\bar{s}e$ because of the link between the epicycle motion and the Sun's motion. The epicycle radius is r_E and the deferent radius is r_D.

because there is no visible object at point f. The center of the epicycle is just an empty point in space. However, because the epicycle motion for a superior planet is linked to the Sun's motion we can calculate the angle θ if we have a measurement of the *time* it took for the planet to move from opposition to quadrature. Once we have θ we can use trigonometry to find the ratio r_E/r_D. The details of the calculation are given in Appendix A.8.

With the periods and relative sizes of the deferent and epicycle circles it seems like Ptolemy's theory would be all set, but it turns out that this basic theory doesn't do a very good job of matching the observed motions of the planets. One problem is that the model we have described so far predicts that the retrograde arcs of each planet (the segments of the planet's motion in which it moves westward, contrary to the order of the signs) will be spaced evenly along the zodiac. It also predicts that these arcs will all be the same size. Neither of these predictions is correct. For example, the retrograde arcs of Mars tend to be spaced close together in one part of the zodiac and farther apart on the opposite part of the zodiac. In the region where the arcs are spaced close together they also tend to be larger, while they are smaller when they are spaced far apart. It is not possible to reproduce this behavior with the simple deferent–epicycle model described so far.[47]

One way to improve the model is to make use of Hipparchus' idea of an eccentric: the center of the deferent circle could be shifted away from the Earth.[48] This modification can help to solve half of the problem. Shifting the center of the deferent away from the Earth means that some of the retrograde loops will be closer to Earth and others farther away. The loops that are closer to Earth will appear larger while those farther away will appear smaller. Unfortunately, the larger loops will also appear to be farther apart. Basically, *all* of the angles get bigger for the retrogrades that are closer to Earth: the size of the retrograde arc gets bigger but so does the angle between one arc and the next. This effect does not fit with what we observe, so placing the deferent off-center from Earth is not enough.

The solution that Ptolemy devised for this problem was to disconnect the center of uniform motion from the geometrical center of the deferent. He proposed that the center of the epicycle would move along the deferent in such a way that its motion would *appear* uniform as seen from a special point that was not at the deferent's center.[49] This point is called the **equant** point because an observer at that point would see the epicycle center move through equal angles in equal times. Ptolemy found that he could match the observed motions of the planets if he shifted the center of the deferent circle away from Earth and then placed the equant point twice as far away in the same direction, so that the center of the deferent lies at the midpoint between the Earth and the equant point.[50]

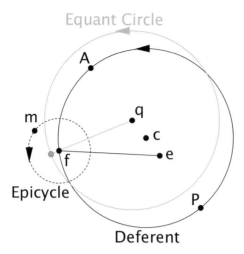

Equant Circle

Epicycle

Deferent

Figure 4.6 Ptolemy's equant. The center of the epicycle f appears to move uniformly as seen from the equant point q. Therefore, it must move faster at perigee P and slower at apogee A. The planet (m) moves uniformly around the epicycle with respect to the line $\bar{qf}$.

Placing the deferent center halfway between the Earth and the equant is known as "bisecting the eccentricity."

The geometry of Ptolemy's model is shown in Figure 4.6. Recall from our discussion of Hipparchus' eccentric that, if an object moves uniformly along a circle, to an observer located off-center the object will appear to move slower when it is farther away and faster when it is closer. However, in Figure 4.6 an observer at the equant point (q) is supposed to see the point f move around the circle in a uniform way. In order to make this work the point f must actually change its speed as it moves around the circle. In particular, it must move faster when it is far from the equant point and slower when it is close to the equant point. Meanwhile the planet itself moves uniformly around the epicycle.[i]

There are two important things to note about Ptolemy's equant. First, it represented a major break with tradition. From the time of Eudoxus on, ancient Greek astronomers had attempted to describe the celestial motions using combinations of uniform circular motions. Ptolemy was the first to introduce nonuniform motion into astronomy, even if he could argue that the motion would appear uniform from the special equant point.[51] Second, because the equant point is on

---

[i] Technically the planet moves uniformly on the epicycle with respect to the line $\bar{qf}$. We could also say that it moves uniformly with respect to the stars, since $\bar{qf}$ rotates uniformly with respect to the stars. However, the planet's motion on the epicycle is *not* uniform with respect to the line $\bar{cf}$ or the line $\bar{ef}$ since each of these lines moves nonuniformly with respect to the stars.

the opposite side of the center from the Earth, when f is farther from the equant it is closer to Earth and vice versa. So we also can say that the point f moves faster when it is close to Earth and slower when it is far from Earth.

The addition of the eccentric and equant to the basic deferent–epicycle theory, as well as the use of Hipparchus' eccentric orbit for the Sun, changes the way the motion of the planets is linked to the Sun.[52] Ptolemy recognized that he could not link a nonuniform motion to a uniform motion. In the basic deferent–epicycle model the motion of a superior planet on its epicycle was synchronized to the motion of the Sun, but Ptolemy had to synchronize that motion to the mean Sun, not the true Sun, because the true Sun appears to speed up and slow down. For the inferior planets the motion of the epicycle center on the deferent was linked to the motion of the Sun, but in the *Almagest* *both* of these motions are nonuniform. Ptolemy solved this problem by linking the uniform motion of the epicycle center as seen from the equant point (i.e. the line $\bar{q}f$ in Figure 4.6) to the uniform motion of the mean Sun.[i]

So far we have been talking about the motions of the planets as though the deferent and epicycle circles lie in the same plane as the Sun's orbit, as in Figures 4.4 and 4.5. However, observations show that we sometimes see planets above or below the ecliptic. One easy way to accommodate these deviations from the ecliptic is to tilt one or both of the circles out of the ecliptic plane. That is exactly what Ptolemy did.[53] For the superior planets he tilted the deferent at a fixed angle such that the plane of the deferent passed through Earth. He also tilted the superior planet epicycles, but he found that he had to make the epicycle wobble back and forth instead of maintaining a fixed tilt angle. For inferior planets the deferent was tilted so that its plane passed through the Earth, but the tilt angle varied, giving the deferent a wobble like that of the superior planet's epicycle. The epicycles of the inferior planets were even more complicated as they wobbled simultaneously in two different directions.

After discussing his theory of latitudes in the *Almagest*, Ptolemy noted that these motions might seem to be overly complicated. He then stated that, while we should try to find simple models to fit the observed motions, if our simple models don't fit the data then we need to find more complicated models that do fit.[54] He also noted that "simplicity" depends on context. If we were to build a mechanical version of Ptolemy's planetary model out of earthly materials it would be hard to make it work. On the other hand, the celestial spheres were

---

[i] For Mercury Ptolemy found that he had to add an extra complication: the center of Mercury's deferent moved around on a small circle with a period of one year. This extra motion brought Mercury closest to Earth at two different points, so Mercury had two perigees (but still only one apogee).

made of ether and what was difficult or impossible for sublunary elements might be accomplished with ease using celestial materials which had a fundamentally different nature. Human intuition for what counts as simple or complicated did not apply to the heavens.[55]

4.3.3   The Planetary Hypotheses

The *Almagest* was not the only book Ptolemy wrote. He also wrote books on optics and musical harmony, as well as an important book on geography in which he specified the latitude and longitude of every place he knew.[i] He also wrote a book called the *Tetrabiblos* (Four Books) in which he laid out a system for casting astrological horoscopes. Ptolemy's astrological work can be viewed as a practical application of his astronomical work. He believed that the positions of the Sun, Moon, and planets along the zodiac at the moment of a person's birth influenced, but did not necessarily determine, that person's characteristics and personality.[57] The calculation of the positions of the Sun, Moon, and planets (using the models of the *Almagest*) could allow the astrologer to explain and predict certain tendencies in human beings.

Ptolemy also wrote another important work on astronomy and cosmology. The *Planetary Hypotheses* was an effort to merge the mathematical models of the *Almagest* with the cosmology of Aristotle.[58] Working from the outside in, he started with a sphere that rotated once per sidereal day around the celestial poles. The motion of this outermost sphere was transmitted to the next sphere down: a sphere that carried the fixed stars and rotated 1° per century around the ecliptic poles, to match the model he had used in the *Almagest* to account for the precession of the equinoxes.

Inside the sphere of fixed stars lay the spheres of the planets. In the *Almagest*, Ptolemy pointed out that astronomers generally agreed about the order of the next three planets down: Saturn, then Jupiter, then Mars. But he stated that there was disagreement about the order of the Sun, Venus, and Mercury. Some said Venus and Mercury must be above the Sun because we never saw them pass below the Sun, unlike the Moon which passes below the Sun during a solar eclipse. Ptolemy noted that this was not a very strong argument because the planets might not be in the ecliptic plane when they crossed below the Sun.[59]

---

[i] Prior authors usually gave latitude in terms of the number of hours of daylight on the summer solstice, which ranges from 12 hours at the equator to 24 hours at the poles. Ptolemy was one of the first to specify latitude in terms of degrees from the equator. He also measured longitude in degrees but with his meridian running down the eastern side of the Atlantic Ocean and passing through a group of islands that were probably the Canary Islands.[56]

Ptolemy's preferred order in the *Almagest* was to put the Sun in the middle of the planets: the Moon, Mercury, and Venus lie below the Sun while Mars, Jupiter, and Saturn are above it. In that way the Sun serves as a natural dividing point between the two types of planets: inferior and superior.[60] In fact, it is from Ptolemy's ordering that we derive the names of these categories: inferior planets are *below* the Sun while superior planets are *above*. In the *Planetary Hypotheses* Ptolemy specified that Mercury lay between the Moon and Venus. He mentioned the erratic movements of the Moon and Mercury as evidence that these planets were influenced by the corrupting nature of the sublunary sphere, which suggested that they were the heavenly bodies closest to the Earth.[61]

It is important to note here that Ptolemy's ordering of the planets was based on Aristotle's physical principles. It was not based on astronomical observations. Ptolemy recognized this problem and admitted that his preferred ordering might be wrong, but he was forced to use non-astronomical arguments to determine the order of the planets because the models in the *Almagest* say nothing at all about the distances of the planets from Earth. Remember that the deferent and epicycle can be made larger or smaller as long as we scale both circles by the same factor. Ptolemy's models for the Sun, Moon, and planets were really separate models, one for each object. There was nothing in the mathematical models themselves to show how the different pieces fit together.

Fitting the pieces together was one of the main goals of Ptolemy's *Planetary Hypotheses*.[62] He started by adding flesh to the skeletal circles given in the *Almagest*. He considered each of those circles to be the equator of a heavenly sphere, or orb.[i] That sounds a bit like the homocentric spheres of Eudoxus, but because of his eccentrics and epicycles Ptolemy had to deal with spheres that were not centered on the Earth. The general structure that Ptolemy proposed for the heavenly spheres is illustrated in Figure 4.7 (although the diagram is taken from a much later source that mimics Ptolemy's proposal). Venus is the black star shape that lies on the equator of the epicycle sphere, depicted as a circle inside the white band labeled B. That band represents the deferent sphere, with its center at point C while the Earth lies at D and the equant point is at H.

Recall that empty space was not allowed in Aristotle's physics. In order to fill the space between the off-center deferent sphere and the properly centered spheres above and below it, Ptolemy proposed that the space in between was

---

[i] Ptolemy also suggested that sections of spheres, similar in shape to a tambourine, could be used. The *Planetary Hypotheses* were probably intended to guide the construction of physical models made of wood or metal, and it would be much easier to build and view those models if only a portion of the sphere was constructed.

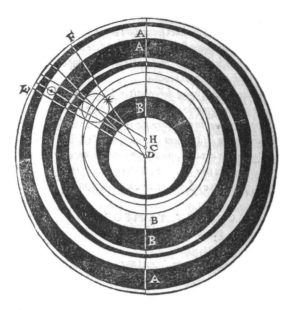

Figure 4.7 Total and partial spheres for Venus and the Sun, from Peurbach's *Theoricae novae planetarum* (1542). Venus is the black star shape on the small epicycle circle inside the white band labeled B. The Sun is represented by the ⊙ symbol inside the white band labeled A. Image courtesy History of Science Collections, University of Oklahoma Libraries.

occupied by filling spheres as shown by the black regions labeled B in Figure 4.7. The figure also shows the spheres for the Sun's eccentric orbit (labeled A).

Each of the eccentric and filling spheres was considered a "partial sphere" while the entire collection of spheres for a planet was the "total sphere" for that planet. These total spheres were nested inside each other just like the homocentric spheres of Eudoxus. If one only paid attention to the total spheres, the system looked just like that of Aristotle (except for the number of spheres involved), but in its details it was more like the models presented in the *Almagest*.

Although he tried to embody the mathematical models of the *Almagest* in the physical motions of spheres, Ptolemy was unable to find a physical mechanism for the equant. He noted that the eccentric sphere whose equator plays the role of the deferent must rotate uniformly, not with respect to its own center, but rather with respect to another point (the equant point). However, Ptolemy offered no suggestions about how that uniform motion could be brought about physically.[63]

We have seen that the astronomical models of the *Almagest* could not be used to find the distances to the planets, but the physical theory of total

Table 4.1 *Minimum and maximum distances to the planets from Ptolemy's* Planetary Hypotheses.

	Minimum distance (Earth radii)	Maximum distance (Earth radii)
Moon	33	64
Mercury	64	166
Venus	166	1,079
Sun	1,160	1,260
Mars	1,260	8,820
Jupiter	8,820	14,187
Saturn	14,187	19,865

and partial spheres gave Ptolemy a way to determine these distances. He determined the thickness of each total sphere that was required to accommodate the eccentric and epicycle from the *Almagest*. He then assumed that the outer surface of one total sphere must coincide with the inner surface of the next total sphere, since it was impossible for there to be any empty space between them. From parallax measurements he had a good estimate of the Moon's minimum distance, which would serve as the distance to the bottom surface of the Moon's total sphere. Working upward from there he determined the minimum and maximum distances for each planet, in units of the Earth's radius. Finally, he placed the sphere of fixed stars just outside Saturn's sphere.

Ptolemy's results are shown in Table 4.1.[64] One thing to note is that the maximum distance for the Sun agrees very well with the distance of 1273 Earth radii we obtained from Aristarchus' method in Section 3.3.1. That is no coincidence: Ptolemy's distances for the Sun were based on his eccentric model for the Sun and his own implementation of Aristarchus' method, not on the principle of nested spheres. But look at how close the maximum distance of Venus (1079 Earth radii) is to the minimum distance of the Sun (1160 Earth radii)! These values differ by only 7%. The slight difference could be fixed by tweaking some of the numbers, although Ptolemy didn't work out those details. The already close agreement of these values provided evidence for the validity of Ptolemy's distance to the Sun, his principle of nested spheres, and his chosen ordering of the planets.[65]

Based on these distances, and measurements of the apparent diameters of the planets and stars, Ptolemy provided estimates for the physical sizes of these objects.[66] He estimated that Mercury's diameter was only 1/27 that of Earth, while the Sun was five and half times Earth's size. His estimates for Jupiter,

Saturn, and the brightest fixed stars were all between four and five times the size of Earth.[i]

Ptolemy also provided a new theory for planetary latitudes in the *Planetary Hypotheses*.[67] After writing the *Almagest* he must have realized that his latitude theories were not only complicated but they were also inaccurate. The problem was that those theories were based on some bad observational data. Ptolemy must have obtained some better data, probably through making his own measurements, because his latitude theory in the *Planetary Hypotheses* was much better than the theory in the *Almagest*. For superior planets he gave the deferent circle a fixed tilt but had the epicycle remain always parallel to the ecliptic plane. For inferior planets he gave a fixed tilt to the epicycle, while the deferents were almost (but not quite) in the ecliptic plane.

The vastly improved latitude theories of the *Planetary Hypotheses* were mostly ignored by later astronomers.[68] From Ptolemy to Copernicus almost all latitude tables were based on the defective models in the *Almagest*. Practicing astronomers were more concerned with determining planetary longitudes, which were important for astrology, than latitudes. However, Ptolemy's merger of his mathematical models with Aristotle's cosmology became the dominant picture of the heavens. That's not to say, though, that it was not subject to criticism by later astronomers.

4.4   Astronomy and cosmology after Ptolemy

After Ptolemy, European astronomy went into a period of decline. Ptolemy's models were so successful that all astronomical work was based on the *Almagest* and other astronomical works were forgotten. The rise of Christianity in Europe led to the neglect of mathematical astronomy, and of the "pagan" knowledge of the ancient Greeks and Romans generally. At the same time, though, astronomy flourished in the Arabic world. Many Greek astronomical and philosophical works, including the *Almagest* and *Planetary Hypotheses*, were translated into Arabic. From the 9th to 14th centuries, many new Arabic astronomical texts were produced. In some cases these texts were presentations of Ptolemaic astronomy with updated values and minor refinements, but in other cases these works criticized Ptolemaic astronomy and suggested significant additions or improvements.[69]

During much of this period the only astronomical knowledge that was widely available in Europe came from medieval encyclopedias. These books were highly

---

[i] We now know that the apparent diameters measured by Ptolemy were incorrect. In fact, the problem of the apparent size of stars was not resolved until long after Newton's death. We will see why in Section 10.3.2.

condensed summaries of ancient knowledge. The astronomical information in these books was usually of low quality because the authors of the encyclopedias were not trained in astronomy.[70] By the 12th century Latin translations of Arabic astronomical texts filtered into Europe. By the late 13th century Latin translations were made from original Greek works, including important works by Aristotle.[71]

With the availability of Latin translations of Aristotle's works came the opportunity to merge Aristotelian thought with Christianity. Commentaries on Aristotle by scholars such as Albertus Magnus and Thomas Aquinas helped to blend Aristotelian philosophy and cosmology with Christian theology to such a great extent that they came to seem inseparable. Latin translations of the *Almagest* helped to familiarize European astronomers with Ptolemaic astronomy and its Arabic modifications. By the 13th century Aristotelian thought became a standard part of the European university curriculum.[72] Likewise, students in European universities learned astronomy from such works as the *Tractatus de Sphaera* of Johannes de Sacrobosco,[73] which covered the two-sphere model described in Chapter 2, and the *Theorica planetarum*, an anonymous work that covered the theory of planetary motions.

By the end of the 15th century, European astronomers had mastered Ptolemaic astronomy. For example, Georg Peurbach lectured on Ptolemaic planetary theory at the University of Vienna. His lectures were eventually published, by Peurbach's student Johannes Müller (also known as Regiomontanus), as *Theoricae novae planetarum* in 1472.[74] Regiomontanus went on to write *Epytoma in almagestu Ptolemei* (Epitome of Ptolemy's *Almagest*), an abridged version of Ptolemy's great work with commentary. The revival of European astronomy allowed writers of Latin to continue the tradition of commentary and criticism that was initiated by writers of Arabic. It is worth taking a look at some of the specific criticisms that were leveled against Ptolemy and Aristotle.

4.4.1   Criticisms of Ptolemy

Medieval scholars criticized Ptolemy most strongly for his failure to follow Aristotle. It was not hard for astronomers to recognize that the alliance between Ptolemy's astronomy and Aristotle's cosmology was an uneasy one. Aristotle proclaimed that the natural motion of the ethereal spheres was a uniform rotation around the center of the universe, which was also the Earth's center. Ptolemy's models included eccentrics and epicycles that were not centered on Earth, as well as equants that produced nonuniform motion.

Critics of the equant included Ibn al-Haytham in the 11th century and Maimonides in the 12th century.[75] Other Arabic writers went further. In the 12th century Ibn Rushd (later called Averroes), proposed a return to the system of homocentric spheres like those of Eudoxus. His contemporary al-Bitruji developed a detailed proposal to account for the motions of the planets using homocentric spheres, but al-Bitruji's models were unable to match the accuracy of the Ptolemaic models.[76] Later, European astronomers took up the cause of homocentric spheres. In the 1530s, treatises on homocentric astronomy were published by the Italians Girolamo Fracastoro and Giovanni Battista Amico,[77] both of whom had studied at the University of Padua.

Other astronomers criticized Ptolemy's failure to match with observations. We have seen that there were problems with Ptolemy's latitude theory in the *Almagest*, but that was not a major source of concern. Ptolemy's longitude theory was generally accurate, such that any discrepancy between Ptolemy's predictions and the observed longitudes of the planets could be attributed to observational error. Larger errors could often be corrected by finding better values for the numerical parameters that were used in the models. In fact, for over a thousand years tables of the planetary positions, such as the *Alfonsine Tables* produced in Spain in the 13th century, were based mostly on the models in the *Almagest*.

However, if the Ptolemaic models really represented what was happening in the heavens they could be used to predict things other than the positions of the planets among the stars. For example, Ptolemy's lunar theory indicated that the distance between the Earth and the Moon changed by about a factor of two during the Moon's motion. That change in distance should cause the Moon's apparent diameter to change by a factor of two and the apparent area of the lunar disk to change by a factor of four. In fact, the Moon's apparent diameter changes by only 14%.

A similar problem seemed to plague Ptolemy's models for the planets. The Moon shines by reflecting sunlight and therefore it goes through a sequence of phases as described in Section 3.1.2. The planets, on the other hand, always appear to the naked eye as tiny, full circles. To explain this fact some writers assumed that the planets (but not the Moon) produced their own light. Others assumed that the planets received light from the Sun, but that the planets were transparent so that they took in the light of the Sun and re-emitted it. Some even suggested that the Moon was partially transparent, which explained why the portion of the Moon facing away from the Sun was sometimes dimly visible. Regardless of the source of the planets' light, there were problems with Ptolemy's theory.[78]

Ptolemy's theory for Venus predicted that the distance from Earth to Venus would change by a factor of more than six, which would produce dramatic changes in its size and brightness, but as we have seen, the appearance of Venus does not change very much. Mars does exhibit noticeable changes in size/brightness, but its apparent diameter seems to change by a factor of about two while Ptolemy's theory predicted that it would change by a factor of five or more.

Some writers after Ptolemy discussed the changing brightness/size of the planets, but in many cases their comments seem to assume the correctness of the Ptolemaic system while ignoring the actual appearance of the planets. For example, Simplicius cited the variation in the appearance of the planets as the primary reason why the Eudoxan homocentric spheres were abandoned in favor of the deferent–epicycle model. As evidence he claims that Venus is seen to be many times larger when it is in the middle of its retrograde arc. This claim is clearly not based on observation because Venus is in conjunction with the Sun, and thus not visible at all, when it is in the middle of retrograde. Simplicius' claim is based on Ptolemaic theory, not on actual measurements.[79] The first person to investigate this problem through careful measurement may have been Levi ben Gerson in the 14th century. His observations showed that the variations in the apparent size of both Venus and Mars were much less than predicted from Ptolemy's theories.[80]

Another area of criticism was the ordering of the planets. Astronomers noted that Ptolemy's models were unable to set a definite order. Nobody doubted that the Moon was closest to Earth, or that from the sphere of the stars inward the order was Saturn, Jupiter, and Mars. But many astronomers noted the uncertainty about the order of the Sun, Mercury, and Venus. There are six possible orderings for these planets and most of them were considered by astronomers in this period. In fact, some writers even proposed that the order might be variable, with Mercury and Venus orbiting around the Sun while the Sun orbits the Earth. Although this idea was proposed by some ancient writers, it was introduced to Latin Europe by Martianus Capella in his 5th-century encyclopedia *De nuptiis Philologiae et Mercurii* (On the Marriage of Philology and Mercury) and later supported by Jean Buridan in the 14th century.[81]

Another oddity, although not an error, in the Ptolemaic theory was the way the motions of the planets (and the Moon) were linked to the Sun. Ptolemy himself noted, in the *Planetary Hypotheses*, that the Sun was stronger than the other celestial bodies, exerting its influence on the planets without being influenced in return. That was why the Sun had such a simple orbit (an eccentric, but no equant or epicycle) while the other planets had more complicated motions, some of which were linked to the Sun.[82]

The link between the motions of the planets and that of the Sun was necessary, in Ptolemy's models, in order to match with observations. Superior planets always retrograde in opposition and inferior planets are always found near the Sun. The only way to achieve these effects with the deferent–epicycle model was to link the deferent motion of inferior planets, and the epicycle motion of superior planets, to the motion of the (mean) Sun. However, Ptolemy's models could not suggest *why* the motions would be linked in that way. As the 15th-century astronomer Georg Peurbach put it, in his *Theoricae Novae Planetarum*, "it is evident that the six[i] planets share something with the sun in their motions and that the motion of the sun is like some common mirror and rule of measurement to their motions."[83] The link between the planets and the Sun was not a natural consequence of the deferent–epicycle model, but rather something that had to be built in. What, then, was the underlying reason for this mysterious link?

4.4.2   Criticism of Aristotle

Ptolemy was not the only ancient authority who was subjected to medieval criticism. Although Aristotle's grand cosmological system was not seriously challenged during the Middle Ages, some of his specific ideas about physics and cosmology were questioned. One notable example is Aristotle's theory of projectile motion. We have seen that Aristotle believed that motion was only possible if one moving object exerted a force to cause motion in another object. How, then, did he explain the continued motion of a thrown object after it left the hand of the thrower? Clearly the hand could no longer exert a force on the projectile, since it was no longer in contact with it. Aristotle argued that the projectile would continue to move because its movement would create an empty space where the projectile had been the previous moment. Since Aristotle thought empty space was impossible, the surrounding medium had to rush in to fill that space instantaneously. The motion of the medium could then exert a force on the projectile and keep the projectile in motion.

Several later writers were critical of this explanation for projectile motion. In the 6th century John Philoponus argued that the hand must impart to the projectile a motive power that stayed within the projectile and kept it moving, at least for a while. This motive power came to be known as **impetus**.[ii] It was endorsed by the influential philosopher Ibn Sina (or Avicenna) in the 11th century, and in the 12th century it was used by al-Bitruji to explain the continued motion of the celestial spheres. The idea was further elaborated by Jean Buridan in the 14th century. Buridan taught that the impetus of a projectile was lost because of the resistance of the medium through which it travelled – exactly the opposite of Aristotle's idea that the medium is what sustains the motion. The impetus idea would ultimately give way, in the 17th century, to an entirely new way of thinking about motion.[84]

---

[i] Peurbach counted the Moon as a planet influenced by the Sun.
[ii] Arabic natural philosophers called it *mail*.

Not all objections to Aristotle were on physical grounds, however. Theologians also criticized Aristotelian ideas, particularly the infinite age of the universe. The notion of the eternal universe directly contradicted the creation account of Christian scripture, and scholars such as Aquinas discarded that idea, claiming that it was not an essential part of Aristotle's philosophy and cosmology.

Most of Aristotle's ideas could be merged with Christian theology without much difficulty, and men like Aquinas helped to establish a widely accepted view of the cosmos that blended Aristotle and the Bible. Some medieval theologians, however, went to great length to assert the predominance of the Bible over Aristotle. In 1277 Étienne Tempier, Bishop of Paris, issued a condemnation of 219 propositions drawn from Aristotle and other non-Christian philosophers. These condemnations did not necessarily deny the correctness of Aristotle's physics and cosmology, but they did insist that God's creative power was not limited by Aristotelian rules. Aristotle's physics and cosmology might happen to be correct, but God could have created a world with a different structure and a different set of rules if He had wished to do so.[85]

The Condemnations of 1277 and other religious critiques of Aristotle opened up space for medieval scholars to think outside of the Aristotelian box and cast doubt on many things that had been considered certain in philosophy. This doubt led some medieval scholars to embrace a form of empiricism, in which all knowledge was derived from sense experience. In the 14th century William of Ockham insisted that scientific theories should be derived from experience and that a simpler theory should be preferred over a more complicated one, but the infinite power of God meant that humans could never be sure about the true causes of phenomena. God could choose to do things in a complicated, rather than simple, way.[86]

These philosophical doubts led scholars to consider alternative explanations for astronomical phenomena. In the 14th century, Nicole Oresme, a scholar at the University of Paris, claimed that there is no way to experimentally determine whether it is the celestial sphere that rotates, or the Earth that rotates (although he eventually sided with the traditional view of a stationary Earth).[87] In the 15th century, Nicholas Cusanus argued that the cosmos was not finite and therefore had no center.[88] The ideas of an infinite universe and a rotating Earth were not widely adopted by their contemporaries, but the writings of Oresme and Cusanus helped to make these ideas worthy of consideration.

4.4.3   New additions

Medieval astronomers didn't just criticize ancient Greek authorities, they also made new contributions of their own. Several Arabic astronomical works introduced new observations and mathematical models that, although they did not fundamentally alter the Aristotelian–Ptolemaic cosmos, had a lasting impact on astronomy.

In the 9th century, astronomers working in the caliphate of al-Ma'mun carefully measured the obliquity of the ecliptic and found it to be 23° 33', noticeably less then Ptolemy's value of 23° 51' 20''. They concluded that the obliquity was gradually decreasing over time.[i] These astronomers also investigated the rate of precession by comparing their own measurements of star positions to those of Ptolemy. They found a rate of 1° in 65 years, noticeably faster than the rate Ptolemy had calculated by comparing his measurements to those of Hipparchus.[89]

The apparent changes in the obliquity and rate of precession were incorporated into a new theory, usually attributed to Thābit ibn Qurra, which came to be known as **trepidation**. In the theory of trepidation the ecliptic wobbled around on the celestial sphere in such a way that both the obliquity and the precession rate went through cyclic changes, with the precession even reversing its direction periodically.[ii] The period of the wobble was just over 4000 years.[90]

Although trepidation was not embraced by many Arabic astronomers, it was adopted by astronomers in Islamic Spain. The theory of trepidation was incorporated into influential astronomical tables such as the *Toledan Tables* of the 11th century.[91] Trepidation was then passed along to Latin writers in Europe. New measurements indicated that precession had advanced farther than the back-and-forth motion of trepidation could account for, so astronomers added a steady precession onto Thābit's trepidation model. This combination of steady precession and trepidation ensured that the stars always drifted eastward through the signs of the zodiac, but with a speed that increased and decreased. In the 13th century this "precession plus trepidation" theory was built into the *Alfonsine Tables*,[iii] which became the standard astronomical tables for European astronomers until the mid-16th century.[92]

In Aristotelian physics every simple body has a single simple motion, so to account for motions like precession and trepidation astronomers had to add new spheres in the heavens. Although the exact number of celestial spheres was often debated, by the late Middle Ages many astronomers had settled on a system of ten spheres.[93] Each of the seven "planets" was assigned a single

---

[i] Although the obliquity is decreasing, it is not decreasing as rapidly as they thought. Ptolemy's value was too great, which made the change appear larger than it actually was.

[ii] Thābit got the idea of an oscillating precession from the writings of Theon of Alexandria, a 4th-century astronomer who wrote commentaries on Ptolemaic astronomy. Theon mentioned the idea of an oscillating precession and attributed it to some ancient astrologers, but he chose to stick with Ptolemy's uniform precession in his own work.

[iii] Except for the addition of trepidation, the planetary models used to construct the *Alfonsine Tables* are standard Ptolemaic models, with updated parameter values.

sphere, which included within it all of the partial spheres needed to account for the planet's motion. At the center of the cosmos lay the partially mixed spheres of earth, water, air, and fire. Working outward from this sublunary region there were celestial spheres for the Moon, Mercury, Venus, the Sun, Mars, Jupiter, and Saturn. Outside the sphere of Saturn lay the eighth sphere, the sphere of the fixed stars (sometimes called the "firmament"), which moved with the motion of trepidation. The ninth sphere, often called the "crystalline sphere" because it was completely transparent, accounted for the steady precession. Finally, the tenth sphere was the *primum mobile*, the "first moved" sphere, which performed the daily rotation about the celestial axis.

By the 12th century, Christian writers such as Anselm of Laon and Peter Lombard had added an eleventh celestial sphere: an immobile, invisible, theological heaven.[94] The eleven sphere system is illustrated in Figure 4.8, a depiction of the cosmos from Peter Apian's 16th-century *Cosmographicus liber*. The words around the outer edge indicate that the eleventh sphere represents "the home of God and all the chosen." Note the crossed circles next to the western (clockwise)

Figure 4.8 The eleven spheres of the cosmos, as illustrated in Peter Apian's *Cosmographicus liber* (1524). Image courtesy History of Science Collections, University of Oklahoma Libraries.

edge of Aries (♈) and Libra (♎) in the ninth (crystalline) sphere. These circles represent the mechanism that produces the trepidation motion in the eighth sphere (the firmament of the fixed stars). That is why there is a slight offset between the constellations in the eighth sphere and those in the ninth sphere (for example, the western edge of Cancer, ♋, in the eighth sphere is slightly east of the same spot in the ninth sphere). There is an even bigger difference between the positions of the constellations in the eighth and ninth spheres and the locations of the corresponding signs in the tenth sphere, because of the cumulative effects of steady precession. As a result, the vernal equinox, at the western edge of the sign of Aries (♈) in the tenth sphere, lies far into the constellation of Pisces (♓) in the eighth sphere.

Trepidation was not the only new addition provided by Arabic astronomers. Concerns about the equant led to the development of new mathematical devices for producing nonuniform motion out of uniform circular motion. In the 13th century, al-Ṭūsī showed that a back-and-forth motion could be produced by combining two uniform circular motions, a mechanism now called a "Ṭūsī couple."[95] In the 14th century, al-Shāṭir showed that the same effect produced by the equant could be generated by using a small epicycle (sometimes called an "epicyclet") instead.[96] He also proposed a model for the Moon that used two epicycles. The lunar theory of al-Shāṭir greatly reduced the problem of the variations in the Moon's apparent size that had plagued Ptolemy's theory.[97] The innovations of al-Ṭūsī and al-Shāṭir did not improve the calculation of planetary positions, so they were not incorporated into practical works like the *Alfonsine Tables*. However, these mathematical devices would play an important role in the work of Copernicus.

Of course, astronomy was not the only area in which new knowledge was generated that went beyond what was known by the ancient Greeks. Ptolemy's *Geography* was overturned by the 1492 voyage of Christopher Columbus to the New World. The discovery of new lands by European navigators may have given European scholars the confidence to challenge ancient authorities and even propose fundamentally new ideas to replace the old traditions. The discovery of the New World was an indisputable piece of new knowledge for Europeans, and it opened the door to the possibility of further discovery.[98]

4.5   Reflections on science (and history of science)

In this chapter we have seen that more than one model can account for the same observational effects. The apparent motion of the Sun can be modeled with an eccentric circle or an epicycle on an Earth-centered deferent. The behavior of Ptolemy's equant can be reproduced using the epicyclet of al-Shāṭir.

Philosophers of science say that scientific theories are **underdetermined** by the available evidence, because there may be many different theories that match all of the data. How, then, are we to ever know what is really going on?

One answer to this question is that we cannot ever know. Perhaps the point of a scientific model is just to provide a compact description of the data we have and to predict the results of future measurements. Maybe we shouldn't worry about what is *really* happening. We will explore this issue in greater depth in the next chapter when we discuss the contrasting points of view known as realism and instrumentalism.

Regardless of the ultimate purpose of science, it is certainly true that astronomers had to build models based on limited data. Because the data were not sufficient to determine the model on its own, astronomers had to make assumptions about how their models should be constructed. For example, some astronomers insisted that all celestial motions must be composed of uniform circular motions. We can never be sure if our assumptions are correct, but we can test the limits of our models as a way of testing those assumptions. In this chapter we saw that Ptolemy was unable to account for the retrograde arcs of Mars using uniform circular motions, and this failure led him to modify the assumption of uniform circular motion by introducing the equant.

In a similar way, the data that seemed to indicate a changing rate of precession guided Thābit ibn Qurra to introduce his theory of trepidation. Unfortunately, in this case Thābit was guided by flawed data and the phenomenon that his model accounted for did not actually exist.

Whatever assumptions we use to build our models, we must then judge whether or not those models are successful. How we judge the success of a model depends on what we care about. Perhaps we just want a plausible explanation of certain qualitative phenomena, such as the Eudoxan model for retrograde motion. Perhaps we want a model that we can build out of real materials, such as the partial and total spheres of Ptolemy's *Planetary Hypotheses*. Maybe we want an accurate fit to detailed observational data, as provided by Ptolemy's *Almagest*. Or we might want to judge our model based on how well it fits with other things we know. For example, the Eudoxan model fit perfectly well with Aristotle's physics and cosmology but the Ptolemaic models did not fit nearly as well.

Our discussion of the Eudoxan model in this chapter indicates that historians face a problem similar to that of scientists. Historians must construct an interpretation of history that is based on limited textual evidence. Inevitably, they must make assumptions in building their interpretation, and they must check their interpretation against other accepted historical ideas. We know of the homocentric spheres of Eudoxus only from a few commentaries, some written hundreds of years after Eudoxus lived. Is Schiaparelli's reconstruction of the Eudoxan model correct? It does fit with the textual evidence that we have and the assumption that Eudoxus' goal was to account for the retrograde motion of the planets, but some scholars question that assumption. We may never know for sure unless new textual evidence is found.

As science advances we tend to demand more and more from our scientific models. Of course our models should fit with the observational data that were used in their construction, but we expect them to give us more. They should provide accurate predictions for our future observations and we may also expect them to fit new types of observations that the model was never intended to address. For example, Ptolemy's lunar theory was designed to match the motions of the Moon on the celestial sphere and it did that well, but some astronomers were disappointed that Ptolemy's lunar theory implied unobserved dramatic changes in the Moon's apparent size.

If we hope for our theories to tell us what is really going on in the natural world, then we might also expect them to provide reasonable explanations for why we see what we see. For example, Ptolemy's model for Mars provided an elegant explanation for why that planet appeared brighter when it was in ret-rograde: when Mars was in retrograde it was on the inner part of its epicycle, closer to Earth. However, Ptolemy's planetary models provided no explanation for the mysterious link between the motion of the planets and the motion of the Sun, nor did they explain why different models were needed for the inferior and superior planets.

With so much expected from our scientific models, it comes as no surprise that our models sometimes fail. Those failures can be interesting, since they point us toward new areas of research that might help us solve those prob-lems and develop better models, but until those better models are developed we tend to stick with what we have. In spite of all of the criticism of the *Almagest*, Ptolemy's models were still used to produce astronomical tables in the 13th century, over one thousand years later. The homocentric spheres of al-Bitruji might fit better into Aristotelian cosmology, but they simply could not match the accuracy of Ptolemy's models.

The overthrow of Ptolemaic astronomy would require a new set of astronom-ical models that could match the accuracy of the *Almagest* while also addressing some of the criticisms raised against Ptolemy's theories. Those astronomical models were supplied by an administrator working in a Catholic cathedral on the Baltic coast of what is now Poland. The name of that church administrator was Nicolaus Copernicus.

5

Moving the Earth: the revolutions of Copernicus

5.1   *On the Revolutions of the Heavenly Spheres*

Nicolaus Copernicus (Figure 5.1) was born in 1473 in Toruń, a town in Royal Prussia (part of the Kingdom of Poland).[1] Copernicus was educated at the University of Kraków, where he had the opportunity to learn astronomy and mathematics from some of the foremost astronomers and astrologers of the time. Later, Copernicus continued his education in Italy. He studied law and medicine at universities in Bologna and Padua, but he also furthered his knowledge of astronomy and astrology. In Bologna he worked for a while as an assistant to the well-known astronomer Domenico Maria Novara.[2] In Padua Copernicus may have learned about the mathematical models developed by al-Ṭūsī and al-Shāṭir, models that he would later use in his own work.[3] He almost certainly became familiar with some of the criticisms of Ptolemy during this time.[4] Eventually Copernicus obtained a doctorate in canon law, or church law, from the University of Ferrara, even though he did not actually attend classes there.

After receiving his doctoral degree in Italy, Copernicus returned to Poland. He worked for about a decade as a personal physician and secretary to his uncle, Lucas Watzenrode, who was then the Bishop of Warmia. It was probably during this time that Copernicus began his work on a new heliocentric (Sun-centered) system of the world, although the only work he published during this time was a translation, from Greek to Latin, of 7th-century Byzantine poems. After his uncle's death, Copernicus settled down to the life of a canon at the Catholic cathedral in Frombork, a small town on the Baltic coast of his homeland. The role of church canon mostly involved administrative work, running the day to day operations of the Frombork cathedral. Although the job was difficult at times,

Figure 5.1 Portait of Nicolaus Copernicus from a copperplate engraving by de Bry. Image courtesy History of Science Collections, University of Oklahoma Libraries.

Copernicus still found opportunities to make new astronomical observations and work on improvements to his heliocentric theory.

At some time before 1514 Copernicus wrote a summary of his heliocentric theory. This summary is now known as the *Commentariolus* (Little Commentary).[5] Copernicus never had the *Commentariolus* printed, but he did make several copies by hand and distributed them to a few of his acquaintances, including some astronomers. It was in this little pamphlet that Copernicus first made known the principal ideas of his new astronomical system: the Earth spins and orbits around the Sun, as do the other planets, while the Sun sits motionless at the center and the stars lie stationary on the outside. Only the Moon retained its motion around the Earth. In the *Commentariolus* Copernicus had worked out many of the details of his new model, deriving his parameters from the *Alfonsine Tables*.[6] Unfortunately, that approach meant that his new model could reproduce the *Alfonsine Tables* but it could not improve upon them. In order to do better, Copernicus had to derive new parameters directly from observations. He spent the next twenty years making observations and performing calculations in order to develop his new heliocentric system into a set of astronomical models that could surpass those of Ptolemy.[7]

Copernicus' new ideas gradually spread through Europe. By 1533 his ideas about the orbiting Earth were explained to Pope Clement VII in Rome.[8] Copernicus received encouragement to publish his work from Cardinal Nicolaus Schoenberg and Bishop Tiedemann Giese, but he was still hesitant to publish. He had run into several technical difficulties in working out his new system and he feared that his radical idea of a moving Earth would be ridiculed by others who could not understand the intricate details of his mathematical models. The *Commentariolus* might have forever remained the only public statement of Copernicus' theory were it not for the intervention of Georg Rheticus, a young professor of mathematics at the University of Wittenberg.[9]

Rheticus, a German Protestant, was given permission to travel to Frombork to learn astronomy from the Catholic canon Copernicus at a time when some Catholic areas were expelling Protestants (and vice versa). Rheticus arrived at Frombork in 1539 and by 1540 he had learned enough to publish his own account of the Copernican system, the *Narratio Prima* (First Account).[10] Both the *Commentariolus* and the *Narratio Prima* provide an overview of the Copernican system, but without all of the mathematical details needed to make the theory useful for professional astronomers.

Positive feedback on the *Narratio Prima* helped Rheticus persuade his teacher to publish the complete theory. Rheticus agreed to oversee the printing, although he was unable to see it through to the end, as we will see shortly. Copernicus' great work, entitled *De revolutionibus orbium coelestium* (On the Revolutions of the Heavenly Spheres), was finally printed in 1543.[11] By this time Copernicus had suffered a stroke and was near death. Some of the printed pages of the book may have been delivered to Copernicus when he was still aware enough to read them, but it is not clear if he ever saw the final printed version of his greatest work.[12]

Copernicus' great work is to some extent a new version of the *Almagest*. Copernicus did not believe that the nonuniform motions introduced by Ptolemy could describe the true motions of the heavens, so he eliminated the equant from his models, using instead the epicyclet devices of al-Ṭūsī and al-Shāṭir which were constructed from uniform circular motions. Even these devices cannot entirely eliminate nonuniform motions from Ptolemy's system. It may have been the desire to eliminate these last traces of nonuniform motion that led Copernicus to consider his heliocentric arrangement.[13] Alternatively, he may have been motivated to find a definite ordering for the planets in order to address a powerful critique of astrology by Giovanni Pico della Mirandola, published in 1496.[14] Whatever his reasons for considering a Sun-centered system, once he did so he liked what he saw, but aside from the issue of nonuniform motions Copernicus believed that Ptolemy's models were accurate.[15] In *De revolutionibus* he did not invent a new astronomy from scratch. Rather, he started from Ptolemy's models

and made what he saw as necessary modifications, including the switch to a
moving Earth. Even the presentation of his astronomical models mimicked the
Almagest. As we will see, this approach led to several oddities in the Copernican
models that are really just artifacts of Ptolemy.

De revolutionibus was destined to ignite a revolution in astronomy, and we will
take some time to examine just what Copernicus claimed in his ground-breaking
work. We won't go into all of the (considerable) mathematical details. Instead,
we will take a look at a simplified version of the Copernican theory in order to
understand its basic structure. Once we have the basics down we will discuss
some of the ways in which the full theory differs from the simplified version.

The most striking thing about Copernicus' theory is that the Earth moves.[16]
Copernicus assigned three motions to the Earth. Two of these motions were
an annual revolution about the Sun and a diurnal (daily) rotation about a set
axis. Today we believe that the Earth really does move in these two ways. Let's
take a look at how Copernicus used these two motions of Earth to explain the
phenomena of the heavens, and why Copernicus felt the need to give Earth an
additional motion.

By assigning the Earth an annual revolution (or orbit) about the stationary
Sun, Copernicus argued that it is the Earth's motion, not the Sun's, that gives
rise to the *apparent* annual journey of the Sun around the ecliptic. His idea is
illustrated in Figure 5.2. As the Earth travels about its orbit, the Sun appears

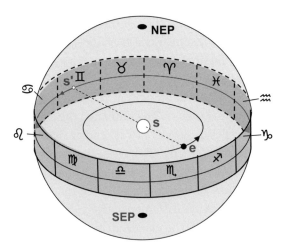

Figure 5.2 The annual orbit of the Earth. The Sun does not physically move along
the ecliptic, but it appears to because we on the Earth move around it. The plane
of the ecliptic is visualized by Copernicus to not be the plane of the Sun's orbit,
but to be the plane of the *Earth's* orbit. The N and S ecliptic poles are shown for
reference.

to move along the ecliptic. The plane of the ecliptic is thus understood in the Copernican model to be the plane of the Earth's orbit, rather than the plane of the Sun's orbit. In Figure 5.2, the Sun s remains stationary while the Earth e moves about it in a circular orbit. From the Earth's point of view the Sun appears to be at point s′ in the constellation Gemini, moving to the east against the sky. In one year, it makes a single trip around the ecliptic; that is, in this time the Earth makes a single trip around its orbit.

Copernicus achieved exactly the same effect with Earth's annual orbit around the Sun that Ptolemy achieved with the Sun's annual orbit around the Earth. So why did Copernicus prefer a moving Earth to a moving Sun? The strongest reason for his preference will be revealed when we examine his theory of the planets. As we will see, the motion of the Earth solved some of the mysteries of the Ptolemaic models (but also created some new mysteries). Copernicus also argued that it was more appropriate for the glorious Sun to be at the center of the universe than the dull Earth.[17]

Copernicus transferred to the Earth not only the Sun's motion, but also the celestial sphere's. In his model it is the Earth and not the celestial sphere that rotates once every 23 hours 56 minutes (see Figure 5.3). In order to account for the appearances, Copernicus set the rotation of the Earth to be west-to-east; this accounted for the *apparent* east-to-west motion of the celestial sphere, which was now stationary. All we can actually see is the motion of the stars *relative to* the Earth. From the appearances alone it is impossible to tell which is really rotating:

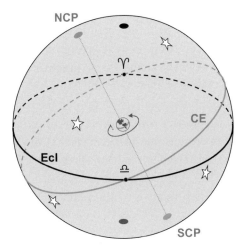

Figure 5.3 The daily rotation of the Earth. The 23-hour-56-minute east-to-west rotation of the celestial sphere is transformed by Copernicus into a west-to-east rotation of the Earth, while the celestial sphere sits stationary. The Earth's axis of rotation is tilted 23.5° relative to the ecliptic axis.

the celestial sphere or the Earth (or both). As Copernicus said it: "when a ship is floating calmly along, the sailors see its motion mirrored in everything outside, while on the other hand they suppose that they are stationary, together with everything on board. In the same way, the motion of the earth can unquestionably produce the impression that the entire universe is rotating."[18]

Again, we can ask why, if things look the same either way, Copernicus chose to spin the Earth and not the celestial sphere. Copernicus' argument for the Earth's rotation hinged on the relative size of the Earth and the celestial sphere.[19] Ptolemy claimed that the Earth was like a point in comparison to the celestial sphere and we will see that for Copernicus the celestial sphere was even larger. Why, Copernicus asked, would you have the enormous celestial sphere rotate and carry the stars around at unimaginable speeds when you could just as easily have the much smaller Earth rotate. True, the surface of the Earth had to move rapidly (about 1000 miles per hour near the equator) in order for the Earth to complete its rotations, but 1000 miles per hour was a snail's pace compared to the speed at which the stars had to move if the celestial sphere were rotating.[i]

There are several comments to be made about Figure 5.3. First, for the sake of demonstrating the Earth's rotation (and not its orbital motion, shown in Figure 5.2), the Sun has been removed from the image. It may be argued that this figure is not relevant because it shows the Earth in the center, which is against the theory of Copernicus. We mentioned above that the Earth is tiny compared to the celestial sphere but, as we will see later, the entire orbit of the Earth is also tiny compared to the celestial sphere. For practical purposes the entire orbit of Earth may as well *be* in the center. It is like setting a hula hoop at the center of the Earth; the hoop is so small compared with the Earth itself that every point on the hoop, although technically offset from the true center of the Earth by a couple of feet, can be considered to be at the center for all practical purposes.

Also, notice that the celestial globe is shown in the same manner as in Figure 5.2, with the ecliptic oriented horizontally. The fainter celestial equator is seen here to be a projection of the Earth's own equator out onto the celestial sphere. This demonstrates that Copernicus not only transferred motions from the heavens to the Earth, he also transferred *orientations*; the Earth is now tilted 23.5° relative to the ecliptic, not the other way around as in the left image of Figure 4.1. Since the Earth is tilted (relative to the plane of its orbit, the

---

[i] Note that to an Aristotelian the idea that it is better to move the small Earth than the large celestial sphere literally "carries no weight." In Aristotle's cosmology the celestial bodies are composed of ether which is weightless, because it has no tendency to move up or down. Instead, ether naturally moves in circles around the center of the universe. For an Aristotelian, not only would the sphere of stars be easier to rotate, but it *must* rotate.

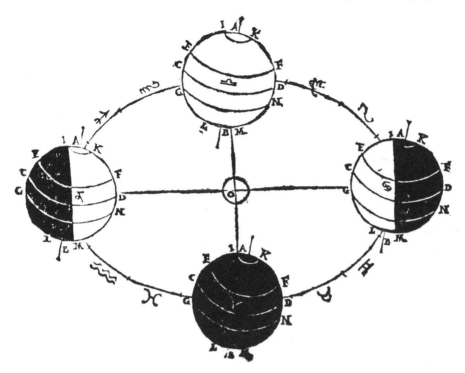

Figure 5.4 Diagram from Galileo's *Dialogo* illustrating the Copernican seasons.
The Earth is shown at four locations in its orbit, corresponding to the celestial
points of the summer solstice (♋, right), winter solstice (♑, left), vernal equinox
(♈, bottom) and autumnal equinox (♎, top). The Sun is represented by the small
circle at the center. Note that on the date of the summer solstice the Earth must
be located at ♑ so that we see the Sun at ♋. Likewise on the date of the winter
solstice Earth is at ♋, on the vernal equinox Earth is at ♎, and on the autumnal
equinox Earth is at ♈. On the summer solstice (Earth at ♑) more than half of each
northern hemisphere latitude line (EF, IK) is in the sunlit portion of the Earth,
while more than half of each southern hemisphere line (LM, GN) is in the dark
portion. The reverse is true on the winter solstice. At the equinoxes all latitude
lines are half lit and half dark. Image courtesy History of Science Collections,
University of Oklahoma Libraries.

ecliptic plane) and rotating, the celestial sphere *appears* to be tilted (relative to
the ecliptic) and rotating.

Copernicus realized that for his system to work the orientation of the Earth's
axis had to remain (nearly) constant, as shown in Figure 5.4 (the figure comes
from Galileo's much later *Dialogo* but illustrates Copernicus' idea). Notice that
as the Earth revolves about the Sun the axis is always leaning to the right.
The constant orientation of Earth's rotational axis will ensure that the north
celestial pole will always be in the same part of the sky (near Polaris). However,

Copernicus, who thought of the Earth being carried around in its orbit on the surface of a giant crystalline sphere, realized that the rotational axis would not maintain a fixed direction as that sphere spins around.

You can easily demonstrate this effect for yourself. Grab a straight stick (or a ruler, etc.). Hold it at arm's length from your body with the stick upright. Then give the stick a little bit of a tilt. Look at where the top part of the stick points. Now turn your body around and watch what happens to the direction the stick points – it changes! The stick represents the rotational axis of Earth and your body (including your arm) is playing the role of the great crystalline orb that carries Earth around in its orbit. Copernicus realized that with this arrangement the rotational axis of Earth would not maintain a fixed orientation relative to the celestial sphere, and thus the north celestial pole would not stay near Polaris throughout the year, even though he knew that it really did.

Copernicus corrected for this problem by adding a third motion, a conical movement of the Earth's axis that turned in the opposite sense from the Earth's orbital motion. In our example above, this would be equivalent to rotating your wrist as you turn your body to ensure that the stick always points in the same direction. The motion of your wrist corresponds to Copernicus' third motion for Earth. With this third motion included, Copernicus ensured that the rotational axis kept a fixed orientation relative to the stars.

The fixed orientation of Earth's axis provides an explanation for the seasons. Although the axis has a fixed orientation relative to the stars, it changes its orientation relative to the Sun as the Earth orbits. When the Earth's northern hemisphere was tilted toward the Sun, as shown by the leftmost Earth in Figure 5.4, then the daily path of the Sun as seen from the northern hemisphere of Earth would be like that labeled Sum in Figure 2.4; the Sun would travel high in the sky and would be up for more than 12 hours, as on the summer solstice. Similarly, when the Earth's northern hemisphere was tilted *away* from the Sun, as shown by the rightmost Earth in the figure, then the daily path of the Sun would be like that labeled Win in Figure 2.4: it would stay low in the sky and would be up less than 12 hours, as on the winter solstice.

Figure 5.4 helps to illustrate these facts by showing several latitude lines on Earth and by shading the part of Earth that is in darkness. When the Earth is in its summer solstice position (left), over half of each latitude line in the northern hemisphere is on the bright side of the Earth, which shows that observers at these latitudes will experience more than 12 hours of daylight. The southern hemisphere latitude lines are mostly in the dark, so southern observers get less than 12 hours of daylight at that time. When Earth is in its winter solstice position (right) this pattern is reversed, and at the equinoxes all latitude lines are split equally between the bright and dark parts of Earth. Keep in mind, though, that the ancients could explain all of these facts just as well with their geocentric theory.

Copernicus gained a slight advantage over the ancients when considering the precession of the equinoxes. We know that in fact the axis of Earth's rotation (or of the celestial sphere's rotation if you are Aristotelian) is *not* fixed. The orientation of this axis changes very slowly; that's precession. Copernicus knew about precession, and he could account for a steady precession in a simple way. All he had to do was to make sure that his third motion (the conical counter-motion of Earth's axis) did not *quite* take place at the same rate as the Earth's annual orbit about the sun. By making these two motions slightly out of sync, the orientation of Earth's axis would be slightly different after each annual revolution. The celestial poles would thus drift slowly through the sky in a big circle, centered on the pole of the ecliptic, completing the circle in 25,816 years.

Using the third motion to explain precession was a neat trick and it allowed Copernicus to account for three known motions (the daily rotation of the celestial sphere, the annual motion of the Sun, and the very slow precession of the equinoxes) by proposing three motions for Earth. Unfortunately, Copernicus also felt the need to account for Thābit ibn Qurra's trepidation effect, which produced an irregular rate of precession and a change in the obliquity of the ecliptic. To account for trepidation Copernicus used the Ṭūsī couple to generate a back-and-forth motion from two uniform circular motions. The end result was that Copernicus' motions for Earth were quite complicated. However, he managed to take the motions that had been assigned to the Sun and three different celestial spheres and put them all into a single, relatively small body: the Earth.

What Copernicus did fit with known observations, but was it *better*? That's not clear. After all, there was no scientific benefit; the appearances of the Sun and stars were reproduced equally well in the Copernican and Ptolemaic systems. His theory worked, but adopting it meant giving up the idea of a stationary Earth, which Copernicus admitted was "almost contrary to common sense."[20] In order to get no better science and arguably no better aesthetics, we have to abandon all the evidence of our senses. If this was all there was to Copernicus' theory, then it would almost certainly have been ignored.

What about Copernicus' theory for the Moon? Like Ptolemy, Copernicus had the Moon orbiting the Earth. For his lunar theory, Copernicus used a double-epicycle model like that of al-Shāṭir.[21] His model could account for the motions of the Moon relative to the stars about as well as Ptolemy's, and it did *not* predict large (and unobserved) changes in the apparent size of the Moon like Ptolemy's theory did. In fact, Copernicus' lunar theory became very popular among astronomers who otherwise thought his Sun-centered system was ridiculous.[22] However, Copernicus' lunar theory could easily be placed within a Ptolemaic, Earth-centered framework. To understand why the idea of a moving Earth was appealing to Copernicus, and eventually to others, we need to look at his theory of the planets.

5.2   Copernican planetary theory

The most amazing part of Copernicus' theory is how he explained the apparent motions of the planets.[23] The general structure of his theory is illustrated in Figure 5.5. We have already seen that Copernicus had Earth move in

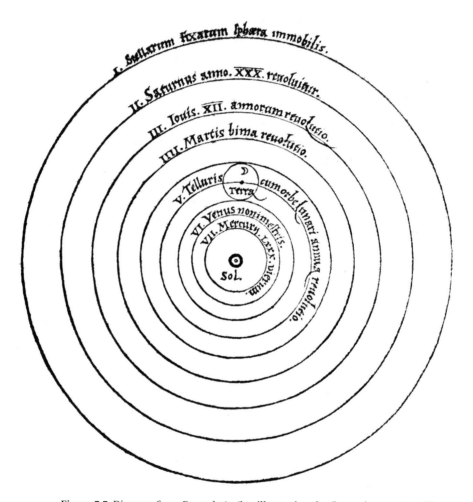

Figure 5.5 Diagram from *De revolutionibus* illustrating the Copernican system. The Sun lies at the center. The Sun-centered spheres of each of the planets, including the Earth with its Moon, are shown in order. The outermost sphere is the immobile sphere of the fixed stars. Image courtesy History of Science Collections, University of Oklahoma Libraries.

a circle with the stationary Sun at the center. Figure 5.5 shows that Copernicus gave the same basic behavior to Mercury, Venus, Mars, Jupiter, and Saturn. Each layer of Figure 5.5 represents a sphere whose rotation carries a planet in its orbit around the stationary Sun at the center. Meanwhile, the Moon's sphere carries it around the Earth while the outermost sphere, which contains the fixed stars, remains immobile. As we will see, this very simple structure explained many aspects of planetary motion.

There are two things to note about the basic framework shown in Figure 5.5. First, it made Earth a planet. Except for the fact that Earth had an orbiting moon, there was nothing in this general arrangement to distinguish Earth from the other planets. In fact, Copernicus' theory redefined the term "planet." For the ancient Greeks, a planet was a celestial object that moved relative to the stars. The Sun and Moon were both planets, along with Mercury, Venus, Mars, Jupiter, and Saturn. Earth was not a planet. For Copernicus a planet was a body that orbited the Sun. So Earth was a planet, but the Sun and Moon were not.

Second, the Copernican theory supplied a natural way to divide the visible planets into two categories: inferior and superior. An inferior planet was a planet whose orbit was contained within Earth's orbit. In other words, an inferior planet was one that lay closer to the Sun than did the Earth. A superior planet was a planet whose orbit lay outside Earth's orbit, or one that was farther from the Sun than was the Earth.

Note how these definitions explained the basic observational distinction between inferior and superior planets that we have already encountered. Inferior planets were inside Earth's orbit, so an observer on Earth had to look *inward*, in the general direction of the Sun, to see an inferior planet. That was why inferior planets were always seen near the Sun. Superior planets, on the other hand, were outside Earth's orbit and could be seen 180 degrees away from the Sun, or in the same part of the sky as the Sun, or anywhere in between. So Mercury and Venus had to be inside Earth's orbit, while Mars, Jupiter, and Saturn were outside Earth's orbit. The basic structure of the Copernican system automatically divided the planets into two groups and gave them exactly the observational characteristics that were seen in the actual planets. In constrast, Ptolemy had to construct his theory in just the right way to produce these different behaviors – they did not follow naturally from his deferent–epicycle model.

Perhaps the greatest triumph of Copernican astronomy was its explanation of retrograde motion. No epicycles were needed, although we will see that Copernicus did use epicycles for other purposes. In the Copernican system, retrograde motion was an optical illusion; planets did not "back up" at any time; there was never any reversal of direction at all. Retrograde motion was merely the effect of observing the planets from an orbiting Earth.

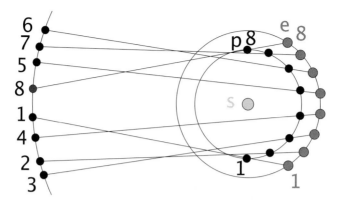

Figure 5.6 Retrograde motion of inferior planets. The diagram shows Earth (e) and an inferior planet (p) at several locations (1 to 8) in their orbits around the Sun (s). The line from Earth to the planet indicates the apparent position of the planet on the celestial sphere. The inferior planet appears to back up when it is passing between the Earth and the Sun.

We will look at the retrograde motions of the inferior planets first. For the moment, let's assume that the time it takes for an inferior planet to orbit (its orbital period) is less than the orbital period of Earth. We will justify this assumption later. In this case the inferior planets should periodically catch up with the Earth and pass it as they travel. Figure 5.6 illustrates this effect. When the inferior planet p is at position 1, the Earth e is at its position 1. A portion of the celestial sphere is shown to the left, and the point labeled 1 on this sphere marks the point among the stars where the planet appears to be, as seen from the small Earth labeled 1. After some time, the Earth moves around the Sun, and so does the planet. The planet appears to move from point 1 to point 2 on the celestial sphere; that is, it seems to have moved eastward (counterclockwise) during this interval. Similarly, for the time between points 2 and 3 the planet has again shifted eastward among the stars. But after another (equal) time interval, the planet has moved to point 4; it appears to back up. The planet then moves westward (clockwise) dramatically, from 4 to 5 and from 5 to 6. Finally, the planet resumes its prograde motion toward the east, from points 6 to 8. So 3 marks the eastern station of the planet, 3–6 mark the planet's retrograde arc, and 6 marks the western station.

Note something very important here: we know from observation that inferior planets always retrograde when they are in conjunction (near the Sun in the sky). In the Copernican system this happens automatically. As the inferior planet is passing Earth it must necessarily be between the Earth and the Sun, so it must be in conjunction as seen from Earth. To make this connection occur in his system,

Ptolemy had to synchronize the deferent motion of his inferior planets to the motion of the Sun. This synchronization is a clear example of a contrivance that was necessary for the Ptolemaic system and that was cleanly disposed of in the Copernican system. So in an aesthetic sense, on this point the Copernican system was better than the geocentric model. However, from the point of view of practical astronomy, Copernicus did no better than Ptolemy; his predictions were no better at matching the observations (or "saving the appearances"), and the aesthetics of the Copernican model were gained at the expense of making the Earth a planet in motion.

The situation is similar for the superior planets: their retrograde motion is also a product of viewing them from a moving Earth. In this case, however, it is the *Earth* that catches up to and passes *them*; since Earth is closer to the Sun than the superior planets, it moves more quickly and regularly overtakes them. As Earth passes a superior planet, the planet's position against the stars of the celestial sphere changes in such a way that it travels eastward, backs up toward the west, and then returns to its regular eastward motion. This effect is depicted in Figure 5.7, where the figure is to be understood in a manner similar to Figure 5.6.

Notice again that the connection between retrograde and opposition for superior planets was resolved nicely, and in a sense more naturally, by Copernicus than by Ptolemy. When Earth laps a superior planet it will lie directly between the planet and the Sun, so the planet will appear opposite the Sun from the point of view of the Earth. For Ptolemy to ensure that superior planets would retrograde at opposition he had to synchronize the motion of the planet on its epicycle to the motion of the Sun. Copernicus did very well here, removing this extra restriction without degrading the accuracy of the system. When a superior planet is lapped by the Earth it should also appear brighter because it is closer to Earth at that time, so Copernicus could explain why Mars was brighter during opposition (just as Ptolemy had before).

According to Copernicus, the motion of the planets relative to the stars as seen in our night sky is really a combination of two motions. One is the actual motion of the planet as it orbits around the Sun. The other is the motion of Earth. Because we are *on* Earth, its motion gets mixed into everything we see. The fact that we are viewing the planets from a moving Earth explains the mysterious connections between the Ptolemaic planetary motions and the apparent motion of the Sun. In the Copernican theory the Sun appears to move because the Earth is orbiting around it. This motion of the Earth gets mixed into the apparent motion of the planets, and the mystery is solved!

Before moving on, let's return to the issue of the orbital periods of the planets. We assumed that inferior planets have periods less than that of Earth, while

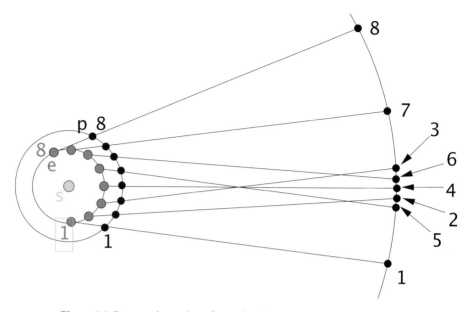

Figure 5.7 Retrograde motion of superior planets. The diagram shows Earth (e) and a superior planet (p) at several locations (1 to 8) in their orbits around the Sun (s). The line from Earth to the planet indicates the apparent position of the planet on the celestial sphere on the right. The superior planet appears to back up when the Earth is passing between it and the Sun.

superior planets have periods greater than that of Earth. Can we justify this assumption? Yes! If an inferior planet had a period longer than one year we would never see it undergo retrograde motion from our viewpoint on the moving Earth. Similarly, a superior planet with a period shorter than one year would never undergo retrograde motion as seen from Earth. So by adopting the Copernican viewpoint we are *forced* to conclude that Mercury and Venus are inferior planets with periods shorter than one year while Mars, Jupiter, and Saturn are superior planets with periods longer than one year. Now let's take a look at how Copernicus determined the periods of the planetary orbits.

In the ancient models of the universe, it was never very difficult to determine the *periods* of the planetary motions. In Ptolemy's theory, as we have seen, each planet had two basic motions: a motion along the epicycle and the motion of the epicycle along the deferent. The period of the planet's motion along the epicycle (measured relative to the deferent – so, from the apogee of the epicycle back to the apogee) was simply the planet's synodic period. The period of the epicycle's motion along the deferent was simply the planet's tropical period (the average time to complete a circuit around the celestial sphere). In Ptolemy's theory, the periods of the planetary motions were taken directly from observation – no calculation was needed.

In the Copernican system we want to measure the period of each planet's orbit. In other words, we want to know how long it takes that planet to complete one trip around the Sun. For one planet, Earth, it is pretty easy to find the period. Since the orbit of Earth is responsible for the apparent motion of the Sun on the celestial sphere, we know that the period of Earth's orbit must be the same as the period of the Sun's apparent motion relative to the fixed stars: one sidereal year. In the Copernican system the Sun was a fixed landmark and the stars were a fixed background, so it was possible to time the Earth's motion by watching the apparent motion of the Sun relative to the stars.

This trick won't work to find the orbital periods of the other planets because neither they, nor we, are fixed landmarks. We are moving observers watching a moving object. How will we know when a planet has circled the Sun once, if we ourselves have been circling the Sun in a different orbit and at a different speed from the planet in question? It turns out that in the Copernican model it's still not hard to do, but it does require a little more careful thought and some calculation. Are the answers the same as in the Ptolemaic system? No! This is because we are working in different systems; *how we measure the planetary periods depends on our theory.*

Let's take a look at how to determine the orbital periods of the planets in the Copernican system.[i] We will have to treat inferior and superior planets separately, so let's start by finding the orbital period of the inferior planet Venus. To determine the orbital period we need first to consider the motion of Venus from one "inferior" (retrograde) conjunction to the next such conjunction. The geometry of this motion is illustrated in Figure 5.8. At time 1, Venus is at point p_1 and the Earth is at point e_1. We will start our clock at this moment. The synodic period of Venus is 584 days, so at that time the planet will once again be in inferior conjunction as seen from Earth. Venus is now at point p_2 while the Earth has moved to point e_2. Since Earth's orbital period is only 365 days, we know that Earth will have completed more than one full orbit, but less than two full orbits in this time. Venus, which we have assumed travels faster than the Earth, must have completed exactly one more full orbit than Earth in order to be lined up between Earth and Sun again.

Figure 5.8 shows us that Venus' synodic period is *not* its orbital period; its orbital period must be less than its synodic period since it traveled more than a

---

[i] Copernicus believed that the planetary orbits were fixed relative to the stars, not relative to the signs of the zodiac, so what we will describe here is how to determine a planet's orbital period measured relative to the stars, or its sidereal orbital period. Because of precession this will differ, but only very slightly, from the planet's tropical orbital period measured relative to the signs of the zodiac.

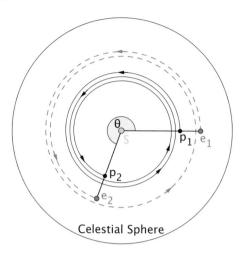

Figure 5.8 Determining the periods of the inferior planets. During the time between successive conjunctions, the Earth (e) moves through some whole number of orbits plus an angle θ. In the same time, the inferior planet (p) journeys through the same angle as the Earth plus one additional complete orbit.

full orbit around the Sun in this time. How, then, will we find the time it takes Venus to complete exactly one orbit around the Sun? We start by figuring out the angle labeled θ in Figure 5.8. We know that in 584 days Earth traveled through one full orbit plus the angle θ. We know that Earth travels all the way around its orbit ($360°$) in 365.25 days, so Earth travels $0°.986$ along its orbit each day. In 584 days it will travel $584 \times 0°.986 = 575°.8$. So the angle θ in Figure 5.8 must be $215°.8$ ($575°.8 - 360°$). Since Venus completed exactly one more full orbit than the Earth did, Venus must have traveled $935°.8$ around its orbit (two full orbits, or $720°$, plus the extra angle $\theta = 215°.8$) in these 584 days. Therefore, Venus travels $1°.6$ each day. Finally, we can find Venus' orbital period by dividing the angle for one full orbit ($360°$) by Venus' angular speed ($1°.6$ per day) to get 225 days. So Venus completes one full orbit of the Sun in 225 days.

This procedure seems a little bit complicated, but it's not too bad and with some mathematical work we can reduce the whole thing to a fairly simple formula (see Appendix A.9). That formula can be used to determine the orbital period of any inferior planet. All we need to know is Earth's orbital period and the planet's synodic period as viewed from Earth. For example, Mercury has a synodic period of 116 days and we can use the procedure described above (or the equation in Appendix A.9) to find that the orbital period of Mercury is 88 days. Note that both Mercury and Venus have shorter orbital periods than Earth, which means that they orbit the Sun more rapidly than Earth does, just as we assumed earlier.

Now that we've figured out the orbital periods of the inferior planets, it's a good idea to step back and compare our results to the *tropical* and *synodic* periods of these planets as determined from direct observation. Looking back at Table 3.1 we find that the tropical periods of both Mercury and Venus are one year. This doesn't match up with the orbital periods we just calculated at all, but in the Copernican system we can understand why these numbers don't match. The tropical period of the inferior planets has *nothing to do* with the motion of those planets, just like the tropical year has nothing to do with the Sun's motion (since, in the Copernican system, the Sun doesn't have any motion). Instead, the tropical period of the inferior planets is nothing more than a reflection of Earth's orbit. Thus, we would expect it to be one year. And it is.

What about the synodic period, which is the time between retrogrades or, as discussed above, the time between successive inferior conjunctions? These don't match up with our calculated orbital periods either. But we saw above that they should not match, according to the Copernican system. The synodic period must be longer than the orbital period. Looking back at Table 3.1 we see that this is the case for both Mercury (synodic period 116 days, orbital period 88 days) and Venus (synodic period 584 days, orbital period 225 days).

This comparison helps us to understand the relationship between the Ptolemaic and Copernican planetary theories.[24] Recall that for inferior planets in the Ptolemaic system, the period of the deferent was just the tropical period of the planet. But we see now that, from the Copernican perspective, this is just the period of Earth's orbit. Thus, the Ptolemaic deferent motion of an inferior planet is nothing more than a reflection of Earth's motion, which explains why Ptolemy had to synchronize the deferent motion to the apparent motion of the Sun. The Ptolemaic epicycle motion of an inferior planet represents that planet's orbit around the Sun. The period of the epicycle motion *relative to the fixed stars* would be the planet's orbital period as determined above, but the period relative to the deferent (which is how Ptolemy thought about it) gives us the synodic period, which is what Ptolemy used for this purpose. We see that Ptolemy's theory is actually intimately connected to the Copernican theory, but whereas Ptolemy viewed the deferent motion of the inferior planets as a real motion, Copernicus viewed it as an apparent motion due solely to the fact that we are riding around on the orbiting Earth.

Now on to the superior planets. To determine the period of a superior planet such as Mars, we must first wait until the planet is in opposition. Figure 5.9 shows Mars in opposition when the Earth is at e_1 and Mars is at p_1; that is, when the Sun, Earth, and Mars all line up and Mars is opposite the Sun from the Earth's point of view. We then wait until the next opposition; this occurs when Earth is at point e_2 and Mars is at point p_2. We already know how long it

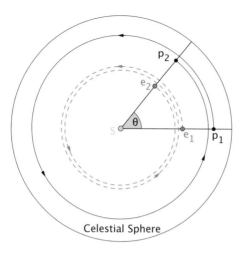

Figure 5.9 Determining the periods of the superior planets. During the time between successive oppositions, the Earth (e) moves through some whole number of orbits plus an angle θ. In the same time, the superior planet (p) journeys through the same angle as the Earth minus one complete orbit.

takes Mars to go from one opposition to the next: it is just Mars' synodic period, 780 days.

Notice that during this time the Earth has orbited the Sun twice plus a bit more, to catch back up to Mars. Earth travels $0°.986$ per day, so in 780 days it will move $769°.1$ around its orbit. This angle is equivalent to two full orbits ($720°$) plus an additional $49°.1$. Figure 5.9 shows us that Mars must have traveled exactly one full orbit less than Earth, for a total angle of $409°.1$ ($769°.1 - 360°$) during this 780-day time span, so it must travel only $0°.524$ per day. We can find Mars' orbital period by dividing the angle for one full orbit ($360°$) by this angular speed ($0°.524$ per day) to get 687 days, or somewhat less than two years.

Just as we did for inferior planets, we can reduce the procedure for finding the orbital period of a superior planet to a simple mathematical equation. This equation is given in Appendix A.10. Again, we only need to know the Earth's orbital period and the planet's synodic period in order to calculate the planet's orbital period. Using our procedure, or the equation in A.10, we find that Jupiter (with a synodic period of 398.9 days) has an orbital period of 4310 days, or just under 12 years. Likewise, Saturn (synodic period of 378.1 days) has an orbital period of 10,634 days, or about 29 years. We see that all three superior planets have orbital periods longer than that of Earth. These planets move around their orbits more slowly than Earth does, as required by the Copernican theory.

As we did with the inferior planets, let's compare the orbital periods of the superior planets to their tropical and synodic periods from Table 3.1. Because

Earth is *inside* the orbits of the superior planets, we might expect that the tropical periods of these planets will be equivalent to their real orbital periods. There may be some slight difference, because in the Copernican system Earth is not exactly at the center. So when the superior planet completes one full orbit, Earth will have changed its location and thus we will view the superior planet from a slightly different angle. But this effect should average out over a large number of orbital cycles. Indeed, we see that the orbital periods calculated above match up pretty well with the tropical periods of Mars, Jupiter, and Saturn in Table 3.1.

The synodic periods, on the other hand, will arise from a combination of the planet's motion and Earth's motion. Because Mars orbits almost as fast as Earth does, it takes Earth more than two years to lap Mars. However, Jupiter and Saturn orbit much more slowly than Earth does so the synodic periods for these two planets are just slightly more than a year. When Earth completes one orbit, Jupiter and Saturn have barely gone anywhere, so it doesn't take long before Earth catches up to them.

Again, this comparison helps us to see the connections between Ptolemy's theory and Copernicus' theory. The Ptolemaic deferent motion of a superior planet is the actual orbit of the planet around the Sun in the Copernican theory. However, the Ptolemaic epicycle motion of a superior planet is just a reflection of Earth's orbit in the Copernican theory, which explains why Ptolemy had to synchronize the motion on the epicycle to the apparent motion of the Sun. The period of the epicycle motion relative to the deferent is the synodic period of the planet, but if we were to measure the period relative to the fixed stars we would find that it is always one year (i.e. Earth's orbital period).

As we have seen in this section, finding the periods of planetary motions was a bit more complicated in the Copernican system than it was in the Ptolemaic. That was because Ptolemy took the apparent motions of the planets to be real motions, so their periods could be taken directly from observation. For Copernicus, the apparent motion of each planet was a mixture of the planet's real motion and the motion of Earth. Disentangling these two effects required a bit of effort and it required information about the planet's apparent motion (its synodic period) and the Earth's motion (its orbital period). Given that information, though, Copernicus could determine the orbital period for each planet.

5.3   The order of the heavens

In addition to determining the periods of the planetary orbits from observational data, Copernicus used his model to figure out the order and sizes of the orbits. It was not the first time someone had used a model of planetary motion to determine the order and sizes of heavenly spheres. As we saw in Section 4.3.3, Ptolemy did it in his *Planetary Hypotheses*, but the sizes of the deferents and epicycles in the Ptolemaic system were not strictly determined by observations. To find their sizes Ptolemy had to invoke an additional principle, the tight packing of celestial spheres, which came from outside of the Ptolemaic model itself. Copernicus was able to use his model of planetary motion to find the sizes of the planetary orbits without any additional assumptions beyond the basic structure of the model.

Strictly speaking, what Copernicus could derive from his model were the *relative* distances to the planets. What this means is that although he did not know the distance between the Earth and the Sun, he *could* find out the size of any other planet's orbit *in terms of* the Sun–Earth distance. To make this relative distance clear, we will define one Astronomical Unit (AU) to be the (average) distance from the Sun to Earth. Copernicus had no way to know what this distance was in more familiar units, such as miles, but if he defined the radius of Earth's orbit to be 1 AU then he could determine the radii of the other planetary orbits in AU.[i]

Recall that Ptolemy could determine the relative size of the epicycle and deferent for a given planet by selecting observations in which the Earth, the planet, and the center of the planet's epicycle formed a right triangle. In that case one side of the triangle was the deferent radius and another was the epicycle radius. If Ptolemy could determine one of the small angles in the triangle, then he could use trigonometry to find the ratio of those two radii. Recall also that Ptolemy had to use a different procedure for inferior planets (using their maximum elongation) and superior planets (using their time from opposition to quadrature). The geometry for the Copernican system is almost identical, so we can use the same approach and the same observational data to find the sizes of the orbits for inferior and superior planets.

First let's look at how to find the size of an inferior planet's orbit in the Copernican system, using an observation of the planet at its maximum elongation. The geometry of this situation is illustrated in Figure 5.10. The figure shows that the elongation ϵ of the inferior planet p is largest when $\angle$eps is equal to $90°$. So the triangle formed by the Earth, the inferior planet, and the Sun is a right triangle. The length of one side (the hypotenuse) of this triangle

---

[i] Copernicus did not, in fact, use the Astronomical Unit in this way, but modern astronomers do. Copernicus typically assigned some large number (say, 10,000) of undefined units to the Earth–Sun distance and then found the radii of the planetary orbits in terms of these units.[25] He used a large number in order to avoid having to deal with fractions – or decimals, which did not come into common use until after the French Revolution.

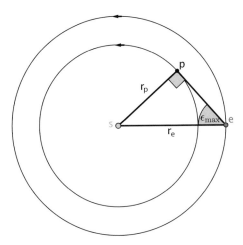

Figure 5.10 Determining the size of an inferior planet's orbit. The diagram shows an inferior planet (p) at maximum elongation (ϵ_{max}) as seen from the Earth (e). The radius of the Earth's orbit is r_e and the radius of the inferior planet's orbit is r_p.

is the radius of Earth's orbit (r_e), while the length of another side is the radius of the inferior planet's orbit (r_p). If we can measure the maximum elongation of the planet, then we can use trigonometry to calculate the ratio r_p/r_e. Since we are using $r_e = 1$ AU, the ratio r_p/r_e is really just the value of r_p in AU. The details of the calculation are discussed in Appendix A.11, but we find that Mercury's orbit has a radius of 0.47 AU while the orbit of Venus has a radius of 0.73 AU.

These numbers should seem familiar: they are exactly the values we got for the epicycle-to-deferent ratios in the Ptolemaic model. In fact, a comparison of Appendices A.7 and A.11 shows that we calculated these values in exactly the same way for the Ptolemaic and Copernican models. This is no coincidence: we have already seen that the Ptolemaic deferent for an inferior planet is really just a mirror image of the Copernican orbit for Earth, while the epicycle is the planet's actual orbit. However, the numbers mean something very different in the two systems. In the Ptolemaic system we were only comparing the size of one of the circles for Mercury (for example) with the size of the other circle for Mercury. There is nothing that tells us how these circles compare to those of any other planet. In the Copernican system, on the other hand, these numbers tell us how the size of an inferior planet's orbit compares to the size of Earth's orbit. The Earth's orbit sets the scale, and we can then determine how all inferior planets fit into that scale.

As was the case for the Ptolemaic model, the sizes of the inferior planet orbits are relatively easy to find, but the superior planets are more challenging. To determine the size of a superior planet's orbit in the Copernican model we can use observations of the planet in two special locations relative to the Sun:

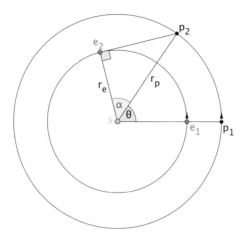

Figure 5.11 Determining the size of a superior planet's orbit. The diagram shows the Earth (e) and a superior planet (p) at opposition (position 1) and quadrature (position 2). The angles θ and α can be calculated from the orbital periods and a measurement of the time between opposition and quadrature. The radius of the Earth's orbit is r_e and the radius of the superior planet's orbit is r_p.

at opposition (opposite the Sun) and at quadrature (90 degrees from the Sun). Figure 5.11 illustrates the arrangement of Earth and a superior planet when the planet is in opposition (Earth at e_1, planet at p_1) and when the planet is in quadrature (Earth at e_2, planet at p_2). The triangle se_2p_2 is a right triangle in which one side corresponds to the radius of Earth's orbit (r_e) while the hypotenuse corresponds to the radius of the superior planet's orbit (r_p). If we can determine the angle α then we can use trigonometry to find the ratio of these two sides.

Unfortunately, the angle α is located at the Sun. That means it can be measured by an observer at the Sun, but it cannot be measured by an observer on Earth.[i] In the case of inferior planets the only angle Copernicus needed to know (the maximum elongation) was an angle that could be measured directly from Earth. He was not so lucky with the superior planets. However, he could still figure out the angle α using what he knew about the motions of the planets. He also needed to measure the time it takes for the superior planet to go from opposition to quadrature, which we will call T_q.

---

[i] Note, however, that the complementary angle, $90° - \alpha$, can be measured by an observer on the superior planet. That angle is just the maximum elongation of Earth as seen from that planet. An alien observer could use that angle to find the ratio of orbital sizes using the procedure for inferior planets because to that observer the Earth *is* an inferior planet.

Let's consider the case of Mars, for which T_q is 106 days (see Table 3.1). Look at the angle θ in Figure 5.11. This angle is just the angle through which Mars moves during the time T_q. Earlier we found that Mars moves along its orbit $0°.524$ per day (360 degrees in 687 days), so in 106 days Mars will travel $55°.5$ around its orbit. Now look at the angle formed by adding together the angles θ and α. This is just the angle that Earth travels on its orbit during the time T_q. We have seen that Earth moves $0°.986$ per day on its orbit, so in 106 days it will travel $104°.5$. So $\theta + \alpha$ is $104°.5$, and θ is $55°.5$ degrees. We can find α by subtracting the value of θ from the value of $\theta + \alpha$. We find that α is $49°$.

That was a lot of work to get the angle α, but with that angle Copernicus could use trigonometry to find the radius of Mars' orbit in AU. The details of the calculation are discussed in Appendix A.12, but the result is that Mars has an orbital radius of 1.52 AU. Note that the reciprocal of this value, $1/1.52 = 0.66$, is equal to the ratio of the epicycle radius to deferent radius for Mars in the Ptolemaic system. Again, this is no coincidence. The Ptolemaic deferent for a superior planet became the orbit of that planet around the Sun in the Copernican system. Similarly, the Ptolemaic epicycle for a superior planet was just the Earth's orbit in the Copernican system. It makes sense that the ratios of the corresponding circles are the same in both models. But once more we see that, while Ptolemy just related one of a planet's circles to another circle for that same planet, Copernicus related the orbits of *all* of the planets to a common reference circle, namely the Earth's orbit.

Appendix A.12 shows how this calculation can be performed for Jupiter and Saturn. The results: in the simplified Copernican system we have shown here Jupiter had an orbital radius of 5.2 AU, while Saturn had an orbital radius of 7.9 AU. The results Copernicus found using his detailed theories were similar for all the planets except Saturn, for which he calculated an orbital radius of 9.3 AU. It is important to note that Copernicus used his model to determine a definite ordering of the planets. Beginning with the Sun, at the center, and working outward toward the fixed stars the order was: Mercury, Venus, Earth (which was now a planet!), Mars, Jupiter, and Saturn. Not only did Copernicus give a definite answer to the question of the order of the planets, for which Ptolemy could not provide a clear answer, but he also found the exact periods of these orbits and their relative sizes.

To summarize our results for the periods and sizes of the planetary orbits, Figure 5.12 indicates the relative sizes of the orbits of the planets, along with the periods we have derived. This figure is just like Figure 5.5 except that the orbits are drawn to scale. Perhaps the best way to get a feel for the full picture of the Copernican system, though, is to read how Copernicus himself described it, from the outside in, in Chapter 10 of Book I in his *De revolutionibus*:[26]

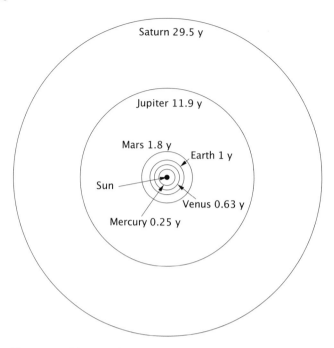

Figure 5.12 Diagram showing the periods and relative sizes of the planetary orbits in the Copernican system. Note that planets with larger orbits have longer periods.

[The sphere of the fixed stars] is followed by the first of the planets, Saturn, which completes its circuit in 30 years. After Saturn, Jupiter accomplishes its revolution in 12 years. Then Mars revolves in 2 years. The annual revolution takes the series' fourth place, which contains the earth … together with the lunar sphere as an epicycle. In the fifth place Venus returns in 9 months. Lastly, the sixth place is held by Mercury, which revolves in a period of 80 days.

At rest, however, in the middle of everything is the sun. For in this most beautiful temple, who would place this lamp in another or better position than that from which it can light up the whole thing at the same time? For, the sun is not inappropriately called by some people the lantern of the universe, its mind by others, and its ruler by still others. [Hermes] the Thrice Greatest labels it a visible god, and Sophocles' Electra, the all-seeing. Thus indeed, as though seated on a royal throne, the sun governs the family of planets revolving around it. Moreover, the earth is not deprived of the moon's attendance. On the contrary, as Aristotle says in a work on animals, the moon has the closest kinship with the earth. Meanwhile the earth has intercourse with the sun, and is impregnated for its yearly parturition.

In this arrangement, therefore, we discover a marvelous symmetry of the universe, and an established harmonious linkage between the motion of the spheres and their size, such as can be found in no other way. For this permits a not inattentive student to perceive why the forward and backward arcs appear greater in Jupiter than in Saturn and smaller than in Mars, and on the other hand greater in Venus than in Mercury. This reversal in direction appears more frequently in Saturn than in Jupiter, and also more rarely in Mars and Venus than in Mercury. Moreover, when Saturn, Jupiter, and Mars rise at sunset, they are nearer to the earth than when they set in the evening or appear at a later hour. But Mars in particular, when it shines all night, seems to equal Jupiter in size, being distinguished only by its reddish color. Yet in the other configurations it is found barely among the stars of the second magnitude, being recognized by those who track it with assiduous observations. All these phenomena proceed from the same cause, which is in the earth's motion.

In the last paragraph, Copernicus is describing how nicely his system accounts for the observed phenomena such as the changes in apparent size of Mars (which looks much larger at opposition, when it rises at sunset, than at other times). He points out how his system accounts for the varying sizes of the retrograde arcs of the planets: planets that are farther from Earth have smaller retrograde arcs. Therefore, Mars has a larger retrograde arc than Jupiter and Venus has a larger arc than Mercury. Copernicus also points out the "harmonious linkage" between the sizes of the planetary orbits and their periods: larger orbits have longer periods. This helps to explain why retrograde happens more frequently for Saturn than for Jupiter (Saturn takes longer to orbit and so it is more frequently lapped by Earth), and also more frequently for Mercury than for Venus (Mercury takes less time to orbit so it laps Earth more frequently).

Let's review what Copernicus accomplished with his theory of the planets. He provided a natural explanation for the division of planets into inferior and superior groups. He supplied a natural (epicycle-free) explanation for the retrograde motion of the planets, as well as a natural connection between retrograde and brightness. He explained the connection between the apparent motion of the planets and the apparent motion of the Sun, which had been one of the great mysteries of the Ptolemaic system. He set a definite ordering for the planets, which Ptolemy could not do. Finally, he uncovered a beautiful link between the size of a planet's orbit and the time it takes the planet to complete its orbit.

All of these remarkable characteristics of the Copernican planetary system point toward one overriding theme that distinguishes the Copernican theory from the Ptolemaic: Copernicus' system was truly a system. In a system, all the parts fit together. You can't mess with any part without messing up the entire system. In the Copernican system, if you wanted to double the size of Mars' orbit you had to double the size of all the other planetary orbits as well or else your theory wouldn't fit with the observations. Recall that this was not the case for Ptolemy. Ptolemy could double the size of Mars' deferent and the only thing he *had* to do to compensate was to double the size of Mars' epicycle. There was nothing that connected the size of Mars' circles to the size of any other planet's circles. Ptolemy's system was not a system at all. It was really a set of models, one for each planet, each of which stood on its own. Not so for Copernicus: his planetary orbits had to stand together as a whole or else they all would fall.

This property of the Copernican system, that all of the pieces were interconnected and had to work together, is called *coherence*. The word "cohere" is derived from the Latin for "stick together." All of the parts of the Copernican system stuck together to form a single unified whole.[27] As Copernicus described it, in his Dedicatory Preface to Pope Paul III:[28]

> Having thus assumed the motions which I ascribe to the earth later on in the volume, by long and intense study I finally found that if the motions of the other planets are correlated with the orbiting of the earth, and are computed for the revolution of each planet, not only do their phenomena follow therefrom but also the order and size of all the planets and spheres, and heaven itself is so linked together that in no portion of it can anything be shifted without disrupting the remaining parts and the universe as a whole.

What is more, Copernicus managed to achieve all of these results without resorting to the use of equants. The elimination of Ptolemy's equant was one of the main motivations for Copernicus' reform of astronomy. Copernicus felt that the equant violated the principal rule of astronomy: that the motions of the heavenly bodies must be constructed out of uniform circular motions. One of his chief goals was to return astronomy to this fundamental rule.

In his Dedicatory Preface he described the problems with the theories proposed by his predecessors:[29]

> I have accordingly no desire to keep it from Your Holiness that I was impelled to consider a different system of deducing the motions of the universe's spheres for no other reason than the realization that astronomers do not agree among themselves in their investigations of this subject. For, in the first place, they are so uncertain about the motion of the sun and moon that they cannot establish and observe a constant length even for the tropical year. Secondly, in determining the motions not only of these bodies but also of the other five planets,

they do not use the same principles, assumptions, and explanations of the apparent revolutions and motions. For while some employ only homocentrics, others utilize eccentrics and epicycles, and yet they do not quite reach their goal. For although those who put their faith in homocentrics showed that some nonuniform motions could be compounded in this way, nevertheless by this means they were unable to obtain any incontrovertible result in absolute agreement with the phenomena. On the other hand, those who devised the eccentrics seem thereby in large measure to have solved the problem of the apparent motions with appropriate calculations. But meanwhile they introduced a good many ideas which apparently contradict the first principles of uniform motion. Nor could they elicit or deduce from the eccentrics the principal consideration, that is, the structure of the universe and the true symmetry of its parts. On the contrary, their experience was just like some one taking from various places hands, feet, a head, and other pieces, very well depicted, it may be, but not for the representation of a single person; since these fragments would not belong to one another at all, a monster rather than a man would be put together from them. Hence in the process of demonstration or "method," as it is called, those who employed eccentrics are found either to have omitted something essential or to have admitted something extraneous and wholly irrelevant. This would not have happened to them, had they followed sound principles. For if the hypotheses assumed by them were not false, everything which follows from their hypotheses would be confirmed beyond any doubt.

Copernicus was pointing out that the homocentric theories (like that of Eudoxus) could display irregular motions such as retrograde, but they could not be made to fit with the observations. Ptolemy's theory, on the other hand, fit the observations pretty well but it violated the first principles of uniform motion by introducing the equant. Moreover, Ptolemy's theory was a "monster" composed of bits and pieces that didn't fit together well (like the creature in Mary Shelley's *Frankenstein*, but 275 years earlier).[30] This description contrasts sharply with the coherence and symmetry of the Copernican system.

5.4   Problems and purpose

As beautiful as it was, the Copernican theory was not without some major problems. To start with, the theory was not as simple and elegant as the preceding description makes it seem. The simplified theory that we have discussed so far does not, in fact, fit very well with the observational data. Copernicus had to modify this basic theory in order to make it match the observations as well as Ptolemy's theory did.[31]

For one thing, the Copernican planetary orbits were not centered on the Sun. The planets all go around the Sun, but the body of the Sun does not lie at the exact center of the planetary orbits. Copernicus shifted the center of Earth's orbit away from the Sun to account for the changing speed of the Sun relative to the stars throughout the year, just as Hipparchus had shifted the center of the Sun's orbit away from the Earth to account for the same effect. Copernicus then used the center of the Earth's orbit, the "mean Sun," rather than the true Sun as his reference point for constructing the other planetary orbits. Even so, the orbits of the other planets were not centered on either the mean Sun or the true Sun.

Copernicus was also forced to use epicycles in his models. Let's consider the epicycles he used in his model for the superior planets. These epicycles were very different from the ones Ptolemy used to produce retrograde motion: they were much smaller and their period (measured relative to the main orbital circle, or deferent) was equal to the orbital period.[i] This construction generated a nearly circular orbit, but one that was shifted relative to the deferent. What is the point of doing that, you might ask, when Copernicus has already shifted the circles around? The point of the epicycle was not really to shift the orbit in space, but to generate nonuniform movement. Remember that Copernicus desperately wanted to get rid of the Ptolemaic equant. However, he still had to produce nonuniform movements in order to match Ptolemy's equant model. Copernicus managed to produce these nonuniform motions with his tiny epicycles. Since the planet moved uniformly on the epicycle, and the epicycle moved uniformly around the deferent, Copernicus could still claim to be following the rule of using only uniform circular motions. In fact, his construction was similar to the epicyclet mechanism of al-Shāṭir.

His theory for Venus was similar, although he reversed the two circles. He placed the smaller circle near the mean Sun. The center of the larger orbital circle moved around on this smaller circle. Oddly, the period of the small circle's motion was equal to half of Earth's orbital period. In spite of all his success in disentangling the apparent motion of the Sun (i.e. the actual motion of Earth) from the apparent motions of the planets, there still remained a mysterious link between the motion of Earth and that of Venus. Copernicus' model for Mercury is even more complicated, and it too retained a link to the motion of Earth.

There were other oddities in Copernicus' theory. One important point is that Copernicus' orbit for Earth is different from the orbits of the other planets. First

---

[i] Measured relative to the fixed stars the period of the epicycle was half of the orbital period.

of all, because Ptolemy did not use an equant in his model for the solar motion, Copernicus did not use an epicyclet to produce nonuniform motions in his model for Earth's orbit. However, Copernicus' orbit for Earth was not simply a circle centered on the Sun. Based on his analysis of ancient observations Copernicus was convinced that the center of Earth's orbit slowly changed its distance and orientation relative to the Sun. To account for the changing distance he had the center of Earth's orbit ride around on a tiny circle, going around once every 3434 years (the same period he used for variations in the obliquity of the ecliptic). To account for the changes in orientation he had the center of that tiny circle move around a somewhat larger circle, centered on the true Sun, with a period of 53,000 years. But even if we ignore these long-period variations, the Copernican orbit for Earth is unique. It lacks the equant-replacing epicyclet and its center serves as the reference point around which the entire system is built.

With all of this work, Copernicus had only explained the apparent motion of the planets along the ecliptic (i.e. their motion in longitude). He still had to explain their motion perpendicular to the ecliptic (their motion in latitude).[32] To explain these motions he had to tilt the planetary orbits relative to the Earth's orbit. He tilted the orbit of each planet in a different way, but since he used the center of Earth's orbit as his reference point all of the orbital planes for the planets intersected at that one common reference point. Because his models were based on the *Almagest*, rather than Ptolemy's later *Planetary Hypotheses*, Copernicus could not keep the planetary orbits at a fixed tilt.[i] Instead he had to make the orbital planes wobble, and these wobbles were synchronized with Earth's orbit. As with the orbits of Venus and Mercury, Copernicus' latitude theory retained a mysterious link between the motion of the planets and that of the Earth.

Many of the features of the Copernican theory that seem strange to us would not have bothered his contemporaries very much. After all, Ptolemy had used eccentrics and epicycles and a weird orbit for Mercury. What would have bothered Copernicus' first readers was not so much the technical astronomy as the idea of a moving Earth and a stationary celestial sphere. The problems with these ideas fell into three general categories: theological, (natural) philosophical, and astronomical. Let's take a look at some problems in each category.

The most obvious theological problem was that the Bible contains several passages that speak of a stationary Earth or a moving Sun.[33] One of the most famous is Joshua 10:12 in which Joshua commands the Sun to stand still and

---

[i] Had he known of the latitude theory from *Planetary Hypotheses* he might have attempted to keep the orbits at a fixed inclination relative to the ecliptic, but because he had the orbital planes intersect at the center of Earth's orbit, not at the Sun, this would not have worked particularly well.

the Sun does so. What could this passage mean if the Sun is always still and it is the Earth that moves? But by the 16th century the Church had a long tradition, dating back at least to Saint Augustine, of re-interpreting Scripture in light of new discoveries in natural philosophy. In fact, one of the first people to write about the Copernican theory was an Augustinian monk (Diego de Zuniga) who in his 1584 commentary on the Book of Job explained how the Copernican theory could be reconciled with the Bible.[34]

Still, there were deeper theological problems with the Copernican theory. In the medieval worldview, Heaven was located outside of the celestial sphere. God played the role of Aristotle's Prime Mover and caused the celestial sphere to spin. It was this movement that served as the cause of all other movements in the universe. A stationary celestial sphere not only removed the source of all motion in the universe, it also removed the one way in which God continued to interact directly with his Creation. Other theological problems bedeviled the Copernican theory: Dante, in the *Inferno* of his *Divina Commedia*, had placed Satan at the lowest point in the Aristotelian universe (the center of the Earth) and now Copernicus was sending Satan riding around the Sun!

One could argue away the theological problems with the Copernican theory by claiming that Christianity and the Bible were not really concerned with the structure of the universe so much as with mankind's salvation, and so these apparent conflicts were not really important. Natural philosophy, on the other hand, was very much about the structure of the universe, and Copernicus' theory did not fit well with Aristotle's physics and cosmology. The removal of the Earth from the center point disrupted the structure imposed by Aristotle's theory of the elements. Making the Earth into a planet, a celestial body, broke down Aristotle's division between the heavens and the sublunary sphere. Copernicus could have no response to these criticisms other than to say that Aristotle was wrong.

Recall that Aristotle's physics required all violent (non-natural) motion to be produced by a mover. What, then, could move the Earth? Copernicus had a partial answer for the Earth's rotation: the Earth (with its surrounding layers of air and fire) was a sphere, and the natural motion of a sphere was rotation.[35] Natural motions didn't need a mover, so the Earth could rotate on its own without anything making it spin. Note the trick Copernicus played here. For Aristotle, circular rotations were natural in the heavens, but not on Earth. Copernicus, though, has moved the Earth into the heavens and claimed that it, too, has a natural circular motion. Copernicus also argued (as noted earlier) that it made more sense to rotate the small Earth than to spin the vastly larger celestial sphere. He considered a state of rest more noble than a state of motion and so claimed it was more suitable for the celestial sphere (and the Sun) to be at rest.[36]

There was also the problem of gravity. What makes objects fall to the Earth? For Aristotle it was the natural tendency of heavy elements to seek their place at the bottom of the universe. Copernicus offered instead the idea that gravity was just the tendency of the parts of a body to stick together:[37]

> For my part I believe that gravity is nothing but a certain natural desire, which the divine providence of the Creator of all things has implanted in parts, to gather as a unity and a whole by combining in the form of a globe. This impulse is present, we may suppose, also in the sun, the moon, and the other brilliant planets, so that through its operation they remain in that spherical shape which they display.

In Copernicus' view, Earth-stuff had a natural tendency to fall toward other Earth-stuff, while Mars-stuff fell toward other Mars-stuff, and so on, which was why all celestial bodies were spherical. He then used this theory to help support his claim that the Earth is a planet (and thus orbits the Sun) since it is spherical like the other planets.

If the Earth was moving, though, why couldn't anyone see or feel its motion? Copernicus brushed away this objection by saying that everything on the Earth shared in the (natural) motion of Earth. Since we can only perceive relative motions, we don't notice the Earth moving because we are moving along with it (as are the clouds, the birds, and everything else in the sublunary sphere).[38]

Believers in Aristotelian physics (which included most academic philosophers and theologians) were unpersuaded by Copernicus' arguments. Even day to day experience seemed to contradict Copernicus' notions about the moving Earth. When you ride a horse you are not automatically carried along with it: if you don't hold on, you will fall off! Copernicus didn't even try very hard to overcome these philosophical problems in *De revolutionibus*. The book was not, after all, primarily about natural philosophy. It was a book about technical astronomy. Unfortunately for Copernicus, there were some astronomical problems with his theory as well.

Some of the astronomical problems were fairly minor. For example, Copernicus' theory predicted that the distance from Earth to Venus would change by a factor of more than six over the course of time and similar changes would occur for Mars. This changing distance would presumably cause the brightness and apparent size of Venus and Mars to change dramatically. In fact, though, Mars changed only a little in brightness/size and the changes of Venus were not noticeable. Ptolemy, though, had the same problem with Venus and Mars, as well as a similar problem with the apparent size of the Moon, so perhaps a difficulty like this could be ignored. Speaking of the Moon, another problem with Copernicus' theory was the inconsistency of having the Moon orbit the Earth while everything else orbits the Sun. Why were there two centers of motion? But since everyone believed the Moon orbited Earth, this one point could hardly be raised as the primary objection to Copernicus.

The biggest astronomical problem was this: just as the motion of Earth shows up in the apparent motions of the planets, so too it should show up as an apparent motion of the stars. One example of this effect was that the celestial poles should move around over the course of a year. To see why this must be, refer to Figure 5.3. That figure is drawn such that the Earth is in the center. If it is truly *at* the center, and not *orbiting* the center at some distance, then the celestial poles (NCP and SCP, the points at which the Earth's axis of rotation "hits" the sphere) should remain in one place for all time. However, suppose that the Earth *did* revolve about the center at some distance from it; in that case we should see something very different, and that something is demonstrated in Figure 5.13. Copernicus claimed that as the Earth orbited the Sun its axis remained at a fixed angle; therefore, over the period of a year, the celestial poles should trace out circles on the celestial sphere. The point about which the stars appeared to rotate would change throughout the year.

Yet the motion of the poles along the dashed circles of Figure 5.13 is not seen. Why not? Two possible explanations immediately come to mind: either Copernicus is wrong, the Earth *is* at the center, and the circles don't exist, or Copernicus is right and the circles do exist, but the celestial sphere on which they are "drawn" is so far away that they are too small to be detected. Copernicus, of course, took the second option:[39]

> Yet none of these phenomena appears in the fixed stars. This proves their immense height, which makes even the sphere of the annual motion, or its reflection, vanish from before our eyes. For, every visible object has some measure of distance beyond which it is no longer seen, as is demonstrated in optics. From Saturn, the highest of the planets, to the sphere of the fixed stars there is an additional gap of the largest size. This is shown by the twinkling lights of the stars. By this token in particular they are distinguished from the planets, for there had to be a very great difference between what moves and what does not move. So vast, without any question, is the divine handiwork of the most excellent Almighty.

Ptolemy thought that the Earth was like a point compared to the celestial sphere, but Copernicus claimed that the Earth's entire orbit (the sphere of the annual motion) around the Sun was like a point compared to the sphere of the stars. He felt that such an enormous universe was fitting for an omnipotent

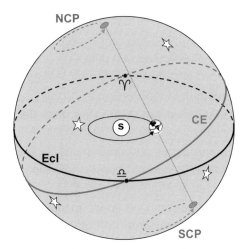

Figure 5.13 If the Earth really revolved around the Sun, we would see the celestial poles trace out annual circles (dashed lines) on the celestial sphere. This effect is not observed.

Creator, but, as we will see, many of his readers felt that it was a tremendous waste of space.

The Copernican theory surely had its problems, but should one really expect an astronomical model to address all of these issues? Wasn't it enough that Copernicus had devised a theory to account for planetary motions without using Ptolemy's equant? When *De revolutionibus* was finally published it began with the following preface addressed to the reader:[40]

> To the Reader concerning the hypotheses of this work:
>
> There have already been widespread reports about the novel hypotheses of this work, which declares that the earth moves whereas the sun is at rest in the center of the universe. Hence certain scholars, I have no doubt, are deeply offended and believe that the liberal arts, which were established long ago on a sound basis, should not be thrown into confusion. But if these men were willing to examine the matter closely, they will find that the author of this work has done nothing blameworthy. For it is the duty of the astronomer to compose the history of the celestial motions through careful and expert study. Then he must conceive and devise the causes of these motions or hypotheses about them. Since he cannot in any way attain to the true causes, he will adopt whatever suppositions enable the motions to be computed correctly from the principles of geometry for the future as well as for the past. The present author has performed both these duties excellently. For these hypotheses need

not be true nor even probable. On the contrary, if they provide a calculus consistent with the observations, that alone is enough. Perhaps there is someone who is so ignorant of geometry and optics that he regards the epicycle of Venus as probable, or thinks that it is the reason why Venus sometimes precedes and sometimes follows the sun by forty degrees and even more. Is there anyone who is not aware that from this assumption it necessarily follows that the diameter of the planet at perigee should appear more than four times, and the body of the planet more than sixteen times, as great as at apogee? Yet this variation is refuted by the experience of every age. In this science there are some other no less important absurdities, which need not be set forth at the moment. For this art, it is quite clear, is completely and absolutely ignorant of the causes of the apparent nonuniform motions. And if any causes are devised by the imagination, as indeed very many are, they are not put forward to convince anyone that they are true, but merely to provide a reliable basis for computation. However, since different hypotheses are sometimes offered for one and the same motion (for example, eccentricity and an epicycle for the sun's motion), the astronomer will take as his first choice that hypothesis which is the easiest to grasp. The philosopher will perhaps rather seek the semblance of the truth. But neither of them will understand or state anything certain, unless it has been divinely revealed to him.

Therefore alongside the ancient hypotheses, which are no more probable, let us permit these new hypotheses also to become known, especially since they are admirable as well as simple and bring with them a huge treasure of very skillful observations. So far as hypotheses are concerned, let no one expect anything certain from astronomy, which cannot furnish it, lest he accept as the truth ideas conceived for another purpose, and depart from this study a greater fool than when he entered it. Farewell.

This preface presents a stark contrast to Copernicus' enthusiasm about harmony and unity. It is hard to believe that Copernicus could have written this preface. In fact, he did not. This preface was written by the Lutheran theologian Andreas Osiander[i] who had been charged with supervising the printing of *De revolutionibus* when Rheticus had to leave to start a new job.[41] Osiander had pleaded with Copernicus to present the system of *De revolutionibus* as what we might

---

[i] Johannes Kepler, whom we will meet later, made this information public more than 50 years later.

now call a "mathematical fiction." He was convinced that this approach would allow Copernicus to avoid the opposition of philosophers and theologians that so worried him. Copernicus refused to follow Osiander's advice, but Osiander added his unauthorized preface during the printing process without Copernicus' knowledge.[42]

Osiander's preface puts forth a view of science known as *instrumentalism*. Instrumentalists think that the goal of science is solely to reproduce and predict observational data: to "save the appearances." It is not the goal, and should not be the goal, of a scientific theory to tell us the *truth* about how things really are. In some ways Osiander's preface sounds very sophisticated, similar to some early 20th-century philosophy of science.

Unlike some instrumentalists of the 20th century, though, people like Osiander were not instrumentalist about everything. Many astronomers, natural philosophers, and theologians were instrumentalists, however, when it came to the heavens. Recall Aristotle's division of the universe into the sublunary sphere and the heavens. Some natural philosophers believed it was possible to discover the truth about the sublunary sphere, but the heavens were the realm of the divine and therefore fundamentally unknowable to mortal humans. Osiander was an instrumentalist about astronomy, but probably not about sublunary physics.[43]

Copernicus made it very clear in his Dedicatory Preface to the Pope that he was NOT an instrumentalist about his astronomy. We have already quoted from this preface where he complained about his predecessors, writing "For if the hypotheses assumed by them were not false, everything which follows from their hypotheses would be confirmed beyond any doubt."[44] He was writing about truth, not about merely saving the appearances. Copernicus was a *realist* about astronomy, believing that the goal of astronomy was to uncover the truth of the heavenly motions.

The expectation for truth in astronomy went against a long-standing tradition that humans could not know the truth about the heavens. Copernicus believed that we could know, and even that we *should* know. In his preface he wrote:[45]

> For a long time, then, I reflected on this confusion in the astronomical traditions concerning the derivation of the motions of the universe's spheres. I began to be annoyed that the movements of the world machine, created for our sake by the best and most systematic Artisan of all, were not understood with greater certainty by the philosophers, who otherwise examined so precisely the most insignificant trifles of this world.

Copernicus used a religious justification for his science: the heavens were made *for us* by a perfect God, and therefore we should be able to figure out what was really going on up there. His expectation that the truth about the heavens should be knowable was what made Copernicus so excited about the harmony and unity of his new system. He thought that the beauty of his system was a sign of its truth; only such a harmonious, unified system would have been created by the "best and most systematic Artisan."

Still, it was one thing for Copernicus to say that it was possible for humans to figure out the heavens, and quite another to say that he himself had figured it out. The harmony and unity of his theory may have convinced Copernicus that he had found the truth, but did it convince anyone else? We will consider the response to Copernicus in the next chapter, but for now let's recap the arguments for and against his theory. Against Copernicus we have the following:

- The Copernican system didn't save the appearances any better than the Ptolemaic system.
- The Copernican system required belief in a moving Earth, which contradicted Aristotelian cosmology as well as Aristotle's theory of gravity. Copernicus had to propose an alternate theory of gravity, claiming substances with similar natures tended to fall together.
- According to Aristotelian physics, the rotation of the Earth should fling us all to the ground, produce violent winds, etc. Copernicus dismissed this problem by claiming that rotation was a natural motion for Earth (and that it was better to spin the small Earth than the huge celestial sphere).
- The Copernican system resulted in an outrageously large universe with lots of empty space between Saturn and the stars.
- The Copernican system led to a few predictions, such as variations in the apparent size of Venus, which were incompatible with observations. It should be noted, however, that Ptolemy's system had similar problems.
- The idea of a moving Earth and stationary Sun seemed to contradict a few passages in the Bible.

In light of these arguments, what astronomer of his day would have believed in Copernicus' theory? Yet Copernicus did (and so did his student Rheticus, and a few others as we will see). What arguments could be made in favor of Copernicus' system? We've seen a few, but let's recap and add a few more:

- Copernicus completely eliminated the large epicycles of Ptolemy's theory. Copernicus showed that one of the two major circles (deferent or epicycle) in Ptolemy's theory of each planet could be viewed as an image of Earth's orbit around the Sun. As a result, in Copernicus' system the planets move in smooth (nearly circular) orbits. In Ptolemy's system the planets move in zig-zag orbits.

- Copernicus replaced the Ptolemaic equant with uniformly rotating epicyclets like those developed by al-Shāṭir. The nonuniform motion produced by the equant had been one of the most heavily criticized aspects of Ptolemy's theory.

- Copernicus provided a natural explanation for the division of the visible planets into two groups: inferior and superior. It all had to do with whether or not the planet was inside Earth's orbit.

- Copernicus explained and (mostly) eliminated the odd link between the motion of the Sun and the motions of the planets in Ptolemy's system.

- In the Ptolemaic system the planets could be arranged in any order. It was impossible to arrange them in order of increasing period because three of them (the Sun, Mercury, and Venus) have equal periods of one year. In the Copernican system the planets had to be placed in order of increasing period in order to make the system work. So what was arbitrary for Ptolemy was essential to Copernicus.

- The relative distances from the Earth to the planets could not be determined in the Ptolemaic system without making additional assumptions, but the relative distances were completely fixed by observation in the Copernican system.

Notice that none of these arguments had to do with saving the appearances. If your goal is to produce a fictitious mathematical system that mimics the movement of the planets and stars in the sky, then why not use equants? What's wrong with zig-zag orbits? And who cares about the *ad hoc* ordering of the planets or the odd connections between the period of the Sun and the other planets? None of these things affect the system's ability to reproduce what we see. As for relative distances, why shouldn't they be arbitrary? Astronomers of the 16th century likely never dreamed that these distances would ever be directly measured. So in the case of distances there were no appearances to be saved. An instrumentalist astronomer had little to gain and much to lose by switching to the Copernican system. But for an astronomer who, like Copernicus, believed that knowledge of the true nature of the heavens was possible the arguments for the Copernican system might be compelling. Fortunately, there were a few such astronomers in the generation after Copernicus.

5.5   Reflections on science

In this chapter we have seen that the result of a "measurement" depends critically on the assumptions made by the measurer. Even the possibility of measurement may depend on these assumptions. In the Ptolemaic theory it is not possible to measure the distance to the planets without making extra assumptions (about tightly packed spheres). In the Copernican system, with its different set of assumptions, the distance of a planet from the Sun can be "measured" directly from observations. Even the meaning of a scientific term may depend on the theory in which it is used. The "period of a planet's motion" means something quite different (and is therefore measured differently, and gives a different result) in the Copernican and Ptolemaic systems. These examples illustrate the ways in which our theories and assumptions color the way we interpret observations.

This dependence of measurement on theory makes it quite difficult to compare two theories. If we start with the Ptolemaic assumptions we obtain a certain set of answers to certain questions. If we start with the Copernican assumptions we obtain a different set of answers to what seem like the same questions, but in a sense the questions we are answering are actually different questions. So how can we compare the two theories? Obviously we can fall back on the requirement that the theories should reproduce and predict the observed positions of the planets and stars, but in this case both theories do that equally well. Even the criterion of simplicity doesn't really help: Copernicus' full theory, with its small epicycles and other complications, is not obviously simpler than Ptolemy's.

If we want to decide between these theories we must apply other criteria, and here things can get tricky. Depending on which criteria we apply we may come down in favor of a different theory. If we focus on common-sense impressions that the Earth doesn't move, or agreement with Aristotelian physics, or agreement with Holy Scripture, then we are likely to favor Ptolemy's theory. On the other hand, if we focus on coherence, unity, and rational appeal (avoiding any *ad hoc* assumptions that aren't part of the basic theory), then we would tend to favor Copernicus.

If, like Osiander, we take an instrumentalist position on astronomy, then these other criteria won't count for much. An instrumentalist does not expect astronomy to reveal the truth about the heavens, and so criteria like harmony and unity are unimportant. From an instrumentalist point of view all that matters is how well the models fit the data, so we are free to choose whichever theory (Ptolemaic or Copernican) we wish. We could even use sometimes one, sometimes the other theory.

Copernicus, however, did believe that astronomy could tell us the truth about the heavens and he was convinced that harmony and unity were the key to finding that truth. Because of this attitude, Copernicus expected his theory to do more than just fit the data: he thought it should *explain* the data. It should not just tell us what we will see, but why the things we see are the way they are. Ptolemy could predict the positions of the planets with his theory, but Copernicus could explain why there were two groups of planets (inferior and superior) and why the apparent motions of these planets seemed linked to the apparent motion of the Sun. Not only could Copernicus explain the observational data in a way that Ptolemy could not, but Copernicus could also answer questions that Ptolemy could not answer. What is the order of the planets? Copernicus provided a definite answer. Why do the planetary motions have the periods they have? For Copernicus, the periods were the result of the "harmonious linkage" between the size of a planet's orbit and its speed along that orbit.

Copernicus' realist attitude toward astronomy played an important role in his innovative work, because it led him to expect things from astronomical theory that others did not expect. It changed the criteria by which he judged astronomical theories, and using the new criteria it became possible to believe in a moving Earth, in spite of the troubling contradictions with Aristotle's physics, because the motion of the Earth explained the mysteries and answered the unanswerable questions of astronomy. Most astronomers were not in a position to appreciate these impressive features of the Copernican theory because they were unexpected: they were answers to questions that most astronomers had not asked. Eventually, though, some astronomers would come to share Copernicus' realist attitude toward astronomy and it is these astronomers who would carry forward what would come to be known as the Copernican Revolution.

6

Instruments of reform: Tycho's restoration of observational astronomy

6.1   The reception of *De revolutionibus*

In the years following the publication of *De revolutionibus* there were very few converts to Copernicus' heliocentric system. So few, in fact, that it is possible to mention all of them. Presumably we can count Copernicus himself, as well as his early advocate Tiedemann Giese. Rheticus, of course, was a convert (although there is some speculation that he may have changed his mind late in life) and we have already seen that de Zuniga advocated the Copernican system in his treatise on the Book of Job (although de Zuniga unequivocally abandoned his Copernicanism in a later work).[1] Rheticus seems to have made converts of two of his physician friends: Achilles Gasser and Georg Vögelin.[2] Gemma Frisius, an astronomer at the University of Leuven in Belgium, published a letter endorsing the Copernican system because of its elegant explanation of the links between the planets and the Sun.[3] These seven men may have been the only advocates for the Copernican theory who were born before the publication of *De revolutionibus* and the death of Copernicus.

Most astronomers who read *De revolutionibus* during the 50 years after it was published did not take Copernicus' claims about a moving Earth seriously.[4] They took the work exactly as Osiander asked them to: as a mathematical tool for prediction rather than as a statement about the true structure of the heavens. This attitude was particularly apparent among the astronomers and mathematicians at the University of Wittenberg (in Upper Saxony, now part of Germany).[5] Wittenberg was the home of Martin Luther, the initiator of the Reformation, who had once dismissed the idea of a moving Earth as ridiculous. It was also home to Philip Melanchthon, an influential Lutheran theologian and founder of the Protestant educational system in Germany. Melanchthon argued against the truth of the Copernican theory, but he also recognized some of the mathematical advantages of the Copernican system and promoted its study, provided it was understood only as a mathematical tool. It was from the University of Wittenberg that Rheticus had come to Copernicus, and it was at the University of Wittenberg that the first astronomical tables based on the Copernican system were developed, by Erasmus Reinhold. Reinhold published his *Prutenicae Tabulae* or *Prutenic Tables* in 1551, but although Reinhold had mastered Copernicus' mathematics it seems that he never believed in the Earth's motion. Nor, it seems, did his successor at Wittenberg, Caspar Peucer. The Wittenberg astronomers recognized some advantages in Copernicus' approach, particularly his technique for avoiding equants, but they seem to have been uninterested in the harmony and order that so impressed Copernicus himself.

The reception of Copernicus' work among his fellow Catholics was no warmer. Although Copernicus had been encouraged to publish his work by high-ranking Catholics such as Cardinal Nicholas Schoenberg and Bishop Tiedemann Giese, his work did not gain widespread support among his co-religionists. But although there may have been some internal objection to the Copernican theory within the Catholic church shortly after the publication of *De revolutionibus*, there was no public condemnation of Copernicus until much later.

The leading Catholic astronomer in the generation after Copernicus, Christopher Clavius, was a staunch defender of the geocentric system of Ptolemy. Clavius was impressed by Copernicus' mathematics but could not stomach the idea of a moving Earth. However, Clavius, who was the primary person responsible for the Gregorian calendar reform of 1582, did take a realist approach to astronomy. He believed that astronomers were as capable of determining the nature of celestial motions as natural philosophers were of figuring out terrestrial physics. So although he never adopted Copernicus' theory, he did share Copernicus' attitude toward astronomy.[6] Other Jesuits, though, were less tolerant of Copernicus and, after some disputes with Galileo (see Chapter 8), by the mid-17th century many Jesuits were vehemently anti-Copernican.[7]

Although there were not many early adopters, Copernicus' ideas did spread across Europe, assisted by the distribution of Reinhold's *Prutenic Tables*. A few astronomers who were born in the generation after Copernicus' death would adopt the Copernican system, and not just as a mathematical tool. One of these early Copernicans was Michael Maestlin, professor of astronomy at the University of Tübingen. Maestlin taught Ptolemaic astronomy in his courses, but he was a believer in the Copernican theory and gave instruction on Copernicus' ideas to some of his students, including Johannes Kepler (see Chapter 7). The Copernican theory was also advocated by Christoph Rothmann, an astronomer in Kassel (Germany), and by Simon Stevin, a mathematician in Holland. All of these men were impressed by the way Copernicus accounted for the links between the solar motion and the motion of the planets.[8]

The most influential proponent of the Copernican model in the English language was the mathematician and astronomer Thomas Digges.[9] His father, Leonard Digges, had published an almanac titled *A Prognostication everlasting*. When Thomas published a new edition of his father's almanac in 1576 he added an appendix titled *A Perfit Description of the Caelestiall Orbes according to the most aunciente doctrine of the Pythagoreans, latelye revived by Copernicus and by Geometricall Demonstrations approved*. This appendix was an English translation of several chapters from Book 1 of *De revolutionibus* and it included the diagram shown in Figure 6.1.

Digges' diagram mimicked the one from *De revolutionibus* shown in Figure 5.5, but he showed stars extending well beyond the sphere of Saturn. Digges' diagram indicated that the stars were not stuck on a thin spherical surface, but instead could be found at a wide variety of distances from the Sun. The possibility of the stars lying at various distances was opened up by the Copernican idea of Earth's rotation. The only reason for placing all of the stars on a spherical surface was to explain their daily motion using the idea of a rotating celestial sphere. With a rotating Earth it was no longer necessary for the stars to move at all, and thus there was no need for a celestial sphere. As long as the stars were far enough away to explain why the celestial poles didn't move, then they could be found at any distance from the Sun, extending even to infinite distances.

Copernicus had recognized that the celestial sphere could extend infinitely in his system, but he had avoided making any definite claim on this point. Digges was not so shy. In his diagram the region of the stars contained the description:

> This orbe of starres fixed infinitely up extendeth hit self in altitude spher-icallye, and therefore, immovable, the Pallace of Foelicitye garnished with prepetual shininge glorious lightes innumerable, farr excellinge our Sonne both in quantitye and qualitye, the very court of coelestiall angelles, devoid of greefe and replenished with perfite endlesse love, the habitacle for the elect.

Digges was saying that the region of stars extends up to infinite distances from the Sun. What is more, the starry heavens were home to the spirits of the saved, so the region of the stars serves the role of the Christian Heaven. In medieval cosmology Heaven had been located outside the starry sphere. Digges extended the stars into this spiritual realm.

Copernicus had made the universe vastly larger than was needed for the Ptolemaic system, but Digges took this process much farther: he made the universe

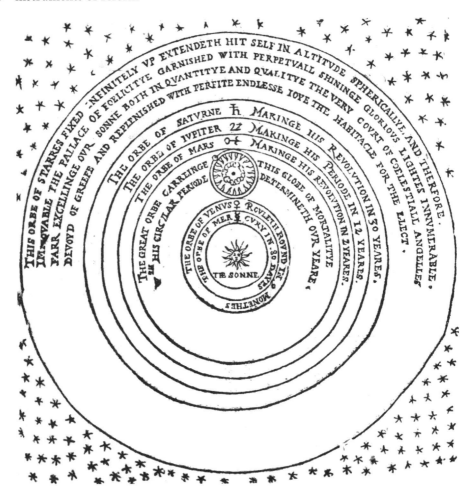

Figure 6.1 Digges' diagram of the Copernican system. The diagram is very
similar to Figure 5.5, but it shows stars extending out well beyond the final sphere
and it also includes some commentary on the theological role of the region of
stars. Image courtesy History of Science Collections, University of Oklahoma
Libraries.

infinite. He also said there were "innumerable" stars. This was a critical point
because it implied that there were stars that cannot be seen by the human eye.
On a clear dark night you can see only about 4500 stars (if you have outstanding
vision). If you could see the stars beneath your feet, those blocked by the Earth
from your view, then you could increase that total to about 9000. Nine thousand
is a lot, but it is not "innumerable." Digges implied that there are many stars
that are so far away we cannot see them. There was no reason to think that there

might be such invisible stars in the Ptolemaic system: all stars are at the same distance from Earth, so if some are visible why wouldn't all of them be visible? Digges' extension of the Copernican system predicts that if we could somehow extend our vision deeper into space, we might see new stars.[i]

The notion of an infinite universe filled with innumerable stars was adopted by another early Copernican, Giordano Bruno.[10] Bruno was a Dominican friar in Italy, but left his order and wandered across Europe speaking and writing about philosophy and theology. He spent some time in England and may have become familiar with Digges' work at that time. Bruno proposed that the universe was infinite and filled not just with stars, but with solar systems. He argued that each star was simply a distant Sun, and that like the Sun each of the stars was surrounded by a system of planets. He even argued that these planets (including those in our solar system) were all inhabited by rational creatures. Along with these radical ideas about the universe, Bruno proposed some radical ideas about theology, denying many of the core teachings of the Church. He was eventually convicted of heresy and burned at the stake in 1600. Although Bruno was not convicted primarily because of his support for the Copernican system, the promotion of that system by a heretic did little to gain support for Copernicus in Italy and other Catholic countries.

The early Copernicans we have mentioned here are almost all of those who adopted the Copernican theory prior to 1600. Most astronomers rejected the truth of the Copernican system, even if they found Copernicus' theory useful as a calculation tool. As we have seen, there were no strong reasons to accept the Copernican system at that time. The evidence was mostly against Copernicus, and although his theory did have a mathematical beauty that was unmatched by the Ptolemaic system, mathematical beauty was not high on the list of important criteria for 16th-century astronomers. Pretty mathematics was certainly not enough to overturn centuries of tradition, or to convince natural philosophers to throw out the only viable physics that was available at the time, that of Aristotle. To really move forward, the Copernican idea had to await two men who believed that beautiful mathematics held the key to understanding the universe: Johannes Kepler and Galileo Galilei. Before we meet these men, we need to discuss the revolution in astronomical *observation* (as opposed to astronomical theory) that took place near the end of the 16th century.

---

[i] Digges idea can also explain why stars differ in brightness. Even if all stars produce the same amount of light, more distant stars will appear dimmer than nearby stars. Stars that are too far away will be so dim that they cannot be seen.

Figure 6.2 Portrait of Tycho Brahe from a line engraving by T. Gemperlin, 1586. Image courtesy History of Science Collections, University of Oklahoma Libraries.

6.2 The noble astronomer

Tyge Brahe (Figure 6.2) was born to a noble family in Denmark. His father, Otte Brahe, served on the Privy Council for the Danish king and his mother, Beate Bille, was also of noble birth. There were some odd circumstances surrounding young Tyge's birth and childhood. He had a twin brother who died at birth, but he did not discover this fact until much later. At the age of two he was taken away to live with his uncle Jørgen Brahe, apparently without the permission of his parents. At the age of twelve he began studies at the University of Copenhagen. While in Copenhagen he developed a keen interest in astronomy, possibly inspired by a solar eclipse in 1560 (a total solar eclipse, but only partial as viewed from Denmark). It was around this time that he began using the Latinized form of his name, Tycho.[11]

After a few years at Copenhagen, Tycho began traveling throughout Europe, visiting universities where he studied the classics. He also studied astronomy and astrology, initially in secret because astronomy was not considered a fit subject of study for a nobleman who needed to focus on the arts of warfare and politics. Soon he was keeping a logbook of his own observations. He made estimates of the positions of the planets relative to the stars by holding a taut string between two reference stars and estimating the location of the planet along that string. He could then determine the approximate location of the planet on a small celestial globe that he owned. Even this crude method for measuring planetary positions was sufficient to show him that ephemerides (tables indicating the positions of the planets at regular time intervals) based on either the *Alfonsine Tables* or the *Prutenic Tables* were in error.[12] These early efforts led to a desire for more accurate measurements in the hope that these might lead to improved tables for the planetary motions.

Later, while studying at the University of Rostock (in Germany), Tycho became embroiled in a duel. He had predicted, on the basis of a lunar eclipse in October 1566, the impending death of the Ottoman sultan Suleiman the Magnificent. Unfortunately for Tycho, it turned out that the sultan had already died in September before Tycho had made his prediction. It may have been ridicule regarding his failed astrological prediction that led Tycho to challenge another Danish student at Rostock to a sword fight. In any case, the result was that Tycho received a disfiguring wound that left him with a scarred forehead and part of his nose missing. For the rest of his life he would wear a prosthetic metal nose.[13]

After returning to live with his parents for a time, Tycho moved to Herrevad Abbey near the residence of another noble uncle, Steen Bille. Uncle Steen helped Tycho to establish an alchemical laboratory at Herrevad, and Tycho also began the construction of new astronomical instruments to make more precise measurements of the heavens. While at Herrevad, Tycho also met the woman who would become his common-law wife. Kirsten Jørgensdatter may have been the daughter of a local pastor, but in any case she was not a noble. Tycho was forbidden to marry her in any formal sense without losing his title, but could live openly with her in a common-law marriage. Kirsten remained with Tycho for the rest of his life and gave birth to Tycho's eight children, none of whom were granted a noble title.[14]

While at Herrevad Tycho saw a new star appear in the heavens in 1572. The following year he published a short book, *De Stella Nova* (On a New Star), describing his observations of the star. These scholarly efforts, combined with his marriage to Kirsten, clearly indicated that Tycho was not planning to follow the traditional path of a noble lord. However, his book won him a degree of fame as an astronomer and the Danish king, Frederick II, wished to keep Tycho in the

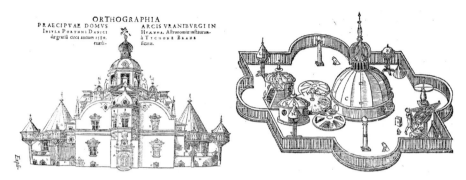

Figure 6.3 Main building of Uraniborg (left) and Stjerneborg (right). Images from Tycho's *Epistolarum astronomicarum*, 1596. Images courtesy History of Science Collections, University of Oklahoma Libraries.

country in order to promote Denmark's reputation in the sciences. To prevent Tycho from emigrating to another country to pursue his astronomy, Frederick offered Tycho control of the island of Hven in the Øresund (the strait that runs between modern Denmark and Sweden).[15]

With his own island and with financial support from the king, Tycho could turn his dreams of a grand observatory into reality. He built a new castle-observatory on Hven, which he named Uraniborg, the castle of Urania, muse of astronomy. Tycho filled his castle with the best instruments he could buy, working closely with the artisans to ensure that each instrument was of high quality. Eventually he would employ his own artisans in a workshop at Uraniborg devoted to producing the biggest and best astronomical tools ever created. Later he wanted to expand his array of instruments and shield them from wind and vibrations, so he built a new, partially underground observatory nearby, which he called Stjerneborg ("star castle").[16] Figure 6.3 shows sketches of Tycho's observatories.

Not only did Tycho have excellent instruments, he also employed teams of students and assistants to use those instruments to measure the stars and planets, night after night, throughout the year. Before Tycho most astronomers focused their planetary observations on special points such as stations or oppositions, rather than continually measuring the positions of the planets throughout their motions across the celestial sphere. Tycho's accurate and regular observations revealed the motions of the planets in unprecedented detail.

Most of Tycho's astronomical instruments were designed to measure angles: the altitude or azimuth of a celestial object, the angle between two objects, or the angle of an object relative to the ecliptic or the celestial equator. Tycho constructed specialized instruments for each type of measurement. Generally

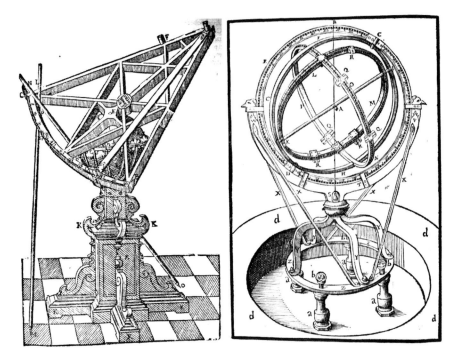

Figure 6.4 Triangular astronomical sextant (left) and zodiacal armillary (right), from Tycho's *Astronomiae instauratae mechanica*, 1602. Images courtesy History of Science Collections, University of Oklahoma Libraries.

the instruments involved a circular arc made of rigid metal and marked with degree markings, plus one or more sights so that the observer could line up the instrument with a particular star or planet. For example, Tycho's triangular sextant (Figure 6.4, left) was designed to measure the angle between two objects. It consisted of a circular arc marked from 0 to 60° (one-sixth of a circle, hence the name "sextant"). There was a fixed sight at the center of the circle and another at one end of the arc so that one side of the sextant could be aligned with a particular star. The body of the sextant could be moved around on its mount in order to line up another object in the second, moveable, sight. Once both objects were properly sighted the angle between the two could be read from the location of the moveable sight among the degree markings on the arc.

To measure angles relative to the ecliptic or equator it was necessary to have an instrument with metal circles that were aligned with the corresponding celestial circle. This was particularly challenging for measurements relative to the ecliptic, because the ecliptic moves around on the sky throughout the night. The metal ring representing the ecliptic and another perpendicular ring for measuring latitude had to be set on rotating axes so that they could be repositioned, as can be seen in the image of Tycho's zodiacal armillary (Figure 6.4, right).

Unfortunately, Tycho's zodiacal armillary proved subject to flexing that rendered the instrument unreliable, as he soon recognized.[17]

Tycho made significant innovations to improve his instruments. He developed a new alidade, or sighting device, that helped the observer precisely align the instrument to the star or planet. He also developed a method of using diagonal lines between the degree markings on the circular arcs that would allow the observer to measure angles to within a small fraction of a degree.[18] Even with this innovation, measuring angles to high precision was hard because even the tiniest movement of the sights might change the angle by a few minutes of arc. The solution to this problem was to make the circular arcs bigger, thereby increasing the distance covered by a single degree of arc. One of Tycho's most productive instruments was also his largest: the great mural quadrant (Figure 6.5). The mural quadrant consisted of a 90° arc (one-fourth of a circle, hence "quadrant") fixed to an east-facing wall inside Uraniborg. The center of the huge circular arc lay at a point on a wall to the south and an opening at that spot allowed the observer to sight a star or planet and thus measure its altitude as it transited across the meridian. Tycho commissioned a mural to be painted above the quadrant, depicting his instruments and assistants with himself (larger than life) directing observations.[19]

In addition to his high-quality instruments, Tycho had his own paper mill and printing press on Hven so that he could publish his work whenever he wished. He also operated alchemical laboratories. Life at Uraniborg was not all work. Tycho was a Danish nobleman and he knew how to hold a feast and to provide entertainment (such as the dwarf named Jeppe whom Tycho thought could see the future) for his guests. There is no doubt, though, that the focus at Uraniborg and Stjerneborg was on the accumulation of observational data. Tycho's observatories were something like a modern research institute, with roughly 100 students, assistants, and artisans employed over the course of twenty years. His efforts represent one of the first "big science" projects in Europe, and the great number of observations he gathered led to one of the first "big data" problems in science.[20]

Tycho's primary goal was the restoration of observational astronomy. He was convinced that he could make more accurate and more complete measurements of celestial phenomena than any of his predecessors, ancient or contemporary, and he believed that improved observations would lead the way to a major reform of astronomical theory. According to a report by Rheticus, Copernicus had claimed that he would be happy if his predictions were accurate to 10′ since the star positions used by Ptolemy were not any more accurate than that.[21] Tycho knew that, with his star castles, his teams of assistants, and the financial support from the king, he could do much better.

Figure 6.5 Mural quadrant, from Tycho's *Astronomiae instauratae mechanica*, 1602. Image courtesy History of Science Collections, University of Oklahoma Libraries.

6.3   Breaking the spheres

The new star, or nova, of 1572 had important implications for understanding the nature of the heavens. In his *De Stella Nova* Tycho described his first glimpse of this remarkable object and its significance.[22]

> Last year [1572], in the month of November, on the eleventh day of that
> month, in the evening, after sunset, when, according to my habit, I was
> contemplating the stars in a clear sky, I noticed that a new and unusual
> star, surpassing the other stars in brilliancy, was shining almost directly
> above my head; and since I had, almost from boyhood, known all the
> stars of the heavens perfectly (there is no great difficulty in attaining
> that knowledge), it was quite evident to me that there had never before
> been any star in that place in the sky, even the smallest, to say nothing
> of a star so conspicuously bright as this. I was so astonished at this sight
> that I was not ashamed to doubt the trustworthiness of my own eyes. ...
>
> For all philosophers agree, and facts clearly prove it to be the case,
> that in the ethereal region of the celestial world no change, in the way
> either of generation or corruption, takes place...

A new star appearing in the sky was a direct contradiction of the Aristotelian
notion that the heavens were perfect and unchanging, but only if that point of
light was, indeed, in the heavens. Tycho set about trying to determine this fact
by measuring the star's parallax.[23] We encountered the basic idea of parallax in
Section 3.3.2, but there we discussed the parallax for observers making simultaneous measurements from two different locations on Earth. Tycho could not be
in two places at once, so that method was unavailable to him. However, from the
Copernican point of view he could allow the rotation of the Earth to carry him
from one location to another. Although Tycho thought that it was the celestial
sphere, not the Earth, that rotated, the geometry is the same. A nearby object
should appear to shift its position relative to the stars as the celestial sphere, or
the Earth, rotates.

He measured the angle between the new star and other stars nearby. The
nova appeared in the constellation Cassiopeia, a circumpolar constellation as
seen from Denmark, so it was always above the horizon. Tycho measured the
star angles when the constellation was high in the sky and again when the
constellation was low to the horizon. If the nova was closer than the Moon these
angles should have changed noticeably from one position to the other, but Tycho
found no evidence of any change at all. Figure 6.6 shows a sketch from a much
later work in which Tycho illustrated the position of the nova among the stars.
If the nova remained in a fixed position relative to the stars, then it must either
be moving with the daily motion of the celestial sphere and also much farther
away than the Moon, or else it must be nearby but have a motion that tracks the
stars *and also* perfectly counteracts parallax. The second possibility was simply
unbelievable, so Tycho concluded that the nova must lie beyond the Moon. Even
six months later he found it to occupy the same spot on the celestial sphere. He
decided that anything that maintained such a fixed location relative to the stars

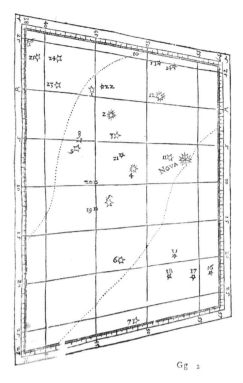

Figure 6.6 A sketch of the 1572 nova in Cassiopeia, from Tycho's
Astronomiae instauratae progymnasmata, 1602. The dotted lines indicate the
boundaries of the Milky Way. Image courtesy History of Science Collections,
University of Oklahoma Libraries.

must lie in the same heavenly sphere as the stars themselves. Eventually, the
nova faded and disappeared.

In his 1573 book on the nova, Tycho dismissed the possibility that the nova
was actually a comet. Comets were known to show tails (which the nova lacked)
and to move relative to the stars (while the nova did not). More importantly,
Aristotle had taught that comets were a meteorological phenomenon taking
place in the fiery sphere below the Moon. Since the nova was far beyond the
Moon it could not be a comet. Tycho expressed the hope that he would have the
chance to observe a comet in order to make his own judgment about whether or
not it lay below the Moon's sphere. He did not have to wait too long.

In 1577 Tycho was out fishing when he noticed what seemed to be a bright
star in the evening twilight. As he continued his observation he saw the star
grow a tail and he knew his wish to see a comet had been granted. As quickly
as he could he set about making a determination of the comet's parallax.[24] The
process was more complicated for the comet than it was for the nova of 1572. For

one thing, the comet was close to the Sun on the sky and therefore only visible for a short time each night. In addition, the comet moved relative to the stars. Tycho could not simply make two measurements separated by a few hours, note a change in position relative to the stars, and declare it a parallax. He had to first determine the comet's actual motion relative to the stars by observing the comet over several days.

After several observations Tycho was able to establish that the comet moved about 2° per day relative to the stars. For observations taken 3 hours apart, the comet would be expected to move by about 15′. Tycho's measurement indicated a shift of about 12′, so the comet's parallax was no more than 3′. For the same interval and angle on the sky the Moon would be expected to show more than 10′ of parallactic shift, so the comet must lie beyond the Moon. Once more Tycho had poked a hole in the Aristotelian picture of the heavens.

Not everyone was convinced by Tycho's observations, however. Some questioned the legitimacy of applying parallax measurements to such unusual objects as a comet or the nova. It is no surprise that natural philosophers were not eager to overturn centuries of tradition and understanding with no new theory of the heavens to take its place. Tycho gathered data from other astronomers, most notably from Michael Maestlin, to establish the accuracy of his results. Maestlin had used cruder instruments than Tycho's but had likewise found no parallax for the comet of 1577. What is more, the observations of astronomers across Europe showed that the comet had appeared at the same spot on the celestial sphere for all observers at any given time, clearly showing that the comet lay far above Earth's surface. In essence, Tycho united the voices of Europe's astronomical community to declare that the comet lay beyond the Moon.

Tycho examined the motion of the comet to try to determine its path through the celestial regions. The comet of 1577 had gradually moved out in front of the Sun and then slowed down while it got progressively dimmer. Tycho also noted that the tail of the comet generally pointed away from the Sun, as had been noted by some previous comet observers. These observations suggested to Tycho that the comet was in orbit around the Sun, outside of the orbit of Venus. When he got his hands on Maestlin's treatise on the comet he found that Maestlin had reached a similar conclusion.

Many years later in 1588, when he published a full account of his comet observations in *De mundi atherei recentioribus phaenomenis* (On recent phenomena in the ethereal world), Tycho presented a diagram showing the comet's orbit (Figure 6.7). Note how the tail of the comet points directly away from Venus, not the Sun, a conclusion that Tycho mistakenly reached after re-analyzing his measurements from 1577.[25] Note also, that Tycho's diagram shows Mercury and Venus orbiting the Sun. We will return to that important point shortly.

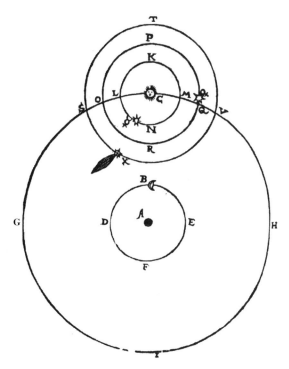

Figure 6.7 Diagram showing the orbit of the comet of 1577, as well as the orbits of the Sun, Moon, Mercury, and Venus, from Tycho's *De mundi atherei recentioribus phaenomenis*, 1588. Image courtesy History of Science Collections, University of Oklahoma Libraries.

Although Tycho's diagram showed the comet of 1577 moving like an inferior planet in the Copernican system, Tycho found that he could not account for the comet's changing position on the celestial sphere using uniform motion on a circle. He considered the possibility that the comet might change its speed or even move in a noncircular, possibly oval, orbit.[26] A comet moving in an oval orbit would pass through the spheres that both Copernicus and Ptolemy thought carried the planets in their motions around the Sun or Earth, but how could a comet pass through a solid, crystalline sphere? After observing the motions of several different comets Tycho was completely convinced that they moved through the traditional celestial spheres. He concluded that no such spheres exist. The heavens were composed of a fluid substance which comets could pass through easily.[i]

---

[i] Tycho was likely influenced in reaching this conclusion by the work of Michael Maestlin and Christoph Rothmann on comets.[27]

The idea of a fluid heaven was not a new one. Indeed, in the Early Middle Ages scholars tended to view the heavens as composed of a fiery, fluid element. After the reintroduction of Aristotle and Ptolemy into Europe, medieval thinkers adopted the notion of heavenly orbs that carried celestial bodies in their motions around the Earth. They considered these orbs to be "solid," but that just meant that they were three-dimensional regions completely full of matter with no empty space within them. Most medieval writers did not address the issue of whether these spheres were hard, or fluid. Even the term "crystalline sphere" usually just referred to the transparency of the sphere, not its hardness. However, it was clear that the orbs for different planets were not supposed to interpenetrate each other, nor was there allowed to be any empty space between the orbs.[28]

By Tycho's day many astronomers regarded the solid spheres of Ptolemy's *Planetary Hypotheses* and Peurbach's *Theoricae novae planetarum* as hard and transparent, like glass or crystal. Certainly, this was Tycho's understanding of the orbs. His determination of the comet's motion led him to claim that these hard orbs did not exist, and that the heavens were fluid. He believed that the stars were denser regions of this celestial fluid, and that a somewhat less dense form of the fluid made up the Milky Way. He even suggested that this fluid could condense to form a nova or a comet.[29] The nova of 1572 had appeared near the edge of the Milky Way (see Figure 6.6) and Tycho thought he could detect a gap in the Milky Way from which the fluid had been drawn to form the new star. As we will see shortly, though, he had even more important motivations for claiming that there were no hard crystalline orbs in the heavens.

Whatever his reasons, Tycho's shattering of the hard celestial spheres was influential. After Tycho astronomers were left searching for a new means of bringing about the planetary motions. If the planets were not carried around by solid (hard) rotating spheres, then what caused the planets to move? Several authors, including Tycho himself, believed that the planets moved through the fluid heavens like birds in the air or fish in the sea, but the motions of the planets were much more regular than the movements of birds and fish. How could the planets follow mathematical rules for their motions without some geometrical structure like solid orbs to dictate their movements?

Tycho would not be the one to answer that question, but the fact that his astronomical observations could even generate that question illustrated the growing power of astronomy in the late 16th century. Questions about the nature of the heavens had previously been reserved for natural philosophers. Copernicus' theory and Tycho's precise observations showed just how powerful the tools of astronomy and mathematics could be for investigating the truth about heavenly motions. The line between natural philosophy and astronomy was becoming blurred.

6.4   Stars against Copernicus

In addition to studying the nova and comets, Tycho carefully measured the positions of the stars on the celestial sphere. He became convinced that the ancient star catalogs that traced back to Hipparchus were inaccurate and that he could do a better job. He devised new methods for determining right ascensions and ecliptic longitudes for stars, and he carried out his measurements using his impressively precise instruments. Star positions were important because the stars served as reference points from which to measure the motions of other objects like comets and planets. At Uraniborg, Tycho constructed an enormous star globe depicting the positions of the stars on the celestial sphere according to his up-to-date measurements.[30]

Tycho's precise star measurements allowed him to investigate the precession of the equinoxes. Copernicus had made the Earth wobble in a complicated way to incorporate the effects of trepidation, but Tycho found little evidence to support trepidation. He measured the rate of precession to be about $1°$ every 71 years, and checking his new star positions against those of Hipparchus he found that the differences could be accounted for by assuming a constant precession rate. He admitted that there might be some long-term variation in the precession rate, and his measurements did indicate a change in the obliquity of the ecliptic since Ptolemy's time, but Tycho's work signaled the demise of trepidation theory.[31]

The star positions that Tycho measured did not seem to change over the course of a night. We have previously discussed the idea of parallax for the Moon and comets, treating the stars as a fixed background, but the stars themselves can show parallax effects. The geometry is a little bit more complicated if we assume that all stars lie on the surface of a celestial sphere, like Tycho did, but as Figure 6.8 illustrates the angle between two stars can appear to change as an observer moves around. Figure 6.8 shows this change for the case of an observer who moves around as a result of the Earth's daily rotation. Tycho would have thought of it as the daily rotation of the sphere of fixed stars around the Earth, but the geometry is the same. Tycho was sure that if the stars had shifted by as much as $1'$ (which corresponds to a parallax angle of $30''$ as defined in Section 3.3.2) he would have detected it.[32] The fact that he never saw any effects from parallax suggested that the distance to the stars was at least 7000 times the radius of Earth.[i]

---

[i] This estimate for the minimum stellar distance makes use of the parallax formula from Appendix A.6, which is not quite correct when dealing with the angle between two stars on the celestial sphere. However, Tycho seems to have been thinking in terms of this type of parallax calculation when he considered the parallax of the stars.

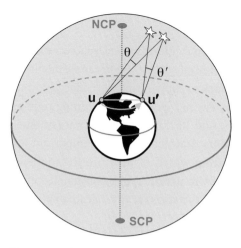

Figure 6.8 Diurnal parallax of the stars. The observer at u measures an angle θ between the two stars. Twelve hours later the observer has been carried by Earth's rotation to u' and measures a different angle, θ', between the stars. If Earth is stationary and the celestial sphere spins, the geometry is identical.

That conclusion came as no surprise. Ptolemy had already declared that the Earth was like a point in comparison to the sphere of the stars. In fact, we have seen that Ptolemy put Saturn at a distance of almost 20,000 Earth radii (see Table 4.1), so it was certainly no surprise that the stars had to be at least that far away. What was more significant was that Tycho found no variation in the postions of stars throughout the year. If the Copernican theory was correct, then the stars should show a parallax effect as the Earth's orbit carried the observer from one side of the Sun to the other over the course of six months. Using Tycho's 30″ value as an upper limit for this annual parallax he found that, *assuming the Earth orbits the Sun*, the distance to the stars must be nearly 7000 times the radius of Earth's orbit around the Sun.

Tycho accepted that the diurnal parallax of the Sun was about 3′,[i] a traditional value that goes back to Aristarchus, Hipparchus, and Ptolemy. Tycho never measured the solar parallax himself.[33] This value for the solar parallax indicated that the Sun was about 1150 Earth radii away. Therefore, the lack of annual parallax indicated that the stars must lie at a distance about 8 million times the radius of Earth. Compared to Ptolemy's distance of 20,000 Earth radii this distance was enormous. What is more, Copernicus' theory placed Saturn about nine times as far from the Sun as the Earth was, so the average distance to Saturn would be about 10,500 Earth radii using Tycho's value for the solar distance. If the

---

[i] The shift of the Sun's position due to parallax over the course of 12 hours would be 6′.

Copernican system were true, then there was a tremendous gap between Saturn and the stars, a situation that Tycho found absurd.[34]

Tycho realized that the great distance to the stars in the Copernican system caused even more problems. Tycho measured the apparent diameters of many stars and found that moderately bright stars (stars of the "third magnitude") had apparent diameters of about $1'$. If such a star was far enough away to explain its undetectable parallax in spite of Earth's orbital motion, then the diameter of that star must be comparable to the diameter of Earth's entire orbit around the Sun. Using Tycho's solar distance, the star would have a diameter 1150 times that of Earth. A brighter star, with a greater apparent diameter, would have to be even larger. In contrast, Tycho found the diameter of the Sun to be five times the Earth's diameter, while the Moon's diameter was less than one-third that of Earth. If Copernicus was correct, then not only were the distances to the stars entirely out of scale with the distances to the planets, but the sizes of the stars were entirely out of scale with the sizes of other celestial bodies.[35]

The problems with the stars were too much for Tycho to accept. These issues contributed significantly to his rejection of Earth's orbital motion.[36] A few of Tycho's fellow astronomers defended the Copernican system against this attack, but their defense was based on religion rather than science. For example, Christoph Rothmann argued that an infinite Creator could easily make stars of such great size and place them at such vast distances.[37] Few of Tycho's contemporaries were diehard supporters of the Copernican theory, though, and most probably accepted Tycho's arguments as further evidence for something they already believed: the Earth does not orbit the Sun.

6.5   The Tychonic system

In many ways Tycho was a fan of the Copernican system, at one point stating that it "expertly and completely circumvents all that is superfluous or discordant in the system of Ptolemy."[38] Like everyone else, he appreciated that Copernicus had eliminated equants. In general, Tycho was impressed with Copernicus' skill as a mathematical astronomer. He was not, however, very impressed by Copernicus' observational ability, and he felt that there were flaws in the way Copernicus had derived the parameters for his system from observational data.[39] Of course, Tycho also could not believe that the Earth truly moved.

Although his star data provided one set of objections to the annual motion of the Earth, Tycho had other reasons to reject Copernicus' theory. The contradictions with Aristotelian physics and with Holy Scripture certainly contributed to his attitude. Furthermore, his analysis of the motion of comets had revealed no annual effect that would arise from Earth's orbital motion.[40] So Tycho had physical, religious, and astronomical reasons for rejecting the idea that the Earth orbited the Sun.[41]

His resistance to the idea of Earth's daily rotation was less severe, but he ultimately decided that this motion didn't exist either. He pointed to arguments, based on Aristotelian physics, that Earth's rotation would alter the paths of moving objects in a noticeable way. Balls dropped from a tower should fall to the west, but this was not observed. Cannonballs fired to the west should travel farther along the ground than those fired to the east, but they don't. He also pointed out that the effects of Earth's rotation on the flight of a cannonball should depend on one's location on the Earth: at the poles the Earth's rotation should produce no movement, but at the equator it should produce a very fast movement. As a result, there should be noticeable differences in the paths of cannonballs fired at different latitudes, but in fact cannonballs seemed to travel the same at all latitudes. Therefore, the Earth must not be rotating.

Tycho's admiration for certain aspects of the Copernican theory, combined with his deep disbelief in the idea of a moving Earth, led Tycho to consider some form of inverted Copernican system that preserved the basic structure of the Copernican theory while keeping the Earth still. We know that he had prepared lectures on that topic as early as 1574.[42] Most likely he first adopted a Capellan system, with Mercury and Venus orbiting the Sun (as in Figure 6.7), but at some point before the 1588 publication of his *De mundi atherei recentioribus phaenomenis* he also decided that the superior planets orbited the Sun.

In that book Tycho laid out the basic idea of his new system: the Moon and Sun orbit the stationary Earth, but the planets (not including Earth) orbit the Sun.[43] Tycho displayed a diagram illustrating his **geoheliocentric** system, shown in Figure 6.9. Note that this system is identical to the Copernican system except that the Earth is held stationary rather than the Sun. Tycho's system also can be viewed as a modified version of the Ptolemaic system, with the Sun's orbit playing the role of the deferent for all of the planets, and the planetary orbits playing the role of the epicycle. For the superior planets that means Tycho had the epicycle larger than the deferent, whereas Ptolemy always had a smaller epicycle, but swapping the two circles didn't change the overall motion.[44]

More importantly, this so-called "Tychonic system" preserved the geometry of the Copernican system and therefore shared some of the pleasing features of that system: a definite order for the planets and definite (relative) sizes for planetary orbits. It also provided an explanation for the inferior/superior distinction, as well as the link between the apparent motions of the planets and the apparent (and actual) motion of the Sun.

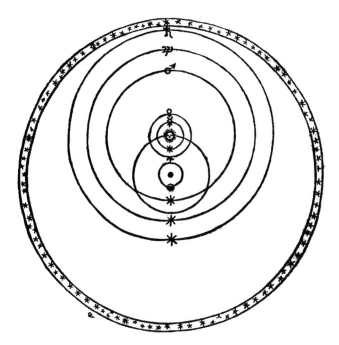

Figure 6.9 Diagram of Tycho's geoheliocentric system, from his *De mundi atherei recentioribus phaenomenis*, 1588. Image courtesy History of Science Collections, University of Oklahoma Libraries.

In fact, the only thing that really got messed up in the transition from Copernicus to Tycho was the nice relationship between period and distance. In Copernicus the relationship was clear: the farther a planet was from the Sun, the longer it took that planet to orbit the Sun. That was still mostly true in the Tychonic system, but there was now a gap between Venus and Mars. What would fit perfectly in that gap in terms of distance was Earth, since its distance from the Sun was between that of Venus and Mars. In terms of period what would fit the gap was the Sun with its one year orbit, but that was an orbit around the Earth rather than around the Sun. So it just didn't work. To anyone who cared about the distance–period relation, the Tychonic system practically screamed to put the Earth in motion and keep the Sun still.[i]

Most astronomers, though, did not care very much about that relation. What they did care about was the way the Copernican system conflicted with accepted science (Aristotle's physics) and theology. The Tychonic system seemed like a perfect compromise. It kept most of the explanatory power of the Copernican

---

[i] To paraphrase a later comment by Kepler.

system, but without introducing the ridiculous idea of a moving Earth. It would have been the perfect solution, except for two problems.

The first problem was that someone else had already proposed it, or something that looked quite like it. That someone was Nicolaus Raimers Baer, also known as Ursus. In 1584 Ursus had been a visitor at Uraniborg and at some point had been caught snooping around Tycho's library. He was found with some Uraniborg papers in his possession and grew upset when those papers were taken from him. Eventually he left Hven, never to return. Then in 1588, just months before Tycho publicly revealed his new system, Ursus published a book called *Fundamentum astronomicum*. In that book he presented a theory just like Tycho's, except that he assigned the daily rotation to the Earth rather than the stars, and his orbit for Mars was slightly different.[45]

The reason Ursus chose a different orbit for Mars was related to the second problem with Tycho's system: Tycho's orbit for Mars intersected his orbit for the Sun, as shown in Figure 6.9. It *had to* do so because in both the Copernican and Tychonic system the sizes of the planetary orbits were fixed relative to the Earth–Sun distance. If you increased the size of Mars' orbit so that it didn't intersect with the Sun, as Ursus did in his book, then it wouldn't fit with the observational data. However, Tycho's contemporaries were bothered by the overlap. If the planets were carried by hard crystalline spheres, then the intersection of the spheres of Mars and the Sun should be impossible.[i] That is where Tycho's comet observations came in. By using his comet observations to eliminate hard celestial spheres, Tycho made space for the acceptance of his new theory.

Although Tycho's elimination of hard spheres made it possible for the orbit of Mars to intersect the orbit of the Sun, it did nothing to show that those orbits actually did intersect. To prove that the orbits overlapped Tycho would need to show that when Mars was in opposition the Earth–Mars distance was less than the Earth–Sun distance. The obvious way to do that would be to measure the diurnal parallax of Mars at opposition. We have seen that Tycho assumed the Sun had a diurnal parallax of 3'. At opposition, if his theory (or Copernicus') was correct, then the diurnal parallax of Mars should be 4' or more.[ii] On the other hand, if Ptolemy or Ursus were correct, then the Martian parallax should

---

[i] Note that there is no worry about the bodies of the Sun and Mars actually colliding because the Sun is always at the center of the Martian orbit, so it can never be at the intersection points of the spheres. Also, note that the orbits of Mercury and Venus also intersect the Sun's orbit. Tycho's contemporaries were not bothered by that fact, presumably because the spheres of those planets could be embedded within a larger sphere that contained the Sun and rotated once per year, somewhat like the way Ptolemy and Peurbach embedded epicycle spheres inside larger Earth-centered spheres.

[ii] So Mars' parallax shift would be 8' or so over 12 hours.

be less than that of the Sun. Tycho's instruments were accurate to about 1′, so a measurement of Martian parallax might be just within his reach. Tycho was well aware that parallax observations could help decide between the Ptolemaic and Copernican/Tychonic arrangements, and he made three separate attempts to measure the diurnal parallax of Mars.[46]

As we have seen, measuring the diurnal parallax of an object that moves relative to the stars is tricky, but Tycho knew how to handle it. By 1582 Uraniborg was up and running and Tycho had his big, high-precision instruments ready to tackle the problem of Mars. During the Martian opposition of late 1582 and early 1583 he used his astronomical sextant (Figure 6.4, left) to measure Mars' position relative to reference stars in the evening and again before sunrise.[47]

When an object is near the horizon, diurnal parallax always shifts the apparent position of the object lower, to a location on the celestial sphere that is nearer the horizon, as compared to where the object would appear on the celestial sphere if it were directly overhead. During opposition, when Mars has just risen above the eastern horizon after sunset, parallax will shift Mars' position eastward on the celestial sphere. When Mars is low on the western horizon just before sunrise, parallax will shift Mars' position westward. Therefore, in comparing the morning position to the previous evening position, parallax will make it seem as though Mars has moved a little bit westward relative to the stars. Of course, Mars is in retrograde at opposition, so during this time it actually *is* moving westward relative to the stars, but parallax effects should make the apparent westward motion of Mars slightly greater than expected.

Tycho's 1582/83 measurements seemed to indicate that the apparent westward shift of Mars, relative to the stars, between evening and morning was slightly *less* than the expected value. The difference was negligible, only about 1′, and could be attributed to imprecision of the sextant, but these results certainly did not indicate that Mars had a measurable parallax. It seemed that the Copernican system, and also the as-yet-not-published Tychonic system, were ruled out. Tycho made another attempt during the Martian opposition of 1585 and again he found the westward motion less than would be expected without parallax.[48] These results made no sense because parallax could only increase the apparent westward motion. Even the Ptolemaic system predicted a 2–3′ increase, while the Copernican and Tychonic systems predicted twice that much. Tycho's results appeared to contradict *all* of the systems! Clearly something was wrong.

By the Martian opposition of 1587 Tycho had determined two possible problems with his parallax measurements. One problem was that wind and vibrations could lead to errors in measurements made with the Uraniborg instruments. With the partially subterranean Stjerneborg observatory available he could avoid these problems. The other problem was refraction.

Refraction is the bending of a light beam when it passes from one substance, or medium, into another. For example, a beam of light bends when it passes from air into a glass lens, and then bends again when it passes out of the glass and back into the air. It is this bending of light that allows properly shaped glass lenses to focus parallel beams of light to (or away from) a common point, which is the basic idea behind eyeglasses and contact lenses. However, refraction also affects the light of celestial bodies as it travels to the Earth. Whether the ether was a hard crystalline substance or a fluid, Tycho was convinced that it was different from air. As the light from Mars traveled from the ether into the air, refraction would bend its path. Tycho knew that when he looked at Mars his eye would trace back the path of the incoming light to determine Mars' location on the celestial sphere. If the path of the light was bent, then the position of Mars he saw would be different from its true position.

If a star or planet is directly overhead, refraction has no effect, but if the object is near the horizon, refraction will cause the object to appear higher above the horizon than it really is. Ptolemy and Copernicus were aware of the problem of refraction, but they tried to avoid it by steering clear of observations near the horizon (hard to do in the case of Mercury, which explains some of the problems with their Mercury data). However, observations near the horizon were exactly what Tycho needed for his parallax measurements. Because refraction would shift Mars' position higher above the horizon, while parallax would shift it lower, the failure to account for refraction could explain why Tycho had found a nonsensical negative parallax in 1582/83 and 1585. Perhaps refraction had shortened the westward shift by a greater angle than parallax had lengthened it. The only way to get at a true measurement of parallax was to figure out how to correct for refraction.

Tycho set out to measure the effects of refraction directly and to do so he used the object whose motion was best understood: the Sun.[49] He made measurements of the Sun's altitude above the horizon and compared them to the predicted values based on solar theory. Crucially, Tycho corrected the predicted position of the Sun for parallax.[50] In other words, he *assumed* that diurnal parallax had shifted the Sun's apparent position downward by 3′. That 3′ had to be added to the measured upward shift of the Sun in order to determine the true effect of refraction. When he compiled a table giving the correction for refraction as a function of the angle above the horizon, his solar parallax value was hidden inside.[i]

---

[i] Incorrect values for the solar obliquity and eccentricity also introduced errors into Tycho's refraction table.[51]

Although Tycho never published the details, it seems likely that he used his solar refraction measurements to correct his observations of the 1587 Mars opposition. By correcting for refraction in this way Tycho effectively built in a parallax for Mars equal to that of the Sun. As a result, he would have found a noticeable parallax for Mars. Because of other, minor problems with his solar refraction values as well as small errors in his observations Tycho would have found a parallactic shift for Mars greater than 5′, in line with predictions based on the Copernican/Tychonic system. Going back and correcting his 1582/83 measurements for refraction in the same way he would have found similar results. It seemed that Mars really was closer than the Sun to Earth. Ptolemy was disproved and the Tychonic system could stand.[52]

It may have been these new results, corrected for refraction, that gave Tycho the confidence to go public with his Tychonic system in 1588, although he did not mention any such measurements in his book. We know that later in 1588 Tycho wrote a letter in which he implied that he had shown Mars to be closer than the Sun, and in early 1589 he wrote to Rothmann suggesting that his 1582/83 observations had proved this result and thus destroyed the Ptolemaic theory. In November 1589 he wrote to the astronomer Thaddeus Hagecius claiming that his 1582/83 Mars observations had shown for certain that Mars had a parallax greater than the Sun's.[53]

We now know that Tycho's parallax measurements were wrong. At opposition Mars is closer than the Sun to Earth, but the diurnal parallax of both the Sun and Mars is too small for Tycho to have detected. The problem was Tycho's assumption that the ancient 3′ solar parallax was correct. By building this value into his solar refraction values he effectively introduced an extra 3′ of fake parallax into his Martian parallax measurements. Tycho may have realized this problem near the very end of 1589 when he investigated the refraction of stars and the planet Jupiter. His refraction angles for the stars were consistently about 4′ less than his solar values. What is more, the values for Jupiter were similar to those of the stars but quite different from the solar values. If Tycho re-evaluated his Mars observations using his results for stellar parallax, which did not incorporate the assumed 3′ solar parallax, he would have found no evidence of parallax for Mars. In any event, he never again claimed to have measured the parallax of Mars.[54]

What he did do was to publish a book of his correspondence with other astronomers. Tycho's 1596 *Epistolae astronomicae* (Astronomical letters) included several letters in which he laid out the plagiarism case against Ursus. It also included his letter to Rothmann claiming that he had shown Mars to be closer than the Sun. Why would he include that letter, when he had come to doubt his claim about Martian parallax? It was probably to help his case against Ursus. By suggesting that he had set out to prove Mars' parallax to be larger than the Sun's in 1582 (rather than 1585 or 1587) he could show that he was already searching for evidence of the Tychonic system, not only before Ursus published his version but before Ursus visited Uraniborg.[55]

The feud between Tycho and Ursus would continue to the ends of their lives. Even after Ursus was dead, Tycho continued to pursue a libel suit against him. More importantly, the astronomical community sided with Tycho. His published letters were sufficient to convince astronomers that the Tychonic system was truly his, and the letter to Rothmann even convinced some astronomers that Tycho had proved that Mars was sometimes closer than the Sun to Earth, thus ruling out the Ptolemaic arrangement.[56]

Shortly after the publication of his *Epistolae astronomicae*, Tycho abandoned his observatories. Frederick II had died in 1588, but at that time his son Christian was only 11 years old. He was not crowned King Christian IV until 1596, but soon after the coronation of the new king the situation changed on Hven. Christian and his advisors were not interested in funding Tycho's astronomical pursuits, and soon they had at hand complaints against Tycho based on his treatment of the locals on Hven and his failure to fulfill his obligation to maintain a chapel at Roskilde Cathedral.[57]

The situation was serious enough that Tycho, his family, and his assistants left Hven to return to Copenhagen. Things went no better in Copenhagen, and it became clear to Tycho that he would be unable to continue his astronomical work in Denmark. In 1597 Tycho packed up his astronomical instruments and left Denmark for good. For about two years he travelled through Europe before finally settling in Prague, where he renewed his astronomical work under the patronage of Holy Roman Emperor Rudolph II.[58]

Once Tycho, his instruments, and his team were settled into Benatky Castle, outside of Prague, Tycho returned to his astronomical work. He was aided by his longtime assistant Christian Sørensen Longomontanus, as well as by a new assistant, a young German named Johannes Kepler, who would use Tycho's data to change the very nature of astronomy.

6.6   Reflections on science

One of the chief things we can learn from Tycho about the nature of science is the importance of precision and accuracy in measurement. Keep in mind that most of the angular measurements we have discussed in this chapter have involved at most a few minutes of arc: apparent diameters of stars and planets, parallax shifts for comets and Mars, etc. Recall that Copernicus believed ancient observations were accurate to no better than 10′ and in fact many of Ptolemy's star positions were off by as much as 1°. On the other hand, modern analysis indicates that Tycho's star positions were correct to about 1′.[59] It is probably not possible to be more accurate than Tycho was without a telescope. Before Tycho developed his impressive instruments and innovative techniques it would have been futile to try to measure changes in the position of a celestial object that amount to only a few minutes of arc.

Precision is important even when trying to measure something that is too small to be measured. Tycho never did find any parallax (diurnal or annual) in the stars. But even that failure provided important information. If he could not detect stellar parallax, that was a strong indication that either there was no parallax or else it was smaller than 1′. In other words, Tycho's observations established a new "upper bound" for stellar parallax of 1′. This was an important result because it meant that either the Earth doesn't orbit the Sun or stars are much farther away, and therefore much bigger, than was previously believed. Most astronomers found it much easier to reject the (already questionable) motion of the Earth than to admit such enormous, and enormously distant, stars.

Precision is also important because it removes one possible excuse for errors in a model. Recall that the parameters of astronomical models are determined from observational data. Inaccurate observations lead to bad parameter values, so that even if the basic structure of the theory is correct you can end up with inaccurate predictions. Errors of about 1° in the *Alfonsine Tables* or *Prutenic Tables* could be blamed on bad observational data rather than on problems with the models used by Ptolemy or Copernicus. By improving the accuracy of the observations that are used to determine the parameters of a model, it becomes possible to really test the limits of that model. If your parameters are derived from observations that are accurate to 1′ but your model makes predictions that are off by 1° it probably indicates a problem with the model itself. In the next chapter we will see that Tycho's precise observations of Mars, even though they failed to reveal the Martian parallax, would show that all three astronomical systems (Ptolemaic, Copernican, and Tychonic) were seriously flawed.

Tycho's career also illustrates the growing power of astronomers in 16th-century Europe. During this period astronomers began to engage in debates that had once been the exclusive realm of natural philosophers or theologians. The accuracy of Tycho's observations made his claim that the nova and comets were celestial, not meteorological, phenomena hard to ignore. Most of Tycho's fellow astronomers were convinced that there could be no solid spheres in the heavens and that the celestial ether was a fluid substance capable of change that could produce new stars or comets. Natural philosophers were slower to accept Tycho's claims, but by the mid-17th century, after further astronomical discoveries by Kepler and Galileo, the notion of solid heavenly spheres was in serious decline. Even more notable is the fact that natural philosophers began to directly address the work of Tycho and others in their writing. Discussions about the true nature of the heavens could no longer ignore the work of astronomers.

The growing influence of astronomy on questions of natural philosophy is well illustrated by Tycho's attempt to measure the parallax of Mars, which serves as an example of a *crucial experiment*. The idea of a crucial experiment is that the outcome of a single experiment or observation can allow us to decide between competing theories. For a while Tycho believed he had eliminated the Ptolemaic theory by showing that the parallax of Mars at opposition was greater than the parallax of the Sun. As we have seen, though, the word "observation" or "measurement" hides a great deal of complication. Tycho's measurement of Mars' parallax involved much more than just sighting Mars in his instruments. It involved calculations and corrections that incorporated many background assumptions, the most important being the $3'$ diurnal parallax of the Sun. Therefore, a crucial experiment does not just test competing theories. It also tests background assumptions. If a theory seems to fail the crucial test, it may be that the background assumptions are really to blame.

Deciding between competing theories is difficult. The evidence is often ambiguous, and sometimes the "evidence" is just plain wrong. Moreover, theories can never be judged in isolation. They are always judged in the context of a whole network of assumptions and beliefs. For a time Tycho felt that his observational data painted a clear picture: his Martian parallax ruled out the Ptolemaic theory, while his measurements of the apparent sizes of stars and failure to find stellar parallax showed that the Copernican theory led to absurd stellar sizes. We now know that his Martian parallax was fictitious and his apparent sizes for the stars were misleading.

In the face of all of these problems, it is not entirely unreasonable to cling to a theory even when the data seem against it. That is particularly true when the data are not well understood or when they are at the limits of our measuring capabilities. A single "crucial experiment" is rarely sufficient to make people abandon a cherished theory. Tycho did not give up on his geoheliocentric world system even when he came to doubt his Martian parallax measurement.

Deciding between competing theories is usually a gradual process, and one that involves many lines of evidence and argument rather than a single crucial experiment. At the start of the 17th century there were strong reasons to believe in the Tychonic model, or one of its semi-Tychonic variants that admitted a rotating Earth. These models incorporated most of the admirable features of the Copernican model, but they avoided the problem of star distances and sizes. More importantly, they avoided conflicts with Scripture and Aristotelian physics. In short, the Tychonic system just seemed to make more sense. To emerge triumphant, the advocates of the Copernican system would need to address the conflicts with the Bible, Aristotle, and star observations, and show that the Sun-centered system could make sense of things that could not be explained in an Earth-centered system.

7

Physical causes: Kepler's new astronomy

7.1   The secret of the universe

Johannes Kepler (Figure 7.1) was a complex man, born into a complex and difficult world, who sought simplicity and harmony throughout his life. He was born to a devout Lutheran family in the Catholic town of Weil der Stadt, which lay within the Lutheran Duchy of Württemberg (now part of Germany). Although Kepler's paternal grandfather served for a time as the mayor of Weil der Stadt, Kepler's father was a soldier-for-hire who would vanish while fighting in a foreign war during Kepler's youth. His mother was the daughter of an innkeeper. Kepler's family was respected, but his social status was far below that of Tycho Brahe.[1]

According to his later recollection, the young Kepler was shown the comet of 1577 by his mother. Not long after that his father took him outside at night to observe a lunar eclipse.[2] Later in life, Kepler would offer the first correct explanation for the reddish tinge of that eclipse. In any event, these moments of wonder in an otherwise difficult childhood inspired Kepler's interest in astronomy.

The young Johannes began his education in the local German school, but his academic gifts soon landed him a spot in the Latin school and placed him on the path toward becoming a Lutheran clergyman. He won a scholarship to study at the University of Tübingen in 1589. There he learned astronomy from perhaps the only astronomy professor in Europe who believed in the Copernican theory, Michael Maestlin.[3] Maestlin taught Ptolemaic astronomy in his general courses, but endorsed the Copernican model in discussions with his more advanced students.[4] Maestlin's unusual beliefs went beyond just an acceptance of the Copernican model: he also believed that the Moon was a body not unlike the Earth, with mountains, valleys, and seas.[5]

Figure 7.1 Portrait of Johannes Kepler (artist unknown). Image courtesy History of Science Collections, University of Oklahoma Libraries.

The mathematical harmonies and explanatory power of the Copernican theory greatly appealed to the young Kepler, and soon he was engaging in debates in which he advocated for the Copernican view. In preparing for one such debate he imagined what the skies would look like for an observer on Maestlin's Earth-like Moon. Kepler realized that the lunar dwellers would believe themselves to be motionless, while they would see each planet weave an intricate dance mixed of the planet's orbit around the Sun, the Earth's annual orbit, and the Moon's monthly orbit around the Earth. The view from the Moon served as a perfect illustration of how the motion of the observer would show up in the apparent motions of other objects, and how that apparent motion could be disentangled from the true motions (just as Copernicus had claimed to do for the planets). Kepler would later expand his notes for this debate into a book, *Somnium* (The Dream), in which a young man makes a journey to the Moon. The *Somnium* may be the first science fiction story ever written.[6]

Although Kepler was an excellent student, there were some reasons for the Tübingen faculty to be skeptical about his success as a Lutheran pastor. Certainly, his defense of the Copernican model indicated Kepler's taste for unconventional ideas. More troubling, though, were Kepler's doubts about some points of Lutheran doctrine and his failure to condemn the rival Calvinists. Whatever the reasons, the faculty at Tübingen had plans for Kepler that did not involve him preaching to a congregation. Instead, shortly before the end of his third year at Tübingen, Kepler was appointed to serve as a math teacher at a boys' school in Graz, Styria (now part of Austria). Kepler was disappointed that his path toward becoming a pastor had been derailed, but he accepted his new assignment and tried to make the best of it.[7]

Kepler did not take to his new job right away. He seems to have been, at best, a mediocre teacher. In addition to his teaching duties he also served as the district mathematician, preparing astrological almanacs for the region. In his first such almanac he made some successful predictions about the weather and the actions of the Turkish army. After that his almanacs were well-received.[8] But it was while teaching that Kepler would have the insight that he thought revealed the secret of the universe.

He was lecturing to his students about "Great Conjunctions" when Jupiter and Saturn appear together in the sky, an appropriate topic for a mathematics teacher to cover since astronomy was considered a branch of mathematics. Kepler drew a diagram (Figure 7.2) to illustrate how the locations of the conjunctions moved around within the zodiac, with each successive conjunction

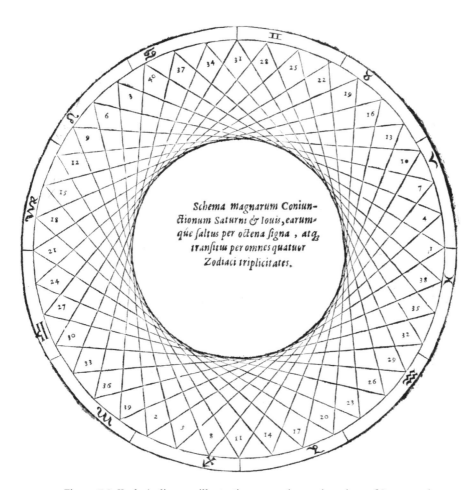

Figure 7.2 Kepler's diagram illustrating successive conjunctions of Saturn and Jupiter, from *Mysterium cosmographicum* (1596). Image courtesy History of Science Collections, University of Oklahoma Libraries.

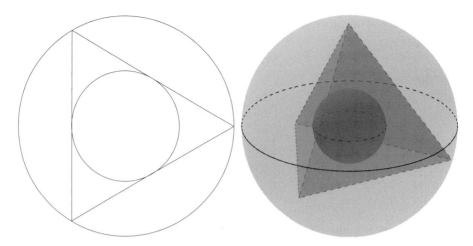

Figure 7.3 Left: circles inscribed within, and circumscribed around, an equilateral triangle. Right: spheres inscribed within, and circumscribed around, a regular tetrahedron.

occurring about 120° from the previous one so that the sequence of conjunctions almost formed an equilateral triangle.[9]

Kepler had been wondering why the orbits of the planets in the Copernican system had the sizes they had. He knew that the relative sizes of the orbits could be determined from observations, but *why* did they have those particular sizes and not others? When he drew his Great Conjunction diagram he realized in a flash that he might have hit on the answer. Perhaps the orbit of Saturn was inscribed around an equilateral triangle, while the orbit of Jupiter was inscribed within the same triangle, as shown on the left in Figure 7.3. If so, the ratio of the orbital sizes would be set by the geometry of the triangle.

It didn't take Kepler long to figure out that this idea didn't work. First of all, the ratio of the two circles shown in Figure 7.3 is two to one, which doesn't fit the ratio of any pair of adjacent planetary orbits. He tried other shapes (squares, pentagons, hexagons) but soon realized that the numbers weren't working out. In any case, there were an infinite number of regular polygons (two-dimensional closed figures with all sides the same length and all angles equal). Perhaps he was thinking in the wrong number of dimensions. The universe was three-dimensional, not two dimensional. Then the light dawned: he should be thinking about spheres rather than circles, and regular polyhedra (three dimensional shapes in which each face is a regular polygon, also called regular solids or Platonic solids) rather than polygons.

Kepler later viewed this insight as something akin to a divine revelation. There were six planets in the Copernican theory (including Earth), so there were five spaces between the planetary orbits, and there were also only five regular polyhedra: the tetrahedron (with four faces), the cube (six faces), the octahedron (eight), the dodecahedron (twelve), and the icosahedron (twenty).[10] By placing the regular polygons between the planetary spheres, Kepler could not only explain the relative sizes of the orbits, he could also explain *why there were exactly six planets!*

He set to work figuring out the details. Did the ratios of the orbits match the ratio for the spheres in and around the regular polyhedra? Kepler immediately ran into the problem of what point to use as the center. Kepler felt that the actual body of the Sun should play the role of the center point. He was partly motivated by his belief that the universe mirrored the structure of the Holy Trinity, with the Sun corresponding to God the Father, the celestial sphere corresponding to the Son, and the intervening space corresponding to the Holy Spirit.[11] He was also motivated by the idea that motions should be referred to physical bodies, rather than to empty points in space.[12]

Recall that Copernicus had used the center of Earth's orbit as his reference point, so all of the values that Copernicus had calculated were measured relative to that point. Kepler needed to recalculate all of the planetary distances (accounting for the eccentricity of the orbits, as well as Copernicus' epicyclets and other complications) from the body of the Sun. That task was beyond his ability so early in his career, so he appealed to his former teacher Maestlin for help. Maestlin obliged and Kepler found the numbers to agree well enough with his idea that he was ready to go public.[13]

Kepler published his novel idea in a short book with a long title that is usually referred to as *Mysterium cosmographicum* (The Secret of the Universe).[14] The book begins with a preface describing the three problems that serve as the focus of the *Mysterium*: explaining the number, sizes, and speeds of the planetary orbits. Since his attempt to answer these questions was grounded in the Copernican theory, Kepler followed this preface with a defense of the Copernican model, citing the usual explanations of retrograde motion and apparent links to the Sun as the primary arguments. Kepler also noted that the Copernican theory explained the pattern of epicycle sizes in Ptolemy: the epicycle of Saturn (relative to its deferent) is smaller than that of Jupiter, which is smaller than that of Mars. Copernicus could explain this pattern because the epicycles of those planets are really all the same circle (Earth's orbit), while the deferents are the true orbits of the planets with Saturn's larger than Jupiter's, etc.[15]

Kepler went on to explain his idea of nesting the planetary spheres (and thus the orbits) within the Platonic solids.[16] Outside the sphere of Mercury lay an octahedron, which was surrounded by the sphere of Venus. Outside of that lay the icosahedron, then the sphere of Earth, then the dodecahedron, then the

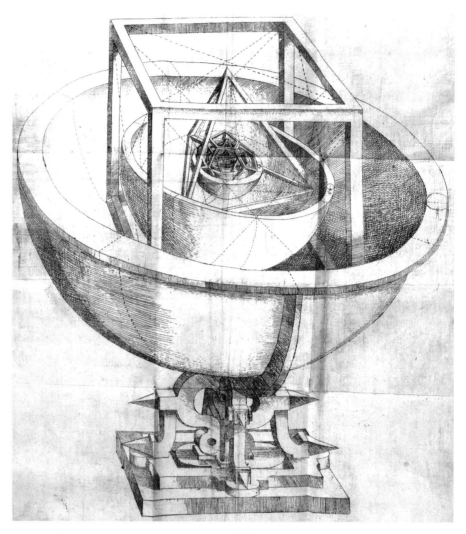

Figure 7.4 Drawing of a physical model illustrating Kepler's regular solids theory, from *Mysterium cosmographicum* (1596). Image courtesy History of Science Collections, University of Oklahoma Libraries.

sphere of Mars. Outside of Mars lay the tetrahedron surrounded by Jupiter's sphere, which in turn was surrounded by a cube and then the sphere of Saturn. Kepler presented an illustration showing a physical model of these nested solids and spheres (Figure 7.4), which he later tried to have made into a metal bowl for serving beverages.

In the *Mysterium* Kepler went to great lengths to explain why the regular solids should have this particular order based on geometrical and astrological considerations, with no account taken for how well this arrangement fit with the actual

Table 7.1 *Distance table from Kepler's Mysterium. The table shows the greatest and least distance of each planet from the center of Earth's orbit (mean Sun) in Copernicus and from the true Sun as calculated by Maestlin. It also gives distances derived from Kepler's regular solids theory in two versions: (I) excluding the Moon's sphere from that of the Earth, and (II) including the Moon's sphere within Earth's sphere. All distances are in Astronomical Units.*[17]

Planet	Height	Mean Sun	True Sun	Solids I	Solids II
Saturn	Highest	9.7	9.99	10.60	11.30
	Lowest	8.65	8.34	8.85	9.44
Jupiter	Highest	5.46	5.49	5.11	5.45
	Lowest	4.98	5.00	4.65	4.96
Mars	Highest	1.67	1.66	1.55	1.65
	Lowest	1.37	1.39	1.31	1.40
Earth	Highest	1.00	1.04	1.04	1.10
	Lowest	1.00	0.958	0.958	0.898
Venus	Highest	0.761	0.741	0.761	0.714
	Lowest	0.678	0.696	0.715	0.671
Mercury	Highest	0.490	0.489	0.506	0.474
	Lowest	0.301	0.233	0.233	0.219
Sun	Highest	0.042	0	0	0
	Lowest	0.032	0	0	0

sizes of the planetary orbits in the Copernican system. Only after presenting this "necessary" ordering did Kepler give Maestlin's computations of the least and greatest distances of each planet from the Sun, along with a comparison of these distances to the predictions of his regular solids theory.[18] Table 7.1 shows his results.

Kepler ran into a few difficulties in working out the details of his regular solids theory. First of all, he could not find any good place for the octahedron. The ratio of the spheres for this polyhedron just didn't seem to fit any of the orbital ratios. However, Kepler found that if he inscribed the inner sphere on the square that lies in the middle of the octahedron, then he got a ratio that worked for the orbits of Venus and Mercury. In this case the inscribed sphere actually poked out from the octahedron a little bit, but Kepler did his best to justify this alternate approach.[19]

Another problem was how to handle the orbit of the Moon around Earth. Either the Moon's orbit could be considered part of Earth's sphere, in which case it would add to the maximum distance and subtract from the minimum distance for that sphere, or else the Moon's orbit could be ignored. After all, the

Moon was a separate body, but Maestlin had argued that it was very similar to Earth. Kepler tried it both ways and both results are presented in Table 7.1. For the superior planets it seemed to work better if the Moon's orbit was included, but for the inferior planets it was better to exclude the Moon's orbit.[20]

Overall, though, the results looked pretty good. The numbers from the regular solids theory matched reasonably well with Maestlin's distances from the true Sun. Kepler pointed out that the Prutenic Tables contained errors of up to 2° in the planetary positions, so perhaps the small discrepancies were due to problems in Copernicus and not in Kepler's regular solids theory.[21] The largest discrepancy was for the Jupiter–Saturn ratio, but these most distant planets would be most subject to errors in the determination of their distances. The other small discrepancies could be blamed on inaccuracies in the determination of the eccentricity or line of apsides (the distance and direction, respectively, of the center of each orbit from the mean Sun) for each orbit.[22] He was particularly concerned about errors in the eccentricity of Earth's orbit, which would affect the calculation of distance from the true Sun for every planet. Only more accurate determinations of the orbital parameters could put his theory to the ultimate test, and decide the issue of how to deal with the Moon's orbit.

After presenting his regular solids theory, Kepler analyzed the speeds of the planetary motions.[23] He noted that the more distant planets had longer orbital periods not just because they had farther to travel, but also because they actually moved at slower speeds. Why, then, did planets farther from the Sun move slower? Kepler suggested that the planets might be propelled by what he called "moving souls," but how did these moving souls act? Kepler stated that[24]

> ... one of two conclusions must be reached: either ... the moving souls are weaker the further they are from the Sun; or, there is ... a single moving soul in the center of all the spheres, that is, in the Sun, and it impels each body more strongly in proportion to how near it is.

Why, though, should the moving soul of a planet be weaker when the planet is far from the Sun? Kepler decided that a single moving soul in the Sun made much more sense, although he did allow the possibility that each planet also had its own moving soul that could steer it closer to, or farther away from, the Sun.

Kepler thought that the Sun's moving soul imparted motion to the planets in an inverse proportion to the planet's distance from the Sun.[i] He then attempted

---

[i] He came to this conclusion by analogy with the spreading of light from a light source, but that was before Kepler understood what is now known as the "inverse square law of light." At this early stage, Kepler thought of light as spreading out in circles rather three-dimensional spheres.

to derive the ratio of orbital sizes from the ratio of speeds, but a mathematical error led him to an invalid result. In spite of this error, Kepler had taken a crucial step in claiming that the motions of the planets were controlled by the Sun.

Kepler supplied additional reasons for believing in the motive power of the Sun. He pointed out that Tycho had removed the hard spheres in the heavens which had previously provided a mechanism for planetary motion. Some new mechanism was needed, and all the evidence pointed to the Sun as the source of motion. Kepler noted that Ptolemy's equant with bisected eccentricity fit with the idea of the Sun's moving power.[25] In the Copernican model bisecting the eccentricity was equivalent to placing the center of the orbit halfway between the equant point and the Sun. For an observer at the equant point the planet would appear, for geometrical reasons, to move faster when it was on the side of the orbit close to the equant, but in fact the planet would actually move slower on that side. The two effects cancel each other and the observer at the equant would see the planet move at its average speed. Likewise, on the other side of the orbit the planet would really move faster but its greater distance from the equant point would reduce its apparent speed back to the average value. The critical point was that the planet moved faster when it was closer to the Sun, slower when it was farther from the Sun.

After opening his book with a defense of Copernicus, Kepler offered some criticisms of Copernican theory near the end. He pointed out that Copernicus gave motions to Venus and Mercury that were inexplicably linked to the motion of Earth. He also pointed out that Ptolemy did not use an equant for the Sun's orbit, nor did Copernicus use his equant-replacing epicyclet for the Earth's orbit. Kepler, who believed that the motion of all the planets was controlled by the same moving spirit from the Sun, considered it doubtful that some planets should be treated differently than others.[26]

Kepler closed the *Mysterium* with an attempt to determine the time of Creation from the arrangement of the planets.[27] He thought that at the moment of Creation the planets were arranged in a perfect line so that the Sun and inferior planets all lay toward the autumnal equinox while the superior planets all lay toward the vernal equinox. He estimated that such a configuration might have occurred about 5572 years earlier.

We now know that Kepler's regular solids theory is wrong.[i] After all, there are more than six planets but there remain only five regular solids. However, in the *Mysterium* Kepler committed himself to several ideas that would prove incredibly fruitful in his later work. First of all, Kepler insisted on asking *why* the heavens were the way they were. Nobody else was asking why there were

---

[i] And so is his estimate for the age of the solar system.

six planets, or why their orbits had certain sizes. There were some speculations about what moved the planets, but they mostly consisted of offhand comments that the planets were moved by angelic souls or else swam through the heavens like fish in the sea. Some thought that humans were not meant to know the ways of the heavens. Kepler, though was a diehard realist. He believed that humans were created by God with the capacity to understand, at least imperfectly, the Creation, which was the visible image of God.[28]

Kepler believed that mathematics was a tool that God had given to humans to allow them to understand the universe.[29] Mathematics could lead to an understanding of the large-scale structure of the universe, as in Kepler's regular solids theory. He also thought mathematics was the key to understanding the detailed motions of the planets. Mathematical analysis could reveal the actual path that the planets followed in their motions through the ether, rather than just a set of circles designed to match the appearances. Furthermore, Kepler believed that mathematical analysis could point the way toward understanding the true causes of planetary motions, and that analysis pointed toward the Sun as the cause.

The idea that the Sun moved the planets would eventually lead Kepler to a completely new type of astronomy, but first he had to get his *Mysterium* in print. He submitted his little book to the faculty at Tübingen to obtain their permission to publish. Maestlin served as Kepler's advocate, although he disapproved of Kepler's attempts to insert physical causes into astronomy. He also felt that Kepler needed to provide more explanation of the Copernican theory, and when Kepler failed to supply that additional material Maestlin added Rheticus' *Narratio Prima* to the book to fill the void.[30]

In his original manuscript Kepler had included arguments to show that the Copernican system was consistent with Holy Scripture. However, the theologians at Tübingen strongly disapproved of Kepler's arguments and urged him to omit that material.[31] Kepler relented and the book was published in 1596.

The *Mysterium* did not make a big splash, which is not surprising for a first book from an unknown math teacher. Some astronomers dismissed the whole thing as wild speculation, while others were impressed by Kepler's regular solids idea but, like Maestlin, disapproved of his attempt to bring physics into astronomy. The *Mysterium* did provide Kepler with an opportunity to make himself known to a few men who played important roles, for better or worse, in his future career. One of these men was Galileo Galilei, whose work we will examine in Chapter 8. Galileo received a copy of Kepler's book and sent him a quick reply before reading it. He expressed his appreciation that Kepler supported the Copernican theory and proclaimed himself an advocate of that theory, though only in private because he feared the reactions of those who did not understand

Copernicus' astronomy. Galileo never followed up with any detailed response to the *Mysterium*.[32]

In 1595, before the *Mysterium* was even published, Kepler sent a letter describing his ideas to Ursus, who at that time was serving as Imperial Mathematician to the Holy Roman Emperor and also deep into his dispute with Tycho. Kepler's letter included much flattery, as might be expected when an obscure math teacher is writing to the Imperial Mathematician. Ursus inserted a copy of Kepler's letter into a book in which he attacked Tycho, without ever asking for Kepler's permission to use the letter in that way.[33]

As it happens, Tycho received a copy of Ursus' book on the same day he received a copy of Kepler's *Mysterium* and a letter from Kepler even more flattering than the one Kepler had sent to Ursus. Tycho wrote back to Kepler to thank him for the book and express interest in his regular solids idea, but also to question him about the letter in Ursus' book. Kepler wrote a quick apology and explained that he was unaware of Ursus' use of his letter and that he certainly did not approve of Ursus' actions. Tycho may have sensed that Kepler could serve as an ally against Ursus, or perhaps he was just impressed by Kepler's mathematical insight. Whatever the reason, Tycho extended an invitation for Kepler to visit him in Prague. Not long after, Kepler had little choice but to accept that invitation.[34]

7.2   A new astronomy from physical causes

While in Graz, Kepler married Barbara Müller, a widow who brought with her a daughter from her previous marriage. Their first child was a son born in 1589, but he survived only two months.[35] Later that year Ferdinand II, Archduke of Inner Austria, decreed that all Lutheran preachers and teachers had to leave Graz. Kepler left, but was later allowed back in.[36] He and Barbara had a daughter in 1599, but she too died soon after birth. Soon the school where Kepler taught was shut down entirely and the rising religious tensions in Graz made it clear to Kepler that he would need to leave.

Kepler's first thought was that he might secure a job at his alma mater, but the faculty at Tübingen were not interested. With few other options left, Kepler set out to visit Tycho in Prague with two great hopes: that Tycho's data could help him prove his regular solids theory, and that Tycho could provide him with a job in a safe place to live.[37]

Tycho welcomed Kepler and set him to work, but did not initially give him a formal position or a definite salary. Nor did Tycho give Kepler full access to his great store of observational data. Instead he assigned Kepler to work as an assistant to Longomontanus, who was working on a model to describe the motions of Mars. Kepler was not pleased to play the role of assistant to the assistant, but soon Tycho recognized that he was not making full use of his talented assistants. He set Longomontanus to work on the theory of the Moon and let Kepler take full control of the problem of Mars.

Kepler was disappointed that he was not given Tycho's eccentricities for the planets, which were needed to verify the regular solids theory, but Kepler eventually realized that Mars, with its large eccentricity, might hold the key to understanding the force that moves the planets around. At first he was confident that he would solve the problem of Mars in a short time.[38] Little did he know that his "war on Mars" would stretch to five years and would yield fruit that Kepler could not have expected when he began.

Before he could really settle into his work on Mars, Kepler had to figure out his income. He returned to Graz to ask his employers there if they might continue to pay him while he worked for Tycho. Perhaps not surprisingly, they refused, suggesting that he should instead do something useful like train to become a physician. About this time, Ferdinand II forced all residents of Graz to convert to Catholicism or leave, and Kepler was forced back to Prague. In the meantime, Longomontanus had returned to Denmark and Tyco, now in dire need of a capable assistant, managed to convince Emperor Rudolph that Kepler should be hired to help compile a new set of astronomical tables to replace the *Prutenic Tables*.[39]

Tycho wanted his new assistant to do more than work on theories of planetary motion. He set Kepler to writing an essay to prove Tycho's priority in developing the geoheliocentric theory and the plagiarism of Ursus, even though by that time Ursus was dead.[40] Kepler hated this task, but complied with his new mentor's wishes. Before the essay was published, though, Tycho died. As Kepler wrote in Tycho's observation log, Tycho had gone to dinner and drank a bit too much. Tycho felt a need to relieve his bladder, but for reasons of etiquette he did not want to excuse himself from the table. By the time he got home he could not urinate. Eventually Tycho became delirious, probably as a result of kidney failure. Kepler reported that on his last night Tycho continuously repeated "Let me not seem to have lived in vain."[41]

Following Tycho's death, Kepler was appointed as the new Imperial Mathematician. He set aside the essay against Ursus and wanted to turn his full attention to the problem of Mars. That was not so easy to do, however. Tycho's heirs claimed ownership of his observational data and they did not trust Kepler to use the data properly (i.e. in a way that would glorify Tycho and make them wealthy). Meanwhile, Kepler was struggling to get the salary promised him by Emperor Rudolph.[42]

After long negotiations Kepler reached an agreement with Tycho's heirs in which he would be allowed access to Tycho's observational data but could not publish any of his work based on that data without the permission of Tycho's son-in-law and former assistant, Frans Tengnagel. Even with free access to the observations, Kepler realized that his work on Mars would be a long haul. To help prove his worth as Imperial Mathematician he published other books while he continued his war on Mars. One of these books was a treatise on astrology, which was likely of particular interest to Emperor Rudolph, who was enthralled by the subject.

Two of the other books written during this period were astronomical in nature. In the *Astronomiae Pars Optica* of 1604 Kepler discussed the functioning of the human eye and the pinhole camera. He described the refraction of light in Earth's atmosphere and (correctly) explained the reddish tinge of an eclipsed Moon as due to light passing through the air around Earth and then bending in toward the otherwise shadowed Moon. He also introduced the inverse square law of light, explaining that the brightness (or intensity) of light falls off inversely as the square of the distance from the light source.[43] In *De Stella Nova* Kepler described his observations of a nova that had appeared in 1604. In that work he rejected the idea, popularized by Giordano Bruno, that stars were just distant suns.[44]

Throughout the period from 1600 to 1605 Kepler continued his work on Mars, and by late 1605 it was completed.[45] Getting Tengnagel's approval to publish was not easy. Tengnagel disliked the book because he felt that it cast doubt on the Tychonic model and promoted the Copernican model instead (as it most certainly did). Eventually he relented and Kepler's work was published in 1609, but with a preface added by Tengnagel warning the reader not to be "moved by Kepler's liberty in certain matters, but especially in his disagreeing with Brahe in physical argumentations ..."[46][i]

Kepler's book on the theory of Mars, like most astronomical works of his day, bore a long and complicated title. It began "NEW ASTRONOMY BASED UPON CAUSES, OR CELESTIAL PHYSICS, Treated by means of Commentaries ON THE MOTIONS OF THE STAR MARS, From the Observations of TYCHO BRAHE ..."[47] We will refer to the book by its shorter, Latin title *Astronomia Nova*, but the longer title makes clear what Kepler was up to: he was attempting to use Tycho's observations to determine not just the motions of Mars, but the physical causes of those motions.[48]

Kepler wrote his book as though it was a historical narrative of his attempts to understand the motions of Mars, although in fact it was a brilliant piece of rhetoric designed to convince his readers that he had been forced to his

---

[i] This preface is oddly reminiscent of Osiander's preface to *De revolutionibus*, though not nearly so well written.

revolutionary conclusions in order to conform with Tycho's accurate observations. Since the book focuses only on the motions of Mars, and the motions of Earth that affect our observations of Mars, Kepler left out any discussion of his regular solids theory, but the entire work was guided by his notion that both Mars and Earth are moved by physical forces from the Sun.

Kepler introduced the *Astronomia Nova* with a defense of the Copernican model and an attack on Aristotle's theory of gravity.[49] Recall that Aristotle believed all objects had a natural tendency to move in certain ways: toward, away from, or around the center of the universe. Gravity, for Aristotle, was just the tendency of heavy objects to move toward their natural place at the center. Kepler, on the other hand, wrote:[50]

> Every corporeal substance, to the extent that it is corporeal, has been so made as to be suited to rest in every place in which it is put by itself, outside the orb of power of a kindred body.
>
> Gravity is a mutual corporeal disposition among kindred bodies to unite or join together; thus, the earth attracts a stone much more than the stone seeks the earth.

For Kepler, there was no such thing as natural motion. All motions required the action of some force. Heavy objects fell because they were made of Earth-stuff and they were attracted by the vast amount of Earth-stuff found in the body of the Earth. Gravity was not specifically connected to the Earth's center or even the Earth itself. Kepler wrote:[51]

> If two stones were set near one another in some place in the world outside the sphere of influence of a third kindred body, these stones, like two magnetic bodies, would come together in an intermediate place, each approaching the other by an interval proportional to the bulk [*moles*][i] of the other.

Kepler rejected the idea of "natural levity" completely, stating that objects rise not because they are light in any absolute sense, but only because they are less dense than the surrounding material. Kepler's new conception of gravity was an important step away from the place-based gravitational theory of Aristotle and toward Newton's later idea of universal gravitation.[52]

---

[i] Kepler uses the Latin term *moles* to represent the "amount of stuff" in an object. Sometimes he uses this term to mean something like our modern defintion for "volume" while at other times he uses it to mean something more like our modern definition of "mass." In this case, if we assume the stones to be made of the same material, then the two uses are equivalent.

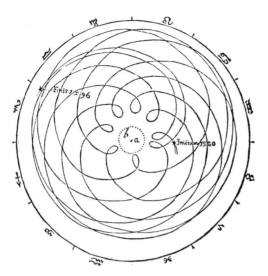

Figure 7.5 Diagram of the path of Mars in the Ptolemaic and Tychonic systems, from Kepler's *Astronomia Nova* (1609). Image courtesy History of Science Collections, University of Oklahoma Libraries.

Kepler also defended the idea of Earth's rotation against common criticisms. People were not knocked down by the Earth's motion, nor did they feel a great east wind, because things on and around the Earth shared in its rotation. Nor was the motion of Earth contrary to Scripture, because the Bible was revealed in terms understandable to common people using everyday language, not in the language of natural philosophy. He noted that the Tychonic system provided a nice compromise, but pointed out how Tycho's model disrupted the beautiful relation between the period and size of a planet's orbit.

In Part 1 of the *Astronomia Nova*, Kepler showed that the Ptolemaic, Copernican, and Tychonic theories for planetary motion are geometrically equivalent.[53] All three theories were equally capable of reproducing the observed motions of the planets. This geometrical equivalency showed that it was necessary to use *physical*, rather than geometrical, reasoning to decide between these theories. Kepler also displayed a picture of the Ptolemaic/Tychonic path of Mars through space (see Figure 7.5). According to the Ptolemaic and Tychonic theories, Mars, like the other planets, wound around like a pretzel. The Copernican paths, on the other hand, were nearly circular and simply repeated the same orbit over and over. The implication was clear: after Tycho's demolition of the hard celestial orbs, only the Copernican orbits made physical sense.

As he had already proposed in the *Mysterium*, Kepler suggested that the Sun emitted a force that drove the planets around in their orbits as the Sun spun on its axis. Since the Sun was the true source of a planet's motion, it was important to use the actual body of the Sun (the "true Sun") as the reference point for determining a planet's orbit. To determine Mars' location *as seen from the Sun* it was necessary to measure the planet's apparent location when the Earth was directly between the Sun and Mars. In other words, the position of Mars must be measured when it is in opposition to the true Sun. All previous astronomers had used oppositions to the mean Sun in constructing their theories of planetary motion, but Kepler, guided by his physical insight, chose to use oppositions to the true Sun instead.[54] With that critical point established, Kepler was ready to begin his attack on Mars.

7.3   The war on Mars

In Part 2 of the *Astronomia Nova* Kepler considered Tycho's attempts to measure Martian parallax. He pointed out that Tycho's data showed no evidence of any detectable parallax, and he questioned the ancient value of 3′ for the diurnal parallax of the Sun. He then demonstrated that if the plane of Mars' orbit is assumed to pass through the body of the Sun, then that orbital plane will have a fixed inclination of just under 2° relative to the ecliptic. In a single move Kepler managed to eliminate the bizarre wobbles of the planetary theories in Copernicus' *De revolutionibus* and Ptolemy's *Almagest*.[55] One extremely important consequence of this result was that the motion of Mars was no longer linked in any way to the motion of Earth. As Kepler says, he had "always opposed this gratuitous connection of diverse orbs as a cause of motion, even before seeing Tycho's observations."[56]

How, though, did Mars move within that fixed plane? Kepler started by assuming that Mars moved in a circle, off-center from the Sun, and that its motion was governed by a Ptolemaic equant. Although Copernicus had objected to the equant because it introduced nonuniform motion, it made sense to Kepler that physical forces might cause a planet to change its speed. Copernicus' epicyclets, on the other hand, made no physical sense.

To determine the orbit for Mars Kepler needed to find four things: the radius of the orbital circle, the distance from the center of the circle to the Sun, the distance of the equant point from the center, and the direction (in longitude) of the *aphelion*. Aphelion was a new term introduced by Kepler to denote the point on a planet's orbit that is farthest from the Sun, just as the apogee was the point farthest from Earth in the Ptolemaic theory. Likewise, Kepler used *perihelion* to denote the point closest to the Sun. Figure 7.6 shows the geometry of Kepler's circular orbit.

Kepler chose not to assume a bisected eccentricity (such that $\bar{q}c = \bar{s}c$ in Figure 7.6) and instead he undertook an extremely difficult set of calculations,

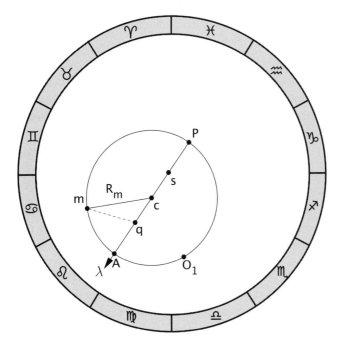

Figure 7.6 Kepler's circular orbit for Mars. In this model Mars m moves counterclockwise on a circle of radius R_m centered at c. The Sun is at s and q is the equant point. The aphelion A lies toward longitude λ while the perihelion P is in the opposite direction. The first octant (O_1) lies 45° from the aphelion as measured from the Sun. The figure is not to scale and the eccentricity is greatly exaggerated.

using four observations of Mars in opposition to the true Sun, to determine all of the orbital parameters. His results produced good agreement with the observed longitudes of Mars, but did not match the observed latitudes.[57] He would later call this orbit his "vicarious hypothesis" because, although it was not the correct orbit for Mars, it did serve to produce accurate longitudes. Note that, in contrast to Ptolemy and Copernicus, Kepler believed that a *single* orbital theory had to account for both the longitude and latitude of the planet.

Kepler found that he could correct the latitude errors by bisecting the eccentricity, but that produced errors in longitude of up to 8′. Specifically, he found that the computed time for the planet to move from aphelion to first octant (where the planet is 45° from aphelion as seen from the Sun – see Figure 7.6) was longer than the observed time.

Kepler pointed out that Ptolemy knew that his observations might have errors as large as 10′, and thus the theory with a bisected eccentricity would have been perfectly satisfactory to him. Not so for Kepler, who had Tycho's much more accurate observations. As Kepler stated it:[58]

Since the divine benevolence has vouchsafed us Tycho Brahe, a most diligent observer, from whose observations the 8′ error of this Ptolemaic computation is shown in Mars, it is fitting that we with thankful mind both acknowledge and honor this favor of God. For it is in this that we shall carry on, to find at length the true form of the celestial motions, supported as we are by these proofs showing our suppositions to be fallacious. In what follows, I shall myself, to the best of my ability, lead the way for others on this road. For if I had thought I could ignore eight minutes of longitude, in bisecting the eccentricity I would already have made enough of a correction in the hypothesis found in Ch. 16. Now, because they could not be ignored, these eight minutes alone will have led the way to the reformation of all of astronomy, and have become the material for a great part of the present work.

Kepler wanted to find the *true* orbit of Mars. He set out to find a simple path, and a simple set of rules for motion along that path, that could reproduce Tycho's observations. Because he knew Tycho's observations were accurate to a few minutes of arc, errors as large as 8′ were simply unacceptable. They showed that he had not yet found the true path. Part 2 ends with a decisive defeat in this particular battle with Mars, but the war was just beginning.

In Part 3, Kepler turned his attention to the orbit of the Earth. He realized that even if he knew the true path for Mars, any errors in his model for Earth's orbit would necessarily produce errors in the predicted positions of Mars as seen from Earth. In any case, the fact the Earth's orbit lacked an equant had bothered Kepler since his *Mysterium* days. He had decided to use an equant for Mars' orbit, and in the Copernican system Earth was just another planet like Mars. If both planets were moved by the same force from the Sun, then Earth had to have an equant as well. So Kepler set out to build a new orbit for Earth.[59]

The best way to show that Earth needed an equant, and to determine the location of the equant point, was to show that Earth moved along its circular orbit at a changing speed. To show this changing speed, Kepler needed to pinpoint Earth's location on its orbit at three different times. Those three points would uniquely define Earth's circular path. Furthermore, Kepler could examine the angles and times between the first location and second, as well as between the second and third, and determine if Earth moved along its orbit at a constant rate. In order to do all of this he first had to find out where Earth was at three different times.

To pinpoint the Earth's location in space, Kepler used the brilliant method of **triangulation**. We can illustrate the method with an example of a ship at sea. The ship's captain knows he is approaching a rocky coast and he wants to

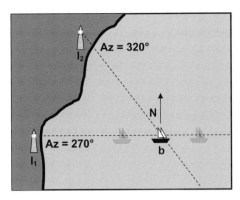

Figure 7.7 An illustration of the triangulation method. Observations of two fixed lighthouses from a ship allow the captain to determine the exact location of the ship at the intersection of the two dashed lines.

know exactly where he is in relation to that coast. He has a map of the coastline which shows the location of two lighthouses, as shown in Figure 7.7. From his ship he sees one lighthouse due west (azimuth 270°), so the captain knows that his ship must lie somewhere along a line running due east from that lighthouse (the horizontal dashed line in Figure 7.7). He sees the other lighthouse 40° west of north (azimuth 320°), so his ship must also lie along the other dashed line in Figure 7.7. Therefore, his ship must lie at the one point where the two dashed lines intersect. Thus, observations of two fixed landmarks allow the captain to determine his exact location relative to those landmarks.

Kepler believed that the Earth served as a ship, carrying him around the celestial sea, but what could he use as his fixed landmarks? As a Copernican, Kepler was certain that the Sun could serve as a fixed landmark, but he needed a second. His brilliant move was to use Mars as that second fixed landmark. Kepler knew that Mars had an orbital period of 687 days. As long as he selected observations that were spaced apart by multiples of 687 days, then Mars would be at the same location in space at each of those times. Therefore, these observations would function just like viewing a fixed landmark.[60]

Kepler found three of Tycho's Mars observations separated by 687-day intervals, and he used Tycho's solar theory to determine the apparent location of the Sun on those days. He also used his vicarious hypothesis to determine the correct longitude of Mars, as seen from the Sun, on those days. With that information he could triangulate the Earth's position on each of the three days, as shown in Figure 7.8.

Once he had identified three locations for the Earth, Kepler used geometry to construct the one and only circle that would pass through those three points. This was the Earth's orbit and it was off-center from the Sun, as expected.

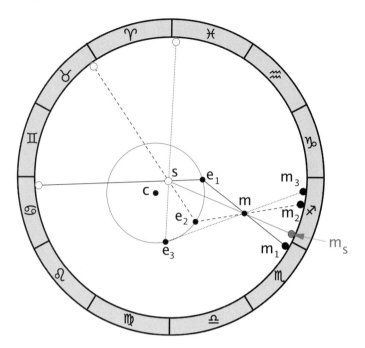

Figure 7.8 Kepler's method for triangulating the position of Earth using observations of Mars and the Sun spaced 687 days apart. For all observations the Sun is at s and Mars is at m (toward longitude m_s as seen from the Sun). Observed longitudes of Mars and the Sun can be used to find Earth's location. Three locations for Earth determine Earth's circular orbit. (Note: diagram is not to scale and the eccentricity of Earth's orbit is greatly exaggerated.)

However, Kepler also found that the Earth changed its speed as it moved along this orbit. The essence of what Kepler found is shown in Figure 7.8. Note that in Figure 7.8 the Earth moves counterclockwise around its circular orbit. Moving from e_1 to e_2 takes 687 days and during that time the Earth completes one full orbit and most of another. On the other hand, in two full years (730 days) the Earth would return to e_1. Therefore it must take 43 days for the Earth to move counterclockwise from location e_2 to e_1. Similarly, it must take 43 days for the Earth to move counterclockwise from e_3 to e_2. Close inspection of Figure 7.8 shows that the *distance* that the Earth moves in going from position e_2 to position e_1 is greater than the distance it moves in going from e_3 to e_2. Therefore, the Earth must travel *faster* between e_2 and e_1 than between e_3 and e_2.

Note that the Earth is closer to the Sun when it is moving from e_2 to e_1 than it is when moving from e_3 to e_2. So what Kepler discovered was that the Earth moves faster when it is closer to the Sun. He found that he could account for this variation in speed by giving the Earth's orbit an equant with a bisected eccentricity (so the equant point is on the other side of the center from the Sun and just as far away). Kepler pointed out that this new orbit for the Earth argued strongly against the Ptolemaic model. Since Ptolemy's epicycles for superior planets had to match the Copernican orbit for the Earth, Ptolemy would have to assign to each superior planet an epicycle with the same bisected eccentricity along the same axis. Why should all of those separate epicycles have these identical characteristics? Only Copernicus (or Tycho) could explain why: because the epicycles are really just mirroring the actual motion of the Earth (or Sun).[61]

Because the motion of the Earth affects our observations of all planets, Kepler could have ended his work at this point and he already would have produced a tremendous advance in astronomy. His new theory for Earth's motion would have dramatically reduced the errors in the Copernican model and also helped to eliminate the odd motions, synchronized with Earth's orbit, that Copernicus had given to Mercury and Venus. For practical purposes there was hardly a need to go further, but Kepler didn't just want accurate calculations, he wanted to determine the true path of Mars and understand the physical causes of planetary motion.

The equant was a geometrical construction, not a physical one, so Kepler set out to find a description of orbital motion that had a more physical character. He showed that in an orbit with bisected eccentricity the speeds at aphelion and perihelion were inversely proportional to the planet's distance from the Sun, and so he concluded that the speed of a planet anywhere along its orbit must be inversely proportional to its distance from the Sun.[62] We will refer to this rule as Kepler's "distance law." Kepler was aware that his distance law was not strictly equivalent to an equant with bisected eccentricity except at aphelion and perihelion, but because the distance law made physical sense he trusted that law more than the geometrical rule of the equant. Throughout the remainder of the *Astronomia Nova* Kepler used the distance law as the rule for governing a planet's motion anywhere along its orbit.

But what was the physical cause that produced this motion according to the distance law? In his *Mysterium* Kepler had suggested that the Sun moved the planets with a force that was inversely proportional to the planet's distance from the Sun and he reintroduced that idea in the *Astronomia Nova*.[i] Kepler stated that the Sun's moving power was transmitted by an immaterial *species* (a Latin term

---

[i] In the *Mysterium* Kepler suggested that the force would be inversely proportional to distance by analogy to the spreading of light, which he did not properly understand at that time. By the time he wrote the *Astronomia Nova* he had discovered that the intensity of light is inversely proportional to the *square* of the distance from the source, but Kepler felt that the moving power of the Sun was effectively confined to the ecliptic plane and therefore it was spread over a circle, rather than over a sphere as was the case for light.

not easily translated into English) that spreads out from the body of the Sun. The Sun rotates on its axis, causing this *species* to sweep around and move the planets. Each planet has its own particular resistance to motion, dependent on its "quantity of matter" (somewhat like what we would now call "mass"). Therefore, all planets would orbit with a period longer than the Sun's rotational period. They could trail behind the *species* but they could not move ahead of it. Kepler thought more distant planets moved slower because the *species* was weaker farther from the Sun.[63]

Kepler was familiar with William Gilbert's *De Magnete*, a treatise on the magnet published in 1600.[64] Magnets seem to exert forces on objects even when they are not in physical contact, a property that came to be known as "action at a distance." Kepler's *species* shared this property, so he suggested that the *species* might be some type of magnetic force. He proposed that a similar magnetic force, this time coming from the Earth, was responsible for moving the Moon in its orbit around the Earth.[i]

The immaterial *species* might explain the motion of the planets *around* the Sun, but Kepler had found that both Mars and the Earth have eccentric orbits, which means that they also get closer to, and farther away from, the Sun. Kepler's *species* could not explain this motion toward or away from the Sun. He rejected the idea that the planets might have some kind of soul, or intelligence, that guided the planet on its eccentric orbit because that intelligence would have to carry out extremely complicated calculations based on measurements of the apparent size of the Sun in order to guide itself along such an orbit. He also completely rejected motion along epicycles as an explanation since there was no physical body inside the epicycle to produce or guide such motion. It seemed that there was no reasonable physical mechanism that could produce an eccentric circular orbit. Midway through the *Astronomia Nova* Kepler dropped a hint that perhaps the orbits of the planets are not, in fact, circular,[65] but he then set the issue aside and focused on investigating the motion of a planet as governed by the distance law.

It turned out that the distance law was impossible to use. Kepler needed to calculate the time it would take for a planet to move from one point on its orbit to another point. With the equant, this was easy. Simply measure the angle between the two points as seen from the equant, and that angle is proportional to the elapsed time. With the distance law there was no way to find this time because the distance, and therefore the speed, of the planet was constantly changing.

Thus, the motive force would be inversely proportional to the distance rather than the square of the distance.

[i] One of the principal conclusions of Gilbert's *De Magnete* was that Earth itself was a magnet.

Kepler tried to account for these changes by dividing the Earth's circular orbit into 360 sectors of $1°$ each, as measured from the Sun. He thought that by adding together the lengths of the lines from the Sun to Earth he could estimate the amount of time the Earth would spend on any part of its orbit. He found that this procedure was too difficult, but he realized that he could approximate the sum of all of the lines by instead calculating the area swept out by the Sun–Earth line as the Earth moved from one point to another.[66]

From then on Kepler approximated his distance law by assuming that the area swept out by the line from the Sun to a planet was proportional to the elapsed time. Another way to say this is that the area swept out by the Sun–planet line is always the same during equal intervals of time, no matter where the planet is in its orbit. It is important to note that in the *Astronomia Nova* Kepler viewed this "area law," which is now known as **Kepler's Second Law of Planetary Motion**, as an approximation to the distance law, which he viewed as the true law governing the motion of the planets.[67]

The area law is illustrated in Figure 7.9. Between time t_1 and $t_1 + \Delta t$ the Sun–planet line sweeps out an area A_1. According to the area law, the ratio of the elapsed time Δt to the orbital period T must be equal to the ratio of the area A_1 to the total area of the circle A. The same is true for the other sectors shown in Figure 7.9, but since Δt is the same for all of the sectors the areas of the three sectors must all be equal ($A_1 = A_2 = A_3$).

Kepler's introduction of the area law closed Part 3 of the *Astronomia Nova*. In Part 4 he returned to the motion of Mars. He used Tycho's observations to construct a new orbit for Mars, this time using his law of areas. This new orbit resulted in errors of $8'$, just like the "Ptolemaic" equant orbit (with bisected eccentricity) he had constructed earlier, except this time the errors were in the opposite direction. Specifically, he found that the computed time for the planet to move from aphelion to first octant was shorter than the observed time.[68]

Kepler realized that his area law pointed him toward the solution of this problem. If the circular orbit was pulled outward along the middle longitudes, midway between aphelion and perihelion, then the area from aphelion to first octant would represent a smaller fraction of the total area (see Figure 7.10), which means the time from aphelion to first octant would be reduced as compared to the circular orbit.[69] That was the opposite of what Kepler wanted, so he realized that the orbit needed to be squeezed in at the middle, just as "if one were to squeeze a fat-bellied sausage at its middle, he would squeeze and squash the ground meat, with which it is stuffed, outwards from the belly towards the two ends, emerging above and below his hand."[70] This squeezed orbit will have an area from aphelion to octant that is a greater fraction of the total area

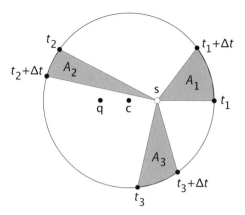

Figure 7.9 Kepler's area law for an eccentric circular orbit. The planet moves along a circle centered on c while the Sun lies at s. The area swept out by the line connecting the Sun and the planet in a time Δt is the same regardless of where the planet is in the orbit, so $A_1 = A_2 = A_3$.

than is the case for the circular orbit. So using the squeezed orbit increases the time from aphelion to first octant, which could correct the error of the circular orbit.

Guided by his area law (which was really just an approximation to his distance law), Kepler abandoned millennia of astronomical tradition. He concluded that the orbit of Mars was not a circle, but rather some kind of oval. But exactly what kind of oval? Kepler knew that oval orbits could be constructed by adding a small epicycle to a circular orbit and his first attempt at such a construction resulted in an orbit he described as egg-shaped. He found that applying his area law to this egg was too difficult, so he approximated the egg as an ellipse (because he knew how to compute the area for a sector of an ellipse). Once more he ran into 8′–9′ errors, but in the opposite direction from the errors in his circular orbit using the area law.[71] It must have seemed like these 8′ errors were simply unavoidable, but Kepler didn't give up. As he stated: "these errors are going to be our path to the truth."[72]

Kepler returned to a consideration of the force that caused Mars to move in and out relative to the Sun. He used his physical intuition to guide the construction of a new oval orbit which he described as "puffy-cheeked," but this orbit (approximated by an ellipse) produced errors similar to that of the egg. Finally, he realized the solution. The circular orbit gave errors of 8′ in one direction. The ellipses used to approximate the egg-shaped and "puffy-cheeked" orbits gave similar errors in the opposite direction. The truth must lie halfway between, but the only thing that lies halfway between a circle and an ellipse is another ellipse![73]

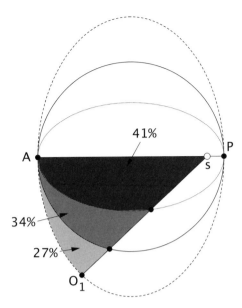

Figure 7.10 The area law points toward an oval that lies inside Kepler's circular "Ptolemaic" orbit. In the circular orbit shown here the line from Sun to planet sweeps out 34% of the total area of the circle as the planet moves from aphelion to first octant, so this motion takes 34% of the total orbital period. If the planet moves along an orbit that is pulled out along the middle longitudes then the area swept out is a smaller percentage of the total area and the planet will reach first octant sooner. If the planet moves along an orbit that is squeezed in along the middle longitudes then the area swept out is a greater percentage of the total area and the planet reaches first octant later. These orbits have a very large eccentricity in order to better illustrate the effect.

Kepler constructed this new ellipse and found that the combination of the elliptical orbit and his area law gave predictions that matched Tycho's observations. In a later work Kepler would state that the Sun lies at one focus of the ellipse, while the other focus plays approximately the role of the Ptolemaic equant point. The fact that planets orbit the Sun in elliptical paths, with the Sun at one focus, is now known as **Kepler's First Law of Planetary Motion** (even though he found it *after* finding his Second Law). Figure 7.11 illustrates the elliptical orbit of Mars. Appendix A.13 provides more details about Kepler's First and Second Laws.

In the final pages of the *Astronomia Nova* Kepler returned to the latitude theory for Mars. He combined his new longitude theory (elliptical orbit plus area law) with his previous conclusion that Mars orbited in a fixed plane passing through the Sun and tilted almost 2° relative to the ecliptic plane. The result gave predictions for the latitude of Mars that matched very well

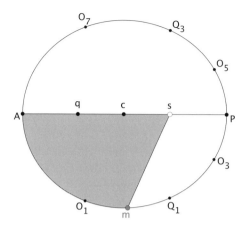

Figure 7.11 Elliptical orbit of Mars. The Sun s lies at one focus while the other, empty focus q serves as an approximate equant point. The planet m moves so that the Sun–planet line sweeps out equal areas in equal times. The aphelion A, perihelion P, quadrants Q, and octants O are labeled. Note: the eccentricity of the ellipse is greatly exaggerated.

with Tycho's observations. Kepler speculated that the same magnetic force that moved the planet in and out relative to the Sun was also responsible for the tilt of the orbital plane. He suggested that the truly fundamental plane of motion was not, in fact, the ecliptic plane but rather the plane of the Sun's equator (i.e. the plane perpendicular to the Sun's rotational axis). Both the Earth and Mars, as well as the other planets, would orbit in planes that were slightly tilted relative to this fundamental plane. By allowing these planes to precess around, while still maintaining a fixed inclination relative to the fundamental plane, Kepler could account for the precession of the equinoxes and a few other astronomical phenomena. However, he completely eliminated the strange wobbles, linked to the Earth's motion, that Copernicus had given to the planetary orbital planes.[74]

In hindsight, Kepler's *Astronomia Nova* was a tremendously important, ground-breaking work. Not only did it introduce elliptical orbits and the area law, two concepts that would prove lasting and important for all future calculations of planetary motion, but it also represented a decisive step toward the merger of astronomy with physics. However, Kepler's contemporaries gave the *Astronomia Nova* a lukewarm reception. Kepler's radical approach of using physical thinking to guide astronomy was too far ahead of its time. Elliptical orbits were too big a break with tradition. Moreover, Kepler's work was difficult for most astronomers to understand. As discussed in Appendix A.13, using Kepler's theory to calculate planetary positions was incredibly difficult and Kepler used old terms in new ways that were hard for astronomers to follow.[75] Still, although the *Astronomia Nova* was not an instant success, it planted a seed that would bear important fruit in the future.

7.4   The harmony of the world

In early 1610, not long after the *Astronomia Nova* went to press, Kepler heard about some amazing astronomical discoveries that had been made by an Italian mathematician using a new optical device. That mathematician was Galileo Galilei, and the device was what we now call a telescope. In the next chapter we will discuss Galileo's discoveries and Kepler's response to them. Galileo's telescope consisted of a tube with two lenses, one convex and one concave. In 1611, Kepler published his *Dioptrice* in which he analyzed the optics of two-lens systems and presented a new telescope design, using two convex lenses, that was generally better than Galileo's for astronomical work. Kepler's design is used in modern refracting (lens-based) telescopes.[76]

Aside from the publication of his *Dioptrice*, 1611 was a difficult year for Kepler. His son Friedrich died from smallpox in February and Kepler's wife, Barbara, died later in the year. Meanwhile Kepler's employment as Imperial Mathematician became increasingly insecure. Rudolph's hold on the Holy Roman Empire was gradually slipping away, and in 1611 his brother Matthias effectively took over. In January 1612 Rudolph died. Matthias, now officially the Emperor, kept Kepler on as Imperial Mathematician but he was not as interested as Rudolph had been in the astrological advice that Kepler could provide. He made it clear that Kepler was free to leave Prague if he wished. Prior to Rudolph's death Kepler had been offered a job as district mathematician in the town of Linz in Upper Austria. After all of the political upheaval, religious clashes, and personal tragedy that Kepler had experienced in Prague he decided to leave for Linz when he had the chance.[77]

Kepler settled into his job in Linz and before long he married again. His second marriage, to Susanna Reuttinger, seems to have been happier than his first, but his move to Linz was not without difficulty. Although Kepler might have expected a warm welcome in a solidly Lutheran town like Linz, in fact he received quite the opposite. Kepler's toleration of Calvinist views was known to the religious authorities in Linz and ultimately Kepler was denied participation in Communion because he refused to agree in writing with the Formula of Concord, which laid out the official Lutheran doctrine.[78]

To make matters worse, in 1615 Kepler's mother was accused of witchcraft by one of her neighbors. These charges against her lingered for several years, and in 1620 she was arrested and imprisoned. Her trial dragged on for a year and

Kepler traveled back to Württemberg to defend her. Eventually Kepler's mother was threatened with torture but still refused to confess. After that, the Duke of Württemberg gave her a pardon, but she died just a few months later.[79]

All of the chaos and turmoil in Kepler's life slowed his astronomical work, particularly his work on the new tables of planetary motion that he had once promised to Emperor Rudolph. When he could work, he sought solace in contemplating the harmony of the heavens. He found himself drawn back to the questions he had probed in his *Mysterium*: what mathematical rules govern the structure and motion of the universe as a whole? In his *Astronomia Nova* he had produced an accurate description of a single planet's motion and a plausible explanation of that motion in terms of physical forces. That explained how the orbits of the planets were maintained, but not how and why they got there in the first place. Kepler still wanted to know why the orbits had the particular sizes and eccentricities that God had apparently given to them.

Kepler's final work on this subject was published in 1619 with the title *Harmonice mundi* (The Harmony of the World).[80] The first two books of the *Harmonice* lay out Kepler's views on the construction of regular geometrical figures and the harmonies (or "congruences") that exist between different figures. In the third book, Kepler applied these ideas of geometrical harmony to develop a theory of musical harmony. In the fourth book, he discussed the role of harmony in astrology. Kepler's astrology focused on "aspects," or the angles between planets, as seen from Earth, measured along the ecliptic. Like Ptolemy, Kepler felt that the planetary aspects at a person's birth could influence (but not fully determine) a person's character and personality.[81]

In the fifth book of the *Harmonice* Kepler applied his notions of harmony to astronomical questions. He reiterated his regular solids theory, stating that it provided the explanation for the number and approximate sizes of the planetary orbits. However, a more detailed understanding of the orbital sizes and their eccentricities could be supplied by Kepler's notion of harmony. In particular, Kepler felt that the maximum and minimum speeds of the planets in their orbits would exhibit harmonic ratios (ratios composed of small integers).

For a single planet, the ratio of its maximum speed (at perihelion) to its minimum speed (at aphelion) had to be a harmonic ratio according to Kepler. Likewise, the maximum and minimum speeds of one planet could be compared to the maximum and minimum speeds of an adjacent planet, and all of the ratios formed from these speeds had to be harmonic. As each planet moved around, it could be thought of as singing a series of notes that changed in pitch as the planet's speed changed. Each planet sang its own particular set of musical notes. Because planets more distant from the Sun moved slower, each planet sang in a different range: Saturn and Jupiter sang bass, Mars sang tenor, Earth and Venus sang alto, and Mercury sang treble.[82]

Kepler went to great lengths to show exactly which harmonic ratios should apply to each planet, and to each set of neighboring planets. From these harmonic ratios Kepler determined the eccentricities of the planetary orbits.[83] To a modern astronomer, Kepler's planetary harmonies may seem like gibberish, but it is important to understand that Kepler was using the mathematical tools available to him to try to understand the heavens. University instruction in Kepler's day included the study of the medieval *quadrivium*: arithmetic, geometry, music, and astronomy. In the *quadrivium*, music was a mathematical subject. For instance, musical harmonies were defined in terms of the ratios of the lengths of two identical strings that, when plucked, play a harmonious sound. Kepler supplied a different definition for musical harmony, but his general goal in the *Harmonice* was to unify the disciplines of the *quadrivium* and use them to explain the construction of the heavens. We may now view his results as incorrect and his methods as too "mystical," but given his educational background and his historical context we can understand why Kepler might have sought connections between geometry, music, astrology, and astronomy.

In any case, there is at least one result in the *Harmonice* that still holds up today. In the fifth book Kepler presented his "harmonic law," now known as **Kepler's Third Law of Planetary Motion**.[84] The Third Law is a mathematical relationship between the period of a planet's orbit and the size of that orbit, something that Kepler had searched for unsuccessfully in his *Mysterium*. As Kepler explained it in the *Harmonice*:[85]

> Again, therefore, a part of my *Secret of the Universe*, put in suspense 22 years ago because it was not yet clear, is to be completed here, and brought in at this point. For when the true distances between the spheres were found, through the observations of Brahe, by continuous toil for a very long time, at last, the genuine proportion of the periodic times to the proportion of the spheres — *only at long last did she look back at him as he lay motionless, But she looked back and after a long time she came;* and if you want the exact moment in time, it was conceived mentally on the 8th of March in this year one thousand six hundred and eighteen, but submitted to calculation in an unlucky way, and therefore rejected as false, and finally returning on the 15th of May and adopting a new line of attack, stormed the darkness of my mind. So strong was the support from the combination of my labor of seventeen years on the observations of Brahe and the present study, which conspired together, that at first I

believed I was dreaming, and assuming my conclusion among my basic premises. But it is absolutely certain and exact that *the proportion between the periodic times of any two planets is precisely the sesquialterate proportion of their mean distances, that is, of the actual spheres,* though with this in mind, that *the arithmetic mean between the two diameters of the elliptical orbit is a little less than the longer diameter.*

In modern terms we would say that the square of a planet's period is proportional to the cube of its average distance from the Sun.[i] The mathematical details of this relationship are discussed in Appendix A.14, but the important thing to know about Kepler's harmonic law is that if you know the period of a planet's orbit it allows you to calculate the mean distance of the planet from the Sun. The orbital periods are directly measurable from observations (see Appendices A.9 and A.10), so Kepler could use this new law to find the mean distances. He could then find the ratio of maximum to minimum distance from his harmonic theory. Using this ratio along with the mean distance Kepler could determine the actual values of the maximum and minimum distance for each planet, as well as the eccentricity of each planet's orbit.

Table 7.2 shows a comparison between the distances Kepler derived from purely astronomical work (using Tycho's observations) and those he found using his harmonic theory. The match is quite good, although one could argue that the agreement is not surprising since Kepler adjusted his harmonic theory until it fit the data. However, Kepler felt that he had provided independent justification for his harmonic theory. The close agreement shown in Table 7.2 served, in Kepler's mind, as evidence that his harmonies were real.

While he was working on his *Harmonice*, Kepler also began writing a textbook that presented his new astronomical theories. Titled *Epitome Astronomiae Copernicanae* (Summary of Copernican Astronomy), the work was published in three parts in 1615, 1620, and 1621. Kepler wrote the *Epitome* as a series of questions that might be asked by a student of astronomy, and in the responses to these questions he described his own astronomical theories (*not* those of Copernicus) including all three of his Laws of Planetary Motion, his regular solids theory, and his speculations on the forces from the Sun.[86]

Although mostly an explanation of work published elsewhere, the *Epitome* also served to introduce some new astronomical material. In the *Astronomia Nova* Kepler had determined that the orbit of Mars was an ellipse governed by the

---

[i] Kepler says the period has the sesquialterate proportion to the mean distance, but that just means the period is proportional to the three-halves power of the distance. That's the same as saying the square of the period is proportional to the cube of the distance.

Table 7.2 *Maximum and minimum distances of each planet from the Sun, in Astronomical Units, as derived from Tycho's observations and from Kepler's harmonic theory.*[87]

Planet	Distance	From Tycho's data	From harmonic theory
Saturn	Maximum	10.052	10.118
	Minimum	8.968	8.994
Jupiter	Maximum	5.451	5.464
	Minimum	4.949	4.948
Mars	Maximum	1.665	1.661
	Minimum	1.382	1.384
Earth	Maximum	1.018	1.017
	Minimum	0.982	0.983
Venus	Maximum	0.729	0.726
	Minimum	0.719	0.716
Mercury	Maximum	0.470	0.476
	Minimum	0.307	0.308

area law (as an approximation to his distance law), and in the *Epitome* he applied that theory to the other planets. He gave a revised version of his distance law, saying that the speed of a planet *in the direction perpendicular to the line from the Sun to the planet* was inversely proportional to the planet's distance from the Sun. This new version of the distance law fit with the area law that Kepler actually used for his calculations.[88] Kepler also provided an explanation of his harmonic law based on "harmonic reasoning" as well as his understanding of the physical force from the Sun (see Appendix A.14 for details). In the *Epitome*, Kepler argued that the Sun was about three times farther away than had previously been thought, which explained why Tycho had been unable to find a diurnal parallax for Mars.[89] Kepler also concluded that, for harmonic reasons, the distance to the stars must have the same ratio to Saturn's orbital size as Saturn's orbital size has to the size of the Sun's body. This conclusion placed the stars at a distance of about 20,000 Astronomical Units, which was far enough away to explain why Tycho had been unable to detect any annual parallax in the stars.[90]

The *Epitome* was Kepler's most influential theoretical work, because it presented all of his important ideas in one convenient package.[91] It was written in a way that was much more accessible than his *Astronomia Nova* or his *Harmonice*, since it was intended to serve as a textbook for university students. But although the *Epitome* helped other astronomers to understand Kepler's work, it did not lead many of them to embrace Kepler's radical ideas.

After his mother's trial and the publication of his *Epitome*, Kepler dedicated himself to compiling the long-overdue tables of planetary motion. In 1614 John Napier had published a book on logarithms, and Kepler made use of Napier's techniques to ease his mathematical computations. The new tables were completed in 1624, but there were some problems with the printing. Meanwhile, Linz had been conquered by Catholic forces and Lutheran preachers and teachers had been expelled. Kepler, as before in Graz, was allowed to stay, but many of his books were confiscated. When a peasant's revolt resulted in the destruction of Linz's printing press, Kepler was given permission to go elsewhere to have his tables printed.[92]

The *Tabulae Rudolphinae* were printed in Ulm, in 1627. In some ways these tables represented the culmination of Kepler's life's work. Kepler's tables were much more accurate than anything that had come before, and it was the accuracy of his tables that ultimately led other astronomers to take Kepler's work seriously. Even those who dismissed his ideas about harmony and solar forces had to respect the accuracy of his elliptical orbits and his area law.[93] The new Holy Roman Emperor, Ferdinand II, who had succeeded to the throne after the death of Matthias in 1619, was impressed with Kepler's work. He offered Kepler a more prestigious job at his court if only he would convert to Catholicism. Kepler, of course, would not and instead found employment with Albrecht von Wallenstein, who served for a time as supreme commander of the Imperial armies during the Thirty Years' War. After his military exploits, Wallenstein had purchased from the Emperor the Duchy of Sagan in what is now Poland. It was there that Kepler spent his final years. In 1630, while traveling to Regensburg, Kepler fell ill and died. He was buried outside the city in a Lutheran cemetery, which was later destroyed during a battle of the Thirty Years' War.[94]

Kepler was a visionary in the best and worst senses of the word. He saw that physical forces held the key to understanding celestial motions and he was able to look beyond the constraints of tradition to see the true patterns revealed by the data. In spite of his own poor eyesight, he was remarkably good at imagining how things would look from a different point of view, whether it was from the Moon (as in his *Somnium*), from Mars (as in his triangulation of Earth's position), or even from the point of view of a Calvinist (which got him in endless trouble with his fellow Lutherans). Kepler also was constantly preoccupied with an imaginary world of geometrical structures and harmonies that, we now know, existed only within his head.

7.5   Reflections on science

In many ways Kepler's work represents a break from astronomical tradition. Obviously, his elliptical orbits represented a major break from the traditional circles. However, it is not just Kepler's results that forge new ground, but also his entire method of approaching astronomy.

Kepler was a realist, in that he believed that the universe had a definite structure, put in place by the Creator, and that humans could discover that structure. Kepler thought that, because humans were created in the image of God, we were capable of comprehending at least some aspects of God's plan in creating the universe. In particular, Kepler believed that God had created the universe using eternal, mathematical principles and that humans were endowed with the ability to recognize and understand those principles. Hence, for Kepler, mathematics held the key to understanding the true nature of the universe.[95]

This confidence that humans could understand the deep structure of the universe led Kepler to ask questions that nobody had asked before. Kepler believed that a true understanding of the universe would *explain* what we see rather than just match with what we see. That is likely what appealed to Kepler about the Copernican model: it explained things such as the link between the motion of the Sun and the motions of the planets that the Ptolemaic model could only match. Kepler, though, was not satisfied with what the Copernican model could explain. He wanted to know why there were exactly six planets, why their orbits had the sizes they had, and why they moved along those orbits the way they did.

Much of Kepler's work was highly speculative. Certainly, we can look back at his regular solids theory and his ideas about planetary harmonies and dismiss these as flights of fancy. But Kepler believed that mathematics could unlock the secrets of the universe, and he made use of the mathematics that was available to him to try to explain the previously unexplained properties of the planets. He did not speculate without testing his ideas. He subjected each of his theories to empirical tests, when possible, and did not try to hide any discrepancies, although he did sometimes blame the discrepancies on bad data.

When he did encounter discrepancies that could not be blamed on poor data, such as the 8′ errors in the *Astronomia Nova*, Kepler worked to modify his theories. He believed that the true theory could not be wrong in any way, so if there were genuine discrepancies then the theory had to be corrected or discarded. Like most modern scientists, Kepler did not discard his theories easily. Rather, he worked to modify them, and the modifications he introduced were guided by the errors he had found.

Kepler was also guided by his belief in the unity and coherence of natural phenomena. He believed that the universe operated according to a small set of mathematical principles, and that these fundamental principles could account for the wide variety of appearances that we see. For example, he believed that all planets should move according to the same rules, so if other planets have an equant then the Earth must have one too. He believed that a single orbital path should explain a planet's observed variations in both longitude and latitude.

Most importantly, he tore down the separation between the physics of the heavens and the physics of Earth, insisting that celestial objects were moved by physical forces just as terrestrial objects are. Kepler was working with a modified version of Aristotelian physics that we now know is incorrect, but sometimes incorrect theories can still be fruitful. Kepler's intuition that the planets were moved by physical forces was a good one, and it led him to his Laws of Planetary Motion even though his detailed theories about the celestial forces were wrong. When the combination of a circular orbit and the area law failed, any other astronomer would have abandoned the area law. The area law made physical sense to Kepler, while an eccentric circular orbit did not, and so Kepler not only stuck with the (correct) area law but he discovered elliptical orbits.

Kepler's extension of terrestrial physics to the heavens was a critical step in the development of astronomy, and we will see later how this idea culminated with the physics of Isaac Newton, but this move by Kepler also illustrates an important principle of modern physical science: the belief that the same set of fundamental physical laws operate throughout the entire universe and at all times (an idea known as "uniformitarianism"). This crucial idea lies at the core of modern physics, and it really started with Kepler. The notion that the universe operates according to physical laws, which can be described mathematically, gave rise to a new metaphor for the universe. For Aristotle, the universe was like a living thing that sought to carry out its purpose and keep all of its parts in their proper place. For Kepler, on the other hand, the universe was a machine designed by God. In a letter to his friend Herwart von Hohenberg in 1603 Kepler wrote:[96]

> I am deeply involved in investigating physical causes. My aim in this is that I describe the celestial machine to be not like a divine being but like a clock (he who believes a clock is alive bestows the artisan's glory upon the work), in that nearly every kind of motion comes from one very simple corporeal magnetic force, just as in a clock all motion comes from a very simple weight. And I show how this physical reason is designated in numbers and geometry ...

For centuries after Kepler's death the clock was used as a metaphor for understanding the functioning of the universe, and to some extent that metaphor still functions today.

One final point to be made about the astronomical discoveries of Kepler is that they could not have been made without the work of his predecessors. The astronomers who came before him gave Kepler the intellectual resources needed to carry out his groundbreaking work. Without the Copernican model, Kepler's regular solids theory and his ideas of a solar motive force make no sense.

Without Tycho's data, and his elimination of hard orbs as a mechanism for celestial motions, Kepler would not have been able to pursue the physical causes of planetary motion and thereby discover the ellipse and area laws. He also relied on Gilbert's investigation of magnets to supply the idea of action at a distance, and Aristotle's physics of violent motion to supply a basic framework for how a physical force from the Sun might move the planets. Kepler's genius was to weave these separate strands into a unified picture of how the heavens function.

Kepler's Laws of Planetary Motion were incredibly accurate. In the 17th century there was little that could be done to improve upon Kepler's astronomy. What remained was to convince other astronomers that Kepler was correct in unifying the heavens and the Earth, and to provide a more solid foundation on which to build a physics of the heavens. Kepler had modified Aristotle's physics by claiming that bodies had a natural tendency to remain at rest in any location, and he had made use of forces acting over a distance, but his understanding of how forces relate to motion was still based mostly on Aristotle. The key to completing the Copernican Revolution was the development of a new understanding of the physics of motion. The first major step toward this new understanding had already been taken, but not yet published, during Kepler's lifetime, by Galileo Galilei.

8

Seeing beyond Aristotle: Galileo's controversies

8.1   Message from the stars

Galileo Galilei (Figure 8.1) was born in Pisa, Italy, in 1564. His father, Vincenzio, was a professional musician from Florence. Vincenzio was not only an accomplished lutenist, he also wrote on the theory of music. Kepler consulted Vincenzio's work during the composition of his *Harmonice mundi*.[1]

Vincenzio wanted his son to be a physician, so when Galileo began attending the University of Pisa he started off studying medicine, but he became increasingly interested in mathematics and physics. He was fascinated by the works of Euclid, whose *Elements* had served as the standard textbook on geometry for nearly two millennia, and Archimedes. Galileo was particularly influenced by Archimedes' work on the law of the lever, centers of gravity, and floating bodies. While Aristotle produced a grand theoretical structure to explain everything in the world, Archimedes, who lived one hundred years later, used mathematics to solve specific, limited, practical problems without much concern for the ultimate causes proposed by Aristotle. Throughout his life Galileo was guided by this Archimedean style.[2]

With his father's permission, Galileo dropped his medical studies and focused on mathematics. Eventually he left the University of Pisa without a degree and began teaching mathematics privately as well as working on the design of precision measuring instruments.[3] In 1589 he was appointed to the chair of mathematics at the University of Pisa where he taught traditional Ptolemaic astronomy, among other things. He also began to develop a new theory of motion, but did not publish his work.[4] (We will come back to Galileo's work on motion later in this chapter.)

Figure 8.1 Portrait of Galileo from *Istoria e dimonstrazioni intorno alle macchie solari* (1613). Image courtesy History of Science Collections, University of Oklahoma Libraries.

While at Pisa, Galileo may have performed the famous experiment of dropping two metal balls, composed of the same material but of different size and weight, from the Leaning Tower. Aristotle had taught that the balls would fall with speeds proportional to their weights, but the experiment would have clearly shown that the two balls fall at nearly the same speed, with the smaller ball hitting the ground just after the larger one. We know of this experiment only from Galileo's recollections in old age, so it is hard to know if it really took place. However, it is certain that Galileo was critical of Aristotelian physics, which made him unpopular among the natural philosophers at Pisa.

Poor relations with the philosophers at Pisa may have encouraged Galileo to move, and in 1592 he took a position at the University of Padua. There he focused on technological problems including the development of his "geometric and military compass," a tool that allowed the user to quickly solve a variety of mathematical problems that arose in practical contexts like firing a cannon.[5] He also continued his research on motion, but in 1609 he heard about a new invention that would postpone his work on motion and change the course of his life.

The new device was what we now call the telescope, and it was invented in Holland, possibly by Hans Lipperhey. When Galileo heard about this optical instrument that could make distant objects appear as though nearby, he quickly worked out the principle of its construction. He produced his own telescope, a tube containing a combination of two lenses: a convex "objective lens" and a concave "ocular lens," or eyepiece. Galileo's first telescope had a magnification of about 8×. He demonstrated his device to the leaders of Venice (Padua was part of the Venetian Republic), who rewarded him by increasing his salary and making his professorship permanent.[6]

Galileo continued to improve his telescope and eventually constructed a device with a magnification of about 20×. He turned this improved telescope toward the skies, recording his first astronomical telescope observation on the first of December, 1609. He continued to observe the heavens whenever possible throughout January and February of 1610, and by March his description of his observations was ready for publication.[7]

Galileo began the *Sidereus Nuncius* (The Starry Messenger, or possibly the Message of the Stars) with a description of his instrument and how he came to make it. He then quickly moved on to his observations of the Moon, writing "the surface of the moon is not smooth, uniform, and precisely spherical as a great number of philosophers believe it (and the other heavenly bodies) to be, but it is uneven, rough, and full of cavities and prominences, being not unlike the face of the earth, relieved by chains of mountains and deep valleys."[8] Although the idea that the Moon's surface was rough and irregular, like that of the Earth, had been anticipated by others (such as Kepler's teacher, Michael Maestlin), Galileo was now able to supply strong evidence to support this view.

Galileo paid particular attention to the appearance of the Moon near the **terminator**, the "line" that separates the dark and bright parts of the Moon. He found that the terminator was not a straight line at all, but a very irregular boundary. There were dark patches on the bright side of the terminator which Galileo took to be depressions or valleys in the Moon's surface. In particular, he noted circular depressions (what we would now call craters) that were dark on the side toward the Sun but brightly lit on the opposite side, just as occurs in valleys ringed by mountains on Earth around sunset or sunrise.[9] The *Sidereus Nuncius* included several sketches of the Moon to illustrate his observations (see Figure 8.2).

Galileo also saw bright spots on the dark side of the terminator, which he took to be mountain peaks that rose up above their surroundings to catch the last (or first) rays of sunlight. He measured how far (in terms of the Moon's diameter)

Figure 8.2 Sketch of the Moon from *Sidereus Nuncius* (1610). Image courtesy History of Science Collections, University of Oklahoma Libraries.

these bright spots were from the terminator and used geometry to find that the mountains rose more than 4 miles above the Moon's surface (see Appendix A.15 for details on his method).[10] With such tall mountains, why did the edge of the Moon look so smooth? Galileo gave two possible reasons:[11] either the combination of many different mountain chains, one behind the other, resulted in a smooth appearance or the smooth "edge" of the Moon was really just the Moon's atmosphere.[i]

Galileo described the dim illumination of the dark part of the Moon by light reflected from Earth, now known as "Earthshine."[ii] Galileo provided a clear explanation of the complementary phases of the Moon as seen from the Earth, and of the Earth as seen from the Moon. When we see a full Moon, the Earth would appear dark ("new") as seen from the Moon and *vice versa*. This complementarity explains why no Earthshine is visible during a total solar eclipse (because no sunlight reflects from Earth to the Moon in that case). Galileo's Moon observations make the Moon seem like Earth, while also making the Earth seem like a celestial body (since it illuminates the Moon).[12] After presenting his Moon observations Galileo promised to write a future book in which he would "prove the earth to be a wandering body surpassing the moon in splendor, and not the sink of all dull refuse of the universe ..."[13] This promise was his first public statement in favor of the Copernican system.

[i] We now know that the Moon has no atmosphere and that Galileo's first explanation was largely correct.

[ii] Earthshine had been described previously by Maestlin and Kepler.

The Moon was not the only thing that Galileo observed. In the *Sidereus Nuncius* he noted that, while the telescope seemed to magnify terrestrial objects and the Moon by the same factor, it did not magnify the stars or planets in the same way. Both fixed and wandering stars appeared smaller than expected through his telescope. Galileo explained this discrepancy by noting that stars and planets, when viewed by the naked eye, appear "as irradiated by a certain fulgor and as fringed with sparkling rays," but that the telescope "removes from the stars their adventitious and accidental rays, and then it enlarges their simple globes" so that they seem to be magnified less than other objects.[14] Galileo's notion of "adventitious rays" suggested that naked eye measurements of the angular size of stars, such as those on which Tycho based his most serious objection to the Copernican theory, might be far too large.

Galileo also used his telescope to investigate the nature of the Milky Way and the nebulae. The hazy whitish appearance of the Milky Way, and the cloudy appearance of the small nebulae, had led Aristotle to propose that these were phenomena in Earth's atmosphere. Others had suggested that the appearance of these nebulous objects might be produced by large numbers of very small stars packed close together. Galileo's telescope resolved the Milky Way and several nebulae into "congeries of innumerable stars"[15] and thus served as another strike against Aristotle. Figure 8.3 shows Galileo's sketch of the stars in the Praesepe nebula (now known as the Beehive Cluster).

NEBVLOSA PRAESEPE.

Figure 8.3 Drawing of the Praesepe nebula (Beehive cluster) from *Sidereus Nuncius* (1610). Image courtesy History of Science Collections, University of Oklahoma Libraries.

Figure 8.4 Diagrams of Jupiter and its moons from *Sidereus Nuncius* (1610). Image courtesy History of Science Collections, University of Oklahoma Libraries.

Galileo's biggest discovery was saved for the end of the *Sidereus Nuncius*.[16] While observing Jupiter he had noticed three little stars that, along with Jupiter, seemed to lie in a straight line along the ecliptic. Returning to observe Jupiter on subsequent nights Galileo found that these stars (as well as a fourth he had not seen on the first night) moved relative to Jupiter, but not in the way he had expected. They seemed to track Jupiter's motion relative to the fixed stars, while at the same time wandering from one side of Jupiter to the other, much like the motions of Venus and Mercury with respect to the Sun (see Figure 8.4).

Galileo concluded that these tiny stars were bodies that were orbiting around Jupiter, just as the Moon orbited the Earth. The discovery of Jupiter's moons helped to diffuse a question that had proved difficult for Copernican astronomers: why did the Moon orbit the Earth if everything else orbits the Sun? Galileo showed that it was not just the Earth that had a companion orbiting around it. With four moons orbiting Jupiter, any system of astronomy would have to accept more than one center of motion. Galileo even found that the more distant the satellite (a term later coined by Kepler) the longer its orbital period, so Jupiter and its four moons appeared like a miniature version of the Copernican system.

Although others had used a telescope to observe the heavens before Galileo (notably Thomas Harriott in England), Galileo was the first to make his observations widely known and his *Sidereus Nuncius* made him instantly famous. His fame did not come solely from the fact that he was the first to publish his observations. Galileo's telescope was so much superior to the other instruments available at the time that he was able to see much more than could anyone else. The superiority of his telescope actually caused some difficulty for him, since others tried to verify his discoveries with inferior telescopes and inevitably failed.[17] Some had difficulty seeing anything even through Galileo's own instrument.[i] A few

---

[i] Inexperienced users often have difficulty looking through a telescope, as we have found in our public outreach work.

Aristotelian philosophers denied Galileo's discoveries and even refused to look through a telescope, claiming that the instrument could not truly show what was in the heavens.

The Jesuit mathematicians at the Collegio Romano were initially skeptical, but eventually they were able to verify Galileo's observations and they celebrated his achievement.[18] Kepler also responded to the *Sidereus Nuncius* with support in a letter to Galileo, which Kepler published in 1610. In his *Dissertatio cum nuncio sidereo* (Conversation with the Starry Messenger), Kepler proclaimed his agreement with Galileo in spite of the fact that he did not have a telescope to make his own observations. He also used Galileo's discoveries as a starting point for several speculations. He suggested that Mars must have two moons and Saturn six or eight, so that the number of moons would form an arithmetic or geometric sequence starting at the Earth and going away from the Sun.[19] He also speculated about life on the Moon and Jupiter, and even interplanetary travel:[20]

> But as soon as somebody demonstrates the art of flying, settlers from our species of man will not be lacking. ... Given ships or sails adapted to the breezes of heaven, there will be those who will not shrink from even that vast expanse.

Galileo was certainly pleased by the support, but probably thought little of Kepler's speculations.

Perhaps the most important response to the *Sidereus Nuncius*, from Galileo's point of view, was that of Cosimo de Medici, Grand Duke of Tuscany. Galileo had dedicated the *Sidereus Nuncius* to the Grand Duke, and had named the moons of Jupiter the "Medicean stars" in honor of Cosimo and his brothers. Galileo hoped to win a position at the Tuscan court, which would free him from teaching responsibilities and supply him with a secure income so that he could focus on his researches. The ploy worked, and in July 1609 he was appointed court mathematician and philosopher to the Grand Duke.[21]

Galileo's telescope observations did not end with his appointment to the Tuscan court. When Saturn became visible Galileo noticed that it consisted of a large central disk with two smaller but substantial companions, one on either side. Later observations would reveal that Saturn's appearance changed over time, sometimes appearing three-bodied, sometimes as just a single disk, and sometimes as a disk adorned with handles or, in Galileo's terms, ears.[22] Even more important were Galileo's observations of the changing appearance of Venus.[23] Through the telescope Venus was seen to go through phases just like those of the Moon.

The phases of Venus were important for three reasons. First, they showed that Venus did not produce its own light. To the naked eye, planets always appear round and thus astronomers had concluded that planets were probably self-luminous (or else they absorbed sunlight and re-emitted it in all directions). Galileo's telescope observations showed that the round appearance of Venus was caused by "adventitious rays." The telescope, by removing these rays, revealed the phases of Venus and proved that this planet, at least, was a dark body that shined by reflecting sunlight. In other words, Venus was just like the Moon and (via the phenomenon of Earthlight) the Earth.

Second, the phases of Venus demolished the traditional Ptolemaic system. The particular sequence of phases shown by Venus only made sense if Venus orbited around the Sun. In the Ptolemaic system the entire epicycle of Venus lies between the Earth and the Sun, and therefore Venus will always display a crescent phase when viewed from Earth (Figure 8.5, top). But if Venus orbits the Sun it will display a full set of phases, including gibbous phases (Figure 8.5, bottom). Galileo's sketches of Venus, shown in Figure 8.6, clearly demonstrate that Venus orbits the Sun. Galileo took these observations as conclusive proof of the Copernican system, but in fact the Tychonic system predicts the same appearances. Galileo objected to the Tychonic system for reasons of physics, but other astronomers felt the phases of Venus ruled in favor of Tycho, not in favor of Copernicus.

The third reason the phases of Venus were important was that they resolved the problems that all three systems had in explaining variations in that planet's brightness as seen by the naked eye. Figures 8.5 and 8.6 show that when Venus is close to Earth, and therefore appears larger, it is in a thin crescent phase. When it is far from Earth, and therefore appears smaller, it is nearly full. Thus the phases offset the changes in distance (or apparent size) and keep the brightness of Venus nearly constant throughout its motion.

Galileo's fame spread across Italy and beyond. In 1611 he was inducted into what may have been the world's first scientific society, the Accademia dei Lincei (Academy of the Lynx-Eyed), founded by Federico Cesi in 1603.[24] Not everyone appreciated Galileo's discoveries as much as the Linceans did, though. Soon Galileo would become involved in a series of controversies that would earn him some powerful enemies.

8.2   Many controversies

Copernicus was a Catholic canon and his new model of the world, while it did not receive enthusiastic support from either Catholics or Protestants, was not officially opposed by any Church authority for seventy years after

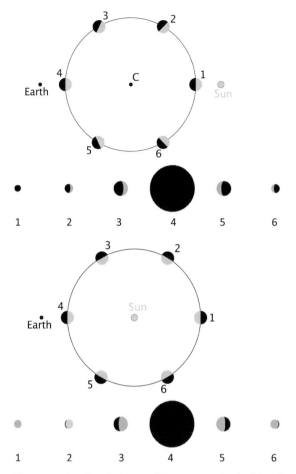

Figure 8.5 Predicted phases of Venus assuming it shines by reflecting sunlight. The top diagram shows the Ptolemaic motion of Venus and the resulting phases. The bottom diagram shows the motions and phases for the Copernican (or Tychonic) system. The diagram is not to scale.

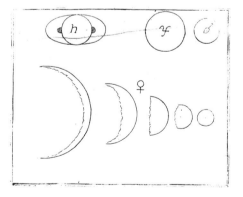

Figure 8.6 Drawings of Saturn (top left), Jupiter (top center), Mars (top right), and the phases of Venus (bottom row) from *Il Saggiatore* (1623). Image courtesy History of Science Collections, University of Oklahoma Libraries.

its publication, but in the 1610s the attitude of the Catholic Church toward Copernicus would suddenly sour. Likewise, Galileo's relationship with Church authorities was initially cordial. As mentioned earlier, the Jesuits in Rome had verified and supported Galileo's telescopic observations, although they did not generally view Galileo's discoveries as a confirmation of the Copernican system. But the 1610s saw Galileo involved in two contentious debates with Jesuit astronomers that led to a dramatic deterioration in his relations with the Jesuit order.

The first of these controversies was over the nature of sunspots.[25] Sunspots had been observed with the naked eye long before, but these observations were fleeting and in some cases they were thought to be transits of inferior planets. A **transit of Mercury or Venus** is when that planet passes across the face of the Sun as seen from Earth, so that it appears as a dark spot on the Sun. Kepler thought he had observed a transit of Mercury in 1607 using a pinhole camera technique, but it is now thought that he saw a sunspot. The first person to observe sunspots with a telescope was Thomas Harriott in December 1610, but as with so much of his work he never published these observations. The first published account of the telescopic observation of sunspots was the *De Maculis in Sole Observatis*, published by Johannes Fabricius in June 1611.[26]

Galileo had shown sunspots to his associates in Rome in early 1611. He only became aware that others had observed sunspots in November of that year. In January 1612 he received a detailed report of sunspot observations from Marc Welser, a banker in Augsburg (Germany).[27] Welser had just published a series of three letters on sunspots written by an author using the pseudonym Apelles. Galileo later found out that Apelles was really Christoph Scheiner,[28] a Jesuit priest in Ingolstadt (Germany).[i]

Scheiner had made several sunspot observations and had realized that the spots appear to move across the face of the Sun. He proposed that the spots could be features on the Sun's surface, which would contradict the perfection of celestial bodies that had been assumed since Aristotle. He also considered that the spots might be opaque bodies orbiting around the Sun, above the Sun's surface. He noticed that the spots seemed to change and the same arrangement of spots did not occur at regular intervals, as would be expected if the spots were

---

[i] Apelles was an ancient Greek artist who reportedly hid near his displayed works in order to hear the commentary of his audience. Scheiner had been cautioned by his Jesuit superiors not to publish his letters, which seemed to contradict established philosophy, under his own name.

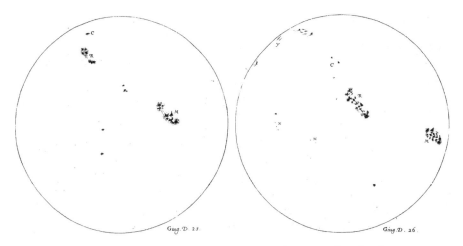

Figure 8.7 Drawings of sunspots on June 23 and 26, 1612, from *Istoria e dimostrazioni intorno alle macchie solari* (1613). Images courtesy History of Science Collections, University of Oklahoma Libraries.

permanent markings on a solid, rotating sphere. Thus, Scheiner concluded that the spots were probably bodies orbiting the Sun well inside Mercury's orbit.[29]

Galileo replied to Welser with two letters written in May and August of 1612.[30] In the second of these letters Galileo included numerous drawings of sunspots made with a technique devised by Galileo's former student Benedetto Castelli. Two of Galileo's sketches are shown in Figure 8.7. Before Castelli astronomers had viewed sunspots by looking at the Sun directly through a telescope, at great risk to their eyesight. Usually, such observations were made when the Sun was low in the sky and perhaps through light cloud cover in order to reduce the risk of damage to vision. Castelli realized that the image of the Sun in the telescope could be projected onto a piece of paper. This not only saved the astronomer's eyes, but it allowed observations to be made at any time of day as long as the weather was clear. Castelli drew circles of predetermined size on his paper and then adjusted the projection of the Sun's image until it matched the circle. Tracing over the image allowed for accurate and consistent representations of the sizes and positions of the spots.[31]

In both of his letters Galileo argued that the spots must be on, or very near, the surface of the Sun. As evidence for his arguments Galileo used Scheiner's observations as well as his own. The spots seemed to move with a collective motion, all following parallel paths across the Sun's face. This collective motion can be seen by comparing the two sunspot drawings in Figure 8.7, which illustrate the

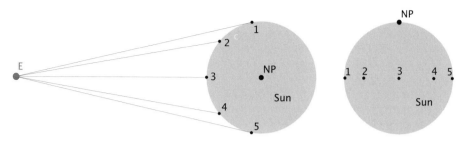

Figure 8.8 The geometry of sunspots. The numbered spots correspond to different positions of a single spot (separated by equal times) or else different spots on the Sun (separated by equal angles). The left part of the diagram shows the view from above the Sun's north pole, while the right part shows the view from Earth.

motion of groups of spots during a period of three days. Such a collective motion was exactly what one would expect for features on a rotating sphere.[32]

Galileo also appealed to projective geometry to show that the spots must lie on the Sun's surface. Spots appeared wider when near the center of the Sun, but narrower (along the direction of motion) when near the edge (or limb) of the Sun's disk. Similarly, spots that were nearby (along the direction of motion) when seen close to the limb were found farther apart when seen near the center of the disk. On the other hand, the sizes and spacings in the direction perpendicular to the motion were constant. Finally, the motion of the spots appeared faster near the center and slower near the limb.[33] All of these features fit with the idea that the spots lie on the surface of a rotating Sun (see Figure 8.8).

On the other hand, Galileo could not ignore the observation that the spots seemed to change over time and that the same pattern of spots did not reappear over and over. He speculated that the spots were something akin to clouds in Earth's atmosphere, which, as Galileo thought, shared in the collective rotation of the Earth but might appear, change shape, merge with other clouds, and disappear over the course of time. Although Galileo could not trace any particular group of spots through a full rotation, he did use the motions that he observed to estimate that the period of the Sun's rotation was about one month.[34]

Scheiner followed up his original series of letters with a new set, published later in 1612.[35] Scheiner agreed with Galileo that the spots moved across the face of the Sun along the ecliptic, but his latest measurements had revealed that "those spots that run along the diameter of the Sun, the ecliptic, remain longer below the Sun than those whose path moves to the south or north of the ecliptic. It is also indubitable (assuming that the Sun is unchangeable and hard, whether we know at present if it rotates or not) that these spots in no way adhere to the

Sun."[36] In other words, Scheiner found that spots at different latitudes moved at different rates, which was entirely inconsistent with the idea that the spots were fixed to the surface of a rotating, solid Sun.

In his new letters, Scheiner maintained his stance that the spots were really orbiting bodies. He explained their changing appearance as being akin to the phases of the Moon or Venus, and also suggested that they might be somewhat translucent.[i] Scheiner thought he had saved the Aristotelian perfection of the Sun, at the price of introducing a host of new bodies orbiting the Sun. He even claimed (wrongly) to have found a fifth moon orbiting Jupiter, which he took as more evidence that there were more bodies in the heavens than had previously been recognized.

Galileo responded to Scheiner's new work in a third letter to Welser written in December 1612.[37] Most of the letter is a detailed geometrical argument demonstrating that if the spots are orbiting around the Sun they must be in extremely tight orbits so that they almost skim the surface of the Sun. Galileo dismissed Scheiner's claim that spots farther from the Sun's equator move faster.[ii] As to the true nature of the spots, Galileo was unsure, and he argued that uncertainty was the most rational position. His statement represents an important aspect of Galileo's approach to the study of nature, which would later be viewed as an integral part of the "scientific method":[38]

> I consider investigating the essence of the nearest elementary substances an undertaking no less impossible and a labor no less vain than that of the most remote and celestial ones. And I seem to be equally ignorant of the substance of the Earth as of that of the Moon, or earthly clouds, and of the spots of the Sun. Nor do I see that, in understanding these nearby substances, we have any advantage other than the abundance of details... But if we want to limit ourselves to learning certain properties, it does not seem to me that we need despair of grasping these in the most remote bodies any more than in the closer ones. At times, indeed, such knowledge will be more exact in the former than in the latter. And who does not know the periods of the motions of the planets better than he does that of the waters of the sundry seas?

[i] Scheiner believed the Moon to be translucent and thought that sunlight, diffused through the body of the Moon, produced the phenomenon that Galileo had attributed to Earthshine.

[ii] In fact, Scheiner was correct that sunspots farther from the solar equator move more slowly. The Sun is a fluid, not a solid, body and different parts of the Sun rotate at different rates.

Galileo's letters were published by the Lincean Academy in March 1613 as *Istoria e dimostrazioni intorno alle macchie solari* (History and demonstration concerning sunspots).[39] Some copies of the book included Scheiner's letters, and all included a preface (written by Angelo de Filiis, a fellow Lincean) that claimed priority for Galileo in the discovery of sunspots and noted that some scholars might not receive Galileo's letters with "openness of mind" because of "pretentions they may have about the discovery of these spots."[40] Scheiner and his fellow Jesuits did not take kindly to this attack.

Near the end of 1612 Galileo heard that the Copernican theory had been attacked as contrary to Scripture by the Dominican priest Nicolò Lorini. Then, in late 1613, Castelli wrote to Galileo that Christina, Grand Duchess of Tuscany, had questioned him during an after-dinner conversation on whether the Copernican theory was not contrary to Holy Scripture. Galileo replied to Castelli with an extended defense of the Copernican system against any charge of heresy.[41] Galileo argued that the Copernican system did not contradict Scripture because Scripture was written in ordinary language, to be understood by common people. It was not written as natural philosophy. When the Bible spoke of the motion of the Sun it was referring to the visual effect of the Sun's motion, not an actual motion through the heavens. Galileo would later expand this letter into a document that came to be known as the *Letter to the Grand Duchess Christina*, eventually published in 1636.[42]

At some point Lorini got a copy of the original letter to Castelli and, dismayed that a layman such as Galileo would dare to make his own interpretations of the Bible, he forwarded the letter to the Holy Office[i] in Rome.[43] A panel of theologians was assembled to rule on the acceptability of the Copernican hypothesis. In February 1616 they ruled the idea that the Sun is stationary at the center of the world to be "formally heretical since it explicitly contradicts in many places the sense of Holy Scripture, according to the literal meaning of the words and according to the common interpretation and understanding of the Holy Fathers and doctors of theology." They also ruled that the idea of a moving Earth was "at least erroneous in faith." Both ideas were judged "foolish and absurd in philosophy."[44] Copernicus' *De revolutionibus* was suspended until corrected, which in practice meant that certain sentences arguing the reality of Earth's motion had to be marked out.[45]

Galileo was not specifically named in the ruling handed down by the Inquisition, but rumors spread that he had been secretly forced to renounce his Copernican beliefs. In fact he had only been told of the ruling and warned not to hold or defend the Copernican theory in the future. Cardinal Roberto Bellarmine

[i] Formally the Congregation for the Doctrine of the Faith, also known as the Inquisition.

later wrote and gave to Galileo a signed statement to this effect. However, an unsigned document in the files of the Inquisition indicated that Galileo was also instructed not to "teach in any way" the Copernican theory. The discrepancy between Bellarmine's account and the one in the Inquisition's files would soon become important for Galileo.[46]

In spite of the ruling of 1616, Galileo still had many powerful friends in the Church, including several Jesuits.[i] However, the appearance of three comets in 1618 led to yet another controversy with a Jesuit that would ruin Galileo's relationship with the Society of Jesus.

It began when the Jesuit astronomer Orazio Grassi proposed that the new comet must orbit the Sun, just as Tycho had claimed for the comet of 1577. Grassi also claimed that the comet was not magnified by the telescope, and thus must be very far away like the fixed stars, which also showed less than expected magnification. Galileo responded through a set of lectures, delivered by Mario Guiducci, a former student of Castelli's and assistant to Galileo. The lectures offered an alternative to Grassi's explanation, suggesting that comets might move in straight lines (as Kepler also believed) and that they might even be an optical effect much like the reflection of sunlight seen on water.[47] The lectures also included a snide remark about "Apelles" and the Jesuits seem to have taken offense at this simultaneous attack on two of their Order.[48]

Grassi, writing under the pseudonym of Lothario Sarsi and possibly with some assistance from Scheiner, responded with a book titled *Libra astronomica ac philosophica* (The astronomical and philosophical balance)[ii] in which he "weighed" Galileo's arguments and found them wanting.[49] Galileo responded in turn with *Il Saggiatore* (The Assayer) in 1623.[50] An assayer is someone who conducts highly accurate, fine-scale measurements of the weights of precious metals, and so Galileo's title poked fun at Grassi's "balance." In general, Galileo offered detailed criticism of Grassi's theory while admitting that his own proposal was uncertain:[51]

> I could illustrate with many more examples Nature's bounty in produc-
> ing her effects, as she employs means we could never think of without
> our senses and our experiences to teach them to us – and sometimes
> even these are insufficient to remedy our lack of understanding. So I
> should not be condemned for being unable to determine precisely the
> way in which comets are produced, especially in view of the fact that I

---

[i] During the 1616 trial it was primarily Dominicans such as Lorini who advocated for a ruling against the Copernican system.

[ii] The title was a pun on the name of the constellation, Libra, in which Grassi thought the comet had first appeared.

have never boasted that I could do this, knowing that they may originate in some manner that is far beyond our power of imagination.

Il Saggiatore is most memorable for its statements on scientific method. Galileo argued that popular opinion, and even the opinion of respected authorities, was not important for deciding scientific questions. What counted in science was observation, reason, and mathematical argument. Galileo particularly emphasized the role of mathematics in science:[52]

> Philosophy is written in this grand book, the universe, which stands continuously open to our gaze. But the book cannot be understood unless one first learns to comprehend the language and read the letters in which it is composed. It is written in the language of mathematics, and its characters are triangles, circles, and other geometric figures without which it is humanly impossible to understand a single word of it; without these, one wanders about in a dark labyrinth.

Galileo also made a distinction between primary and secondary qualities of objects:[53]

> Now I say that whenever I conceive of any material or corporeal substance, I immediately feel the need to think of it as bounded, and as having this or that shape; as being large or small in relation to other things, and in some specific place at any given time; as being in motion or at rest; as touching or not touching some other body; and as being one in number, or few, or many. From these conditions I cannot separate such a substance by any stretch of my imagination. But that it must be white or red, bitter or sweet, noisy or silent, and of sweet or foul odor, my mind does not feel compelled to bring in as necessary accompaniments. Without the senses as our guides, reason or imagination unaided would probably never arrive at qualities like these. Hence I think that tastes, odors, colors, and so on are no more than mere names so far as the object in which we place them is concerned, and that they reside only in the consciousness. Hence if the living creature were removed, all these qualities would be wiped away and annihilated.

The view that scientific explanations should involve only "objective" properties like the position, size, shape, and motion of objects, without reference to "subjective" properties like color and taste, was later championed by the philosophers René Descartes and John Locke.

While Galileo may have gained some enemies among the Jesuits because of the controversy over the comets of 1618, he soon gained a powerful friend in the

Church. In 1623, Maffeo Barberini, a long-time admirer of Galileo, was elected Pope Urban VIII. Galileo obtained an audience with the new Pope, who agreed that Galileo could discuss the Copernican system provided that he treated it only as a hypothesis, not as truth.[54] The door was now open for Galileo to engage once more in the battle between the competing systems of the world.

8.3   Moving beyond Aristotle

The main scientific objections to the Copernican theory, other than the star size problem that Galileo felt he had eliminated with his telescope, were based on Aristotelian physics. The Copernican model was simply incompatible with Aristotle's ideas about motion and gravity, while the Tychonic theory with its stationary Earth fit reasonably well. To overcome the objections to Copernicus, Aristotle's theories of motion would need to be replaced.

As mentioned in Section 4.4.2, medieval scholars had already altered Aristotelian physics by introducing the concept of impetus. In the 14th century, a group of scholars in England known as the Oxford Calculators also investigated different types of motion, including "uniformly difform" motion or what we would now call motion with constant acceleration. They defined uniformly difform motion as motion in which the speed of an object increases with time at a constant rate. The Oxford Calculators derived the "mean speed theorem," which states that the distance covered by a body moving with constant acceleration during a certain interval of time is the same as if that body moved with a constant speed equal to its speed at the midpoint of the time interval (or the average of the initial and final speeds). In France, Nicole Oresme provided an alternate proof of the mean speed theorem using a graphical method that resembles the plots of distance versus time encountered in a present-day physics course.[55]

Although these 14th-century scholars made important advances in understanding possible types of motion, they did not apply their ideas to describe motions in the real world. Galileo, on the other hand, investigated real motions throughout his entire career.[56] He investigated the motion and impact of falling bodies, motions along inclined planes, and motions along circular arcs (such as that of a swinging pendulum). In developing his ideas about motion, Galileo used geometry and careful reasoning much like his medieval predecessors, but he also conducted experiments to see whether or not his theories matched the motion of real objects in controlled situations. In addition, Galileo used what he learned about motion in one context to guide his investigation of motion in other contexts.

One example of this interconnected investigation of two types of motion was Galileo's study of the relationship between the motion of a falling body and that of a swinging pendulum. Early in his career Galileo believed that falling objects experienced a brief period of acceleration and then fell with a constant speed after that, but sometime around 1603 he concluded that falling objects continued to speed up throughout their fall.[57] But exactly how did the speed of a falling body change as it fell?

It was impossible to measure the instantaneous speed of a falling object as it fell, and in fact the concept of instantaneous speed (the speed of an object at a particular moment of time) was unfamiliar to Galileo. We would now calculate a speed as a distance divided by a time, but the mathematics of the 17th century did not permit ratios of unlike quantities so the idea of dividing distance by time made no sense. Instead, mathematicians thought in terms of proportionality. If an object is moving at constant speed then the ratio of the distances covered in two different times is equal to the ratio of the times. If an object is speeding up, then the ratio of the distance covered at a later time to the distance covered at an earlier time will be greater than the ratio of the times. Galileo wanted to test these ratios for a falling body.

Falling bodies fall too fast, so Galileo instead investigated the changing speed of a ball rolling down an inclined plane. He built a flat wooden plane, with a straight groove to guide a rolling ball, and inclined it at a very slight angle (about 2°). In the path of the ball he tied a few taut strings so that when the ball hit the strings it would produce a sound. By repeatedly releasing the ball from the top of the plane, Galileo could adjust the positions of the strings until these noises were equally spaced in time. Galileo's father was a professional musician and Galileo himself was an experienced lutenist, so his sense of musical rhythm allowed him to accurately determine when the noises were "on beat." Modern reconstructions of Galileo's experiment indicate that his timing was accurate to about 1/64th of a second.[58]

When Galileo had the timing right he measured the distances between the strings, and what he found was a remarkable mathematical pattern. The distances between consecutive strings increased like the odd numbers: 1, 3, 5, 7, ... These results clearly showed that the speed of the rolling ball increased continually as it descended the inclined plane. What is more, it showed Galileo that the increase in speed followed a mathematical law.

Galileo's study of pendulum motion had revealed that the period of a pendulum's swing depends only on its length, and he was convinced that the period was connected to the time for an object to fall along the length of the pendulum.[59] To further investigate the changing speed of a falling object, Galileo sought to carefully measure the relationship between pendulum motion and falling. To that end, Galileo timed falling objects and swinging pendulums using a primitive form of stopwatch. He used a setup in which he could start and stop a steady flow of water into a container. By measuring the weight of the water in the container he could get a proportional measure of the time between starting and stopping the flow.[60]

Galileo's results showed that the length of a pendulum was proportional to the square of its period of oscillation. Since he was convinced that the behavior of a pendulum was connected to the behavior of a falling body, this pendulum result suggested that the distance covered by a falling body might be proportional to the square of its time of fall. Returning to his inclined plane results he realized that if he added together the odd-number distances between the strings, the result was the series of perfect squares ($1 + 3 = 4 = 2^2$, $1 + 3 + 5 = 9 = 3^2$, $1 + 3 + 5 + 7 = 16 = 4^2$, etc.).[61] It seemed that the distance covered by a ball rolling on an inclined plane, and therefore the distance travelled by a falling object, was indeed proportional to the square of the elapsed time.

Although Galileo struggled with the notion of instantaneous speed, he eventually came to view speed as a quantity that could change continuously, like distance or time.[62] If acceleration was constant, then the proportionality between distance and the square of time implied a proportionality between instantaneous speed and time. Constant acceleration also implied a proportionality between instantaneous speed and the square root of distance fallen. In fact, Galileo concluded that the speed of an object depended only on how far it had fallen, whether that fall was straight down, along an inclined plane, or along a circular arc such as the swing of a pendulum. Likewise, an object moving upwards would lose speed according to how far up it moved. Putting these ideas together indicated that a pendulum released from rest would swing back to its initial height, as shown in Figure 8.9.[63] Similarly, a ball released from rest would roll down a ramp and then back up another ramp until it reached its starting height.[64]

If an object's speed depended only on how far it had fallen, then an object moving horizontally wouldn't change its speed at all. A ball rolling down an inclined plane would speed up; a ball rolling up an inclined plane would slow down; but, without any interference from the air or friction with the plane, a ball rolling on a perfectly horizontal plane would maintain a constant speed.[i] In this case the ball would maintain its motion without any force to sustain it.[65] Similarly, a spinning sphere would keep spinning on its own without anything

[i] In practice, such a ball would slow down because of friction and air resistance, but Galileo believed that without these impediments the ball would roll forever at a constant speed.

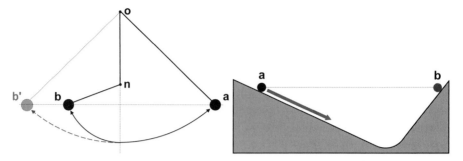

Figure 8.9 A pendulum will swing back up to its initial height, even if a nail is placed in the way of the string. Likewise, a ball rolling down a ramp will roll back up another ramp until it reaches its initial height.

pushing on it because the rotation would not cause the sphere's mass to move either up or down. Galileo came to view these cases as examples of **neutral motions**. Galileo recognized that something had to get these motions started, but once started these neutral motions would sustain themselves forever as long as nothing interfered.[66]

The idea of neutral motion represented an important step beyond Aristotle (who thought that there could be no violent motion without an active force to produce it) and the medieval impetus theory (which stated that an object could sustain its motion for a while, but that eventually the impetus would run out). However, it is important to recognize that Galileo's concept of neutral motion was closely connected to his understanding of gravity. He believed that objects sped up as they approached the center of the Earth and slowed down as they moved away from that center.[i] A neutral motion was a motion in which the object's distance from the Earth's center did not change. As such, it was a *circular* motion. A ball rolling on a perfectly horizontal plane, if it kept going long enough, would encircle the Earth.[67]

Galileo's concept of neutral horizontal motion was initially applied to balls rolling along a horizontal plane, but he later applied it to the study of projectile motion.[68] Galileo conducted experiments to test whether projectiles moved horizontally with constant speed even as they fell with a constant acceleration. He rolled balls down inclined planes and then launched them horizontally from a table. By knowing the relative heights from which the balls were released, Galileo could use his "speed is proportional to the square root of

---

[i] Galileo recognized that very light objects actually speed up as they move upward, but he believed that this occurred only because those objects were lighter than the surrounding air. If there were no air, even these light objects would fall and speed up as they fell.

distance fallen" rule to determine the relative speeds of different projectiles when they shot off the table. If the horizontal motion of the projectile was uniform then the horizontal distance traveled before the ball hit the floor should be proportional to the speed (because the time of fall would be the same for all projectiles launched horizontally from the same table height). He found that his results matched fairly well with that expectation. This experiment not only helped to confirm that horizontal motion was neutral, and that the uniform horizontal motion was independent of the vertical accelerated motion, but it also revealed to Galileo the true shape of a projectile's path: a parabola.[i] More details about Galileo's ideas on falling bodies and projectiles can be found in Appendix A.16.

Although Galileo wrote a few treatises detailing his discoveries about motion, he did not publish them or make them widely available at first. After the election of Pope Urban VIII, Galileo felt free to write a new book in which he could present his ideas about motion in support of the Copernican theory. The result was the *Dialogo sopra i due massimi sistemi del mondo* (Dialogue Concerning the Two Chief World Systems), published in 1632.[69] It was written (in Italian, not in the Latin of scholarly work) as a dialogue spanning four days among three participants: Salviati is Galileo's spokesperson, Simplicio is an Aristotelian, and Sagredo is a supposedly neutral participant. As the title suggests, the discussion deals with the pros and cons of the Ptolemaic and Copernican world systems. Galileo completely ignored the Tychonic system in the *Dialogo*.

On the First Day, Galileo had Salviati demolish arguments based on the dichotomy between the sublunary and celestial regions. Sunspots, mountains on the Moon, Earthlight, comets, and novae were all presented as evidence that the Earth was not fundamentally different from the celestial bodies. Thus, circular motion, whether a rotation or an orbit about the Sun, was just as appropriate for Earth as for the Sun.

The dialogue of the Second Day addressed physical arguments against Copernicus. Here Galileo brought his new theories of motion into play. The rotation of the Earth was self-sustaining because it is a neutral motion. Objects on the

---

[i] Interestingly, Galileo's tests of projectile motion revealed a problem in his theories. Based on his study of falling bodies, he believed that an object dropped through a height h and then projected horizontally from a table of height h should travel a horizontal distance $2h$ before striking the floor. When he performed the experiment he found that the ball traveled a much shorter horizontal distance. He never figured out the problem, but we now know that some of the energy the ball gains as it rolls down the inclined plane goes into *spinning* the ball rather than increasing its speed along the incline. The concept of energy was not developed until long after Galileo. If he had used sliding pieces of ice rather than rolling balls he would have obtained the result he expected.

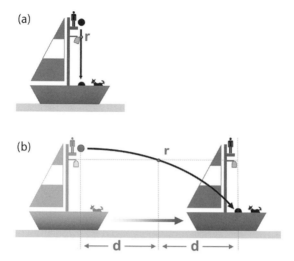

Figure 8.10 A ball dropped from the mast of a ship rings a bell on its way down and lands near the base of the mast. This is true whether the ship is stationary, as in (a), or moving at a uniform speed, as in (b). In case (b) the ball is already moving toward the right (with the ship) before it is dropped, and that motion toward the right continues unchanged as the ball falls.

Earth were not left behind by Earth's rotation because they were already moving in a circle around Earth and that motion was a self-sustaining neutral motion. Objects dropped from a tower would continue to move with Earth's rotation and would land at the base of the tower just as, claims Salviati, a weight dropped from the mast of a moving ship would fall at the base of the mast whether the ship was sitting still or moving uniformly (see Figure 8.10).[70]

In the *Dialogo*, Galileo made the case that there was no way to tell from experiments whether the Earth was rotating or not. Either the Earth was not rotating and bodies fell straight down with constant acceleration, or else the Earth was rotating and bodies fell in a path that combined uniform horizontal motion with vertical acceleration downward. Either way the motion of the object relative to the Earth was the same, so there was no way to tell which is really happening. Galileo/Salviati even suggested that the true motion of a falling object might be along the semicircle connecting its starting point to the Earth's center, so that perhaps *all* natural motions were really circular whether they were in the heavens or on Earth.[71] Later he backed away from that claim, but he did insist that there was no way that the Earth's rotation could fling anything off Earth's surface. Gravity, no matter how small, would always be sufficient to keep objects firmly attached to the Earth.[i][72]

---

[i] This claim of Galileo's is not true, but as Huygens and Newton would later show the actual gravity of Earth is sufficient to keep objects from flying off.

For the Third Day, Galileo returned to astronomical observations, beginning with the undetectable parallax of novae.[i] Galileo then had Salviati use the phases of Venus to guide his listeners to accept that Mercury and Venus must orbit the Sun. Using the brightness of the superior planets at opposition, he achieved the same result for Mars, Jupiter, and Saturn.[ii] Thus, all the planets orbit the Sun! He addressed astronomical criticisms of Copernicus such as the minimal change in the brightness of Venus (explained by the phases), the problem of the Moon orbiting the Earth (eased by the fact that Jupiter had four moons), and the star size argument (the telescope removes the adventitious rays and shows us that apparent diameters of stars are much smaller than previously thought, no more than 5″).[73] He also dismissed the problem of the empty space between Saturn and the stars, saying that we should not be quick to judge such empty space as useless or vain.

To add to his astronomical case for Copernicus, Galileo introduced a new argument based on the apparent motions of sunspots.[74] He pointed out that the Sun's equator was actually tilted relative to the ecliptic.[iii] That tilt produced variations in the apparent path of sunspots across the face of the Sun. If Earth orbited the Sun, then those changes in path could be explained by giving the Sun's rotational axis a fixed orientation relative to the stars. But if the Sun orbited the stationary Earth, then the Sun's rotational axis had to wobble around with a period of one year. Galileo implied that these complicated motions could be avoided if the motion of the Earth was accepted.

Of course, if Earth did orbit the Sun then its axis must wobble around in a similar way, as Copernicus claimed. Galileo argued that the Earth's axis would naturally remain in a fixed orientation, citing as evidence that a ball floating in a rotating bowl of water would spin, relative to the bowl, in the opposite direction. The rotation of the ball relative to the bowl was really an illusion – in fact, the ball was just maintaining itself motionless relative to everything else. In a like manner, the Earth's rotation axis stayed fixed relative to the stars even as the Earth moved around in its orbit. Galileo did *not* go on to point out that the same logic could be applied to explain the apparent wobble in the Sun's rotational axis if one assumed that the Sun orbited the Earth, nor did he supply an explanation

---

[i] Galileo's discussion of this topic is an impressive example of what we would now call "error analysis."

[ii] Though we have seen that these changes in brightness are far from obvious and, in any case, they can be reproduced in the Ptolemaic system.

[iii] The approximately 7° tilt of the solar equator relative to the ecliptic was first discovered by Christoph Scheiner and published in his *Rosa Ursina*.[75] Galileo did not mention Scheiner but instead presented the tilt as his own discovery, based on the curving path that sunspots sometimes take across the Sun (when the Sun's axis is tilted toward or away from Earth).

for the precession of the equinoxes, which Copernicus had explained using the wobble of the Earth's axis.[76]

In the Fourth Day of his dialogue, Galileo presented his theory of the tides. In the first three days of the *Dialogo*, Galileo had shown that the arguments against Copernicus were inconclusive, but he had not supplied conclusive arguments in favor of the Copernican theory. He believed that his theory of the tides provided just such an argument, because his explanation of the tides required both the rotation and the orbital motion of the Earth.

Pick a point on Earth's surface. Sometimes the motion of that point due to Earth's rotation is in the same direction as the Earth's orbital motion. Twelve hours later the motion of that point due to rotation will be in the direction opposite that of Earth's orbital motion. The overall motion of the point through space will be a combination of the rotational and orbital motion, and because of these changing directions the overall motion will go through periodic changes every day. Although a uniform motion would not produce any noticeable effect on the waters of Earth, a changing motion might be able to produce the tides.

The changes in the overall motion of a point on Earth took place over a 24-hour cycle, but Galileo knew that the tides of the Mediterranean followed a 12-hour cycle. He thought that the 12-hour cycle was a result of water sloshing back and forth, east to west and *vice versa*, in the Mediterranean basin. Galileo thought that the period of the tides in a given body of water was determined mostly by its length east to west, so, although the Mediterranean had a 12-hour cycle, other seas or oceans might have different periods. He dismissed Kepler's idea that the attraction of the Moon produced the tides, saying,[77] "that concept is completely repugnant to my mind; for seeing how this movement of the oceans is a local and sensible one, made in an immense bulk of water, I cannot bring myself to give credence to such causes as lights, warm temperatures, predominances of occult qualities, and similar idle imaginings." We now know that Kepler was closer to the truth and that the attraction of the Moon is the primary cause of the tides (though not quite in the way Kepler envisioned), but the idea that attractions are "occult qualities" was a common objection that would later plague Isaac Newton.

Galileo ended his dialogue by having his spokesman Salviati admit that the theory of the tides "may very easily turn out to be a most foolish hallucination and a majestic paradox." The Aristotelian Simplicio then states:

> keeping always before my mind's eye a most solid doctrine that I once heard from a most eminent and learned person, and before which one must fall silent, I know that if asked whether God in His infinite power

and wisdom could have conferred upon the watery element its observed
reciprocating motion using some other means than moving its containing vessels, both of you would reply that He could have, and that He
would have known how to do this in many ways which are unthinkable
to our minds.[78]

Galileo had heard this argument from the Pope (before whom one must fall
silent). Salviati replies:

An admirable and angelic doctrine, and well in accord with another one,
also Divine, which, while it grants to us the right to argue about the
constitution of the universe ... adds that we cannot discover the work of
His hands.

By ending his *Dialogo* in this way Galileo hoped to avoid the charge that he was
arguing for the truth of the Copernican system, even though there could be little
doubt that he was doing just that.

Pope Urban VIII, after finding his words put into the mouth of the simpleton
Simplicio, and perhaps after also discovering the Vatican document that forbade
Galileo to teach the Copernican theory "in any way," was most displeased. He
may have felt betrayed by a man he admired and considered a friend. Heresy
charges were brought against Galileo and in 1633 he stood trial in Rome. In
his defense, Galileo produced the document from Cardinal Bellarmine, which
indicated no restriction on teaching the Copernican theory as a hypothesis. Nevertheless, Galileo was found guilty of "vehement suspicion of heresy" and forced
to publicly "abjure, curse, and detest" his formerly held opinions on the Earth's
motion. He was given a life sentence of house arrest, which he served first in
Rome, then in Siena at the home of Archbishop Ascanio Piccolomini, and finally
at his own home in Arcetri, near Florence.[79]

During the final years of his life Galileo compiled his work on motion for
publication. The *Discorsi e dimostrazioni matematiche intorno a due nuove scienze* (Dis-
courses and mathematical demonstrations relating to two new sciences) was
published in 1638.[80] Galileo could not get permission to publish in Italy, even
though the book does not deal with the Copernican theory. Instead, the book
was published in Leiden, South Holland. It detailed Galileo's mature ideas about
motion, including uniform horizontal motions, accelerated motions in free fall
or along an inclined plane, pendulum motion, and projectile motion, as well as
several other topics.

During his house arrest Galileo's vision deteriorated, and by the time he
received printed copies of his new book he was unable to see it.[81] He died
in January 1642 and was buried in an unmarked tomb in the Basilica di Santa

Croce in Florence, which also houses the tombs of Michelangelo and Machiavelli. Thanks to the efforts of Galileo's student Vincenzo Viviani, a new and impressive tomb was erected for Galileo at the same church, though not until in 1737.[82]

8.4   Astronomy after Galileo

Most of the new observations that could be made with a small telescope were made by Galileo, even if he was not always the first person to make them. Later telescopes were of the Keplerian design, with two convex lenses that displayed an upside-down image. There were some attempts to make these telescopes larger, but the production of large refracting telescopes was difficult. Large lenses were hard to grind accurately and too heavy to mount easily within a telescope tube. These early telescopes also produced fuzzy, rainbow-tinged images, a problem known as "chromatic aberration." Reducing the curvature of the lens seemed to help, but that resulted in very long and unwieldy telescopes.

For all of these reasons the rapid pace of astronomical discovery during the period immediately after 1609 could not be sustained. However, astronomers and mathematicians still made important contributions between the time of Galileo and that of Newton. A brief review of some of those contributions is in order.

In France, Pierre Gassendi observed a transit of Mercury in November 1631. Kepler had predicted that a transit of Mercury would be visible on November 7, 1631, but pointed out that the difficulty of observing Mercury might have led to errors in his theory so the exact timing might be a bit off. As it turned out, the transit occurred on the day predicted. Several astronomers attempted to observe this transit. Only Gassendi was successful enough to publish his results. Initially Gassendi thought he was only seeing a sunspot because he expected Mercury to have a much larger apparent size than the spot he saw, but the spot moved across the Sun faster than any sunspot and Gassendi recognized that he was, indeed, seeing Mercury. He found the apparent diameter of Mercury to be about 20″, much smaller than was previously thought.[83]

Gassendi also did important work on motion, following in Galileo's footsteps. In fact, he actually performed the experiment shown in Figure 8.10, finding that a stone dropped from the mast of a moving ship fell right at the base of the mast and not toward the stern of the ship. Gassendi even gave a description of inertial motion that we now view as correct: bodies in motion will continue moving in a straight line at constant speed unless some force acts on them to alter their motion. This statement represents a significant departure from Galileo, because Gassendi thought of gravity as an external force that caused objects to deviate from their straight-line inertial motion in contrast to Galileo's conception of neutral circular motions. However, Gassendi made other claims that contradicted his notion of straight-line inertia and it's not clear that his idea was very influential.[84]

A much clearer presentation of linear inertia was found in the work of René Descartes. He spelled out the idea of inertial motion in *Le Monde* (The World), which he finished writing in 1633.[85] Descartes' "World" was heliocentric, though, and after the trial of Galileo he decided not to publish it.[86] He rewrote the book and published it in 1644 as *Principia philosophiae* (Principles of Philosophy). There he presented a view of the solar system that was still heliocentric, but that managed to tiptoe around the prohibition against claiming that the Earth moves.

Descartes' solar system was filled with an invisible fluid and that fluid swirled around the Sun in a "vortex." The planets (including Earth) did not move relative to the fluid around them and thus could, in some sense, be considered stationary, but the fluid carried the planets along in its motion around the Sun. Each planet could have its own sub-vortex, such as those that carried Earth's moon or the moons of Jupiter. Comets moved along wandering paths that could escape the Sun's vortex, perhaps to fall into another vortex, since Descartes viewed the stars as other suns, possibly surrounded by their own system of planets, scattered through a universe of indefinite size. Figure 8.11 shows the Cartesian vortices.[87]

Descartes' *Principia philosophiae* supplied a comprehensive cosmology that was capable of replacing the traditional Aristotelian cosmos, and it accomplished that task while accounting for the new observations and ideas of the past century. It was not a mathematical text, nor were Descartes' conclusions derived directly from observation or experience. Instead it was a virtuoso display of reasoning, starting from a set of assumed principles (for instance, that all space is filled with some kind of matter). However, Descartes was an excellent mathematician and his *La Géométrie* (1637) proposed a method of coordinate geometry that allowed mathematicians to use the newly developed techniques of algebra to solve problems in geometry.[i] This algebraic approach made it possible for mathematicians to consider ratios of unlike quantities, which allowed them to define concepts such as speed (distance traveled divided by elapsed time) or acceleration (change in velocity divided by elapsed time) the way we do now.[88]

While Descartes presented a heliocentric theory that he hoped would not raise the ire of Church censors, Giovanni Battista Riccioli decisively rejected the Copernican theory. His decision was clearly indicated in the frontispiece of his 1651

---

[i] You can blame Descartes for the fact that x and y are always used as coordinate variables in mathematics courses.

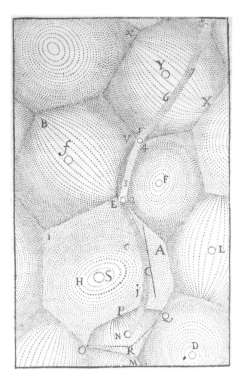

Figure 8.11 Path of a comet through the Cartesian vortices as illustrated in Descartes' *Principia philosophiae* (1644). Image courtesy History of Science Collections, University of Oklahoma Libraries.

Almagestum Novum (New Almagest), shown in Figure 8.12. The image shows the muse Urania weighing a semi-Tychonic system against the Copernican system and finding the semi-Tychonic one heavier (truer) while the discarded Ptolemaic system lies on the ground. Although Riccioli was a Jesuit priest, his opposition to heliocentrism was not primarily for religious reasons. He presented 49 arguments in favor of Earth's motion and 77 against it. He dismissed several of these arguments as either silly or inconclusive (including the two religious arguments against Copernicus), but found that there were a few scientific arguments against the motion of the Earth that seemed conclusive. Both of these arguments were updated versions of the ones that had convinced Tycho of the Earth's stability (see Sections 6.4 and 6.5).[89]

One of these arguments had to do with star sizes. Galileo's telescope showed that Tycho had vastly overestimated the apparent diameters of the fixed stars. Riccioli used a telescope to carefully measure star diameters and found values ranging from about 4″ (for a sixth magnitude star) to 18″ (for Sirius, the brightest star in the heavens). This was a significant reduction in size compared to Tycho's

Figure 8.12 Frontispiece of Riccioli's *Almagestum Novum* (1651). Image courtesy History of Science Collections, University of Oklahoma Libraries.

measurements, but the telescope also allowed for more precise measurements of annual parallax, yet no such parallax was seen. Riccioli felt the annual parallax of a star could not be greater than 10″ or it would have been detected. That smaller parallax value meant that the stars had to be farther away than even the Copernicans had previously believed. At that greater distance, even with the reduced apparent diameters, the stars would still be large enough to fill up Earth's entire Copernican orbit around the Sun.[90]

Riccioli felt that the star size argument ruled out the Earth's annual orbit. A different argument seemed to rule out the Earth's daily rotation. Riccioli pointed out that, while the motion of a ship might not be detectable as in Figure 8.10, the motion of a spinning Earth might still be detectable. Riccioli noted that points on the Earth far from the equator move slower, as a result of the supposed rotation of Earth, than points near the equator. This difference in the speed of Earth's surface as one moves north or south could show up as an apparent deflection of a cannonball, fired north or south, relative to Earth's surface.[91]

In fact, we now know cannonballs fired toward the north are deflected to the east in the Northern Hemisphere. This deflection arises because of the "Coriolis effect," a direct outcome of Earth's rotation. Riccioli was correct that the rotation of the Earth could be detected in the deflection of projectiles. Riccioli believed that the artillery of his time was extremely accurate and that any such deflection would have been detected. In fact, 17th-century artillery was not as accurate as he believed, and the deflections produced are so small that they went unnoticed. Riccioli's intuition was qualitatively correct, but because he could not produce a quantitative prediction of the extent of the deflection he was misled into believing that experience disproved the Earth's rotation.

Riccioli did, however, make accurate quantitative measurements of the acceleration of falling bodies. By dropping balls from the Asinelli tower in Bologna, Riccioli verified that heavy objects do seem to fall with constant acceleration such that their motions in successive equal time intervals follow the pattern of odd numbers as found by Galileo. Riccioli also investigated the effects of air resistance on falling bodies, finding that if he dropped two balls of the same size but different weight from the same height, the lighter ball would hit the ground later.[92]

Riccioli's fellow Italian, Giovanni Borelli, proposed a new explanation of planetary motion based on physical causes, although he was careful to apply his ideas only to the motion of Jupiter's moons so as to avoid declaring the truth of the Copernican system. Borelli proposed that orbiting bodies were subject to two main forces: an attraction toward the center and a centrifugal ("center fleeing") force that pushed them outward because of their motion. If the body moved too rapidly it would fly away from the center, if too slowly it would fall inward. If the centrifugal and attractive forces balanced, however, the body could maintain a stable orbit. These two forces could suffice to produce a circular orbit, but to explain elliptical orbits (which Borelli accepted as correct) he introduced a circulatory force akin to Kepler's *species*.[93]

Although Borelli conceptualized circular motion as a balance between a centrifugal force and an inward attractive force, a detailed understanding of the nature of centrifugal forces was not supplied until the work of the Dutch astronomer Christiaan Huygens. He found that the centrifugal force on an object moving in a circle at a constant speed must be proportional to the square of the object's speed and inversely proportional to the radius of the circle. Huygens also accurately measured the acceleration of a falling object and used his result,

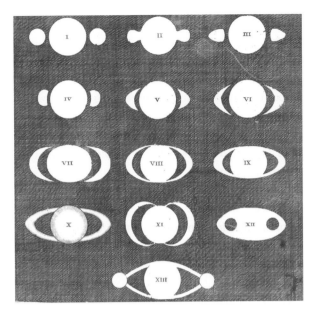

Figure 8.13 Drawings of Saturn from Huygens' *Systema Saturnium* (1659). Image courtesy History of Science Collections, University of Oklahoma Libraries.

along with his new understanding of centrifugal force, to show that an object's weight is far more than is necessary to keep it from being flung off of the Earth by centrifugal force. Huygens did not, however, apply his rule for centrifugal force to the motion of the planets. He was more concerned with applying it to pendulum motion as he worked to develop the first accurate pendulum clock.[94]

Even though Huygens did not investigate the physics of planetary motion, he was a well-known observational astronomer. His most famous discovery was that of Saturn's rings. Huygens observed Saturn over an extended period, making careful sketches of what he saw (Figure 8.13). He realized that the changing appearance of Saturn's "ears" could be explained if Saturn was surrounded by a flat ring that was tilted relative to the ecliptic plane, with a gap between the spherical body of the planet and the inner portion of the ring. As Earth and Saturn moved around in their orbits, we would see Saturn from different perspectives. Sometimes we would see the rings from above, sometimes from below. When our line of sight lay within the plane of the rings, the rings would seem to disappear because they were too thin to be visible.[95]

Huygens also discovered a moon orbiting Saturn and found that the apparent diameters of stars, as seen through a telescope, were reduced if the lens of the telescope was blackened with smoke. The latter observation suggested that even the telescope failed to show the true apparent sizes of stars.[96] That suggestion was confirmed by the English astronomer Jeremiah Horrocks who, with his friend William Crabtree, observed a lunar occultation of the Pleiades star cluster. In a lunar occultation, the Moon moves in front of a star or planet and blocks it from sight. If Riccioli's measurements were correct then the apparent diameter of the star Merope (a fourth magnitude star in the Pleiades) should be about 6″. It takes about 11 seconds for the Moon to move 6″ relative to the stars, so during a lunar occultation Merope should be seen to gradually fade over that period of time. Instead, Horrocks and Crabtree found that the stars of the Pleiades vanished in the blink of an eye behind the Moon.[97] Clearly the apparent diameters that correspond to the true bodies of the stars were much smaller than those measured by Riccioli.

Horrocks and Crabtree were also the only people to observe the transit of Venus in 1639, and Horrocks used his measurements of the transit to make corrections to Kepler's tables of planetary motion. Horrocks was one of the few people in the period immediately after Kepler's death who really mastered Kepler's elliptical astronomy. He demonstrated the elliptical orbit of the Moon and proposed that comets also moved in elliptical orbits. Unfortunately, Horrocks died in 1641, at the young age of 22, and his book on the transit of Venus and other observations was not published until 1672, when the Danzig astronomer Johannes Hevelius paid to have it printed by the Royal Society of London.[98] Hevelius himself was an accomplished astronomer best known for his detailed map of the Moon's surface. He also built and used incredibly long telescopes, including one 46 meters in length, to avoid the problem of chromatic aberration.

Another way around chromatic aberration was to use a mirror, rather than a lens, to focus the light. The Scottish mathematician James Gregory came up with a design for a reflecting (mirror-based) telescope in 1663, but he was a mathematician and not an artisan so he was unable to build a prototype. The first Gregorian telescope was not built until ten years later by Robert Hooke, but by that time Isaac Newton had already constructed the first usable reflecting telescope based on his own new design.[99] Telescope design was not the only activity in which Hooke and Newton were rivals, and in some respects their rivalry helped produce a monumental advance in physics that won the day for the Copernican theory (as modified by Kepler). Their story will be the subject of our next chapter.

8.5   Reflections on science

Looking back at Galileo's telescopic discoveries it might be easy to ridicule those who denied his observations and distrusted what they saw in the telescope. However, it is important to understand how radical was the introduction of the telescope into astronomy. The telescope was the first device to significantly extend human sensory experience. It was, in modern terms, a "disruptive" technology. It was entirely reasonable for people to be skeptical of this new technology that promised to make distant things appear nearby.[100]

The newness of the telescope was not the only challenge to accepting Galileo's observations. Galileo had no theory of optics to explain how the telescope worked. That would be supplied later by Kepler. His telescope was also very difficult to use. It would have been hard for an inexperienced user to see Jupiter's moons, or detailed surface features on the Moon. And there *were* things seen in the telescope that were not to be trusted. Galileo argued that the telescope removed the spurious rays of light around the fixed and wandering stars, but Horrocks' lunar occultation observation showed that the angular diameters of fixed stars measured with the telescope were still far too large.

How, then, were these problems resolved? First, by replication. Once enough telescopes became available (made by Galileo or by others), people could make their own observations. Dedicated observers gained experience with the instrument and soon found that their observations matched those of Galileo. Galileo's subjective claims took a step toward becoming objective facts. The other thing that led to the acceptance of Galileo's observations was that they fit within a sensible theoretical framework. Most of his telescope observations served to show that the heavens were not that dissimilar to Earth. Bright spots on the dark part of the Moon were mountain peaks catching sunlight. Dark spots on the Sun were like the clouds on Earth. Tiny points of light moving back and forth across Jupiter were moons orbiting that planet just as the Moon orbited the Earth. Venus displayed phases just like our Moon's (and like the Earth's as seen from the Moon).

These phenomena seemed even more objective because they could be subjected to mathematical analysis. Geometrical arguments were critical for interpreting the height of lunar mountains, the motion of sunspots, and the phases of Venus. Soon the moons of Jupiter were found to obey Kepler's Third Law. It might not be hard to believe that a new optical device could produce false images, but it was pretty far-fetched to think that the device could produce a whole series of false images that followed strict mathematical laws.

Seeing these new phenomena in the telescope was something like being handed a new piece to a jigsaw puzzle. At first the piece may look odd and may not seem to fit with the rest of the puzzle. Maybe it's just a funny-shaped piece of cardboard. But once you find that the piece snaps in perfectly with a few other new pieces, and maybe even a few older pieces, you start to have confidence that it really is a part of the puzzle. If you hear from other puzzle-solvers that they have found similar pieces, and incorporated those pieces into their puzzles, then the legitimacy of the new piece becomes nearly certain.

Of course, these new puzzle pieces can cause trouble if you thought you already had the whole puzzle solved. That was the problem faced by Aristotelian natural philosophers when confronted with Galileo's observations, as well as his new ideas about motion. Aristotle's cosmology had provided a comprehensive understanding of natural phenomena, but it could not account for the new telescopic observations. Galileo's new observations and ideas threatened to pull apart this comprehensive cosmology, without offering a similarly comprehensive theory to replace it.

In fact, Galileo saw it as an advantage that he didn't try to explain *everything*. His approach to science was to focus on a limited problem and try to solve that problem rigorously and completely. How do objects fall? What path does a projectile follow? He ignored the question of "essences" and ultimate causes, and focused instead on well-defined questions that could be answered through measurement, logical reasoning, and mathematical analysis. In *Il Saggiatore* he said, of natural philosophy, that "the more this partakes of perfection the smaller the number of propositions it will promise to teach, and fewer yet will it conclusively prove."[101] However, Galileo realized that a few *conclusively proven* answers to limited problems could be put together to lead to more general conclusions.

To "conclusively prove" his ideas, Galileo relied heavily on mathematics. He was confident that Nature followed mathematical laws. The trick was to figure out which of the many possible mathematical relationships Nature actually used. This was no easy task, and it required Galileo to develop new ways of applying mathematics to the physical world. Prior to Galileo the speed of an object was viewed as a measure of the object's change in position during some period of time. It was therefore impossible to think about the speed of an object at a single instant of time. If no time passed, then the object would not move, and therefore it made no sense to talk about its speed, but Galileo came to think of speed as a property or state of an object. In that case it was sensible to talk about the speed of an object at a particular instant of time. Only then could he understand that the "instantaneous speed" of a falling object increases proportionally to the time of fall.

Galileo's use of mathematics allowed him to think about motion in new ways, but it also allowed him to measure quantities and test ideas that he otherwise would not have been able to measure or test. For example, Galileo could not directly measure the height of a mountain on the Moon, but if he assumed that a bright spot on the dark side of the terminator was a mountain peak catching sunlight, then he could use geometry to determine that mountain's height. Likewise, he could not measure instantaneous speeds in order to test whether the speed of a falling body was proportional to the time of fall. However, he could use mathematics to rigorously derive consequences of that relation between speed and time (as the Oxford Calculators had done long before, but without applying it to real motions). In particular, Galileo could show that this relation necessarily led to another one: that distance is proportional to the square of time. Even that relation was difficult to test for falling bodies, but Galileo could demonstrate that it held for balls rolling down an inclined plane, which made it plausible that it was also true for falling bodies (as Riccioli's experiments later showed).

Although Galileo's approach sounds a lot like what would later be called the "hypothetico-deductive method," it is not quite the same. In the hypothetico-deductive method the scientist is supposed to formulate a hypothesis, deduce consequences that must follow if the hypothesis is true, and then test those consequences through measurement or experiment. It is true that Galileo followed those steps, but according to the hypothetico-deductive method one is supposed to reject the hypothesis if the experimental results don't match the predicted consequences. Galileo did not take such a simplistic approach to determining the success or failure of his hypotheses. Rather, he recognized that complicating factors might produce small deviations from the expected behavior. That didn't mean that the hypothesis was wrong, it just meant that the hypothesis did not tell the whole story.

For example, Galileo's notion of horizontal (circular) neutral motion indicated that a ball rolling on a horizontal plane would continue rolling forever at constant speed, circling the Earth over and over. In practice, such a ball won't roll very far before it begins to slow down noticeably. Galileo knew that, but he didn't toss out his idea of horizontal neutral motion. Instead, he recognized that in practical situations there were things that interfered with the neutral motion: rolling friction between the ball and the plane, air resistance, a not-perfectly-horizontal plane, etc. These factors caused deviations from the neutral motion, but Galileo viewed the neutral motion as the most fundamental thing. The other factors were just extras that had to be taken into account in practice. He was confident that in the *ideal* case, with a perfectly spherical ball rolling on a perfectly horizontal and frictionless plane in a vacuum, the behavior would be just what he expected.

Likewise, Galileo knew that if you dropped different objects from a tower they probably would not land at the same time, but he attributed the difference to air resistance, and argued that without air the two objects really would land at exactly the same time. Nor was this just a verbal ploy. Galileo could point out that if you dropped the objects in water the difference in time of fall was much greater because water was denser than air.[102] Thus it was reasonable to think that a medium less dense than air would produce a smaller difference in time of fall, and a medium of zero density (vacuum) would produce zero difference.[103]

Idealization plays an important role in science because it allows us to uncover important relationships that might otherwise be hidden within the details of our observations. In physics, in particular, idealization allows us to uncover simple mathematical laws that govern the behavior of ideal physical systems.[104] Often the physical systems we deal with in practice don't show this ideal behavior, but by learning how an ideal system behaves we can gain important understanding about more realistic systems. From there we can work to improve our knowledge of the factors that cause deviations from the ideal case. We can also devise experimental setups that should demonstrate the ideal behavior. The invention of the air pump in the decade after Galileo's death would eventually allow experimenters to demonstrate that a feather and a stone really do fall at the same rate in a vacuum.

So what was the outcome of Galileo's work, and that of his immediate successors? First and foremost, Galileo raised serious questions about Aristotelian cosmology and physics. The celestial/terrestrial divide, already weakened by Tycho, was further undermined by Galileo's telescope discoveries. Galileo's new theories of motion revealed flaws in Aristotle's physics and illustrated the power of the mathematical approach to understanding nature that Aristotle had largely rejected. Galileo's work helped to bring the Aristotelian synthesis crashing down, which created an opening for a new "system of the world." In fact, it did not take long for Descartes to supply an alternative to the Aristotelian synthesis.

But did Galileo prove that Earth rotates on its axis and orbits the Sun? He thought he had, with his theory of the tides. Later investigators found many flaws in his theory, so that case was open to doubt. The phases of Venus had demolished the traditional Ptolemaic arrangement of the planets, but they fit just fine with the Tychonic system. Indeed, Galileo's discussion of moving bodies on the Earth had suggested that the only motion we can observe is *relative* motion. If you ignored the seemingly undetectable annual parallax of fixed stars, the relative motions of the celestial bodies were the same in both the Copernican and Tychonic systems. Galileo had supplied many persuasive arguments in favor of the Copernican theory, but everything except his theory of the tides could be seen as an argument for Tycho instead.

Barring a successful parallax measurement it seemed impossible to make a conclusive case for the Copernican theory based solely on astronomical observations. But the observations did reveal that the heavens were not so different from the Earth, and thus might follow the same physical laws. Kepler had attempted to apply Aristotle's physics of terrestrial motions to the planets while developing his new astronomy. Now, Galileo had undermined Aristotle's ideas about terrestrial motions and he and his successors had made the first steps toward a new physics. It would be the unification of this new physics with Kepler's astronomy that would clinch the argument for Copernicus.

9

The system of the world: Newton's universal physics

9.1   Curious characters: Newton and Hooke

9.1.1   *Isaac Newton*

Isaac Newton[1] (Figure 9.1) was born at Woolsthorpe Manor in the county of Lincolnshire on December 25, 1642, according to the Julian calendar still used in England at the time.[i] Although he was born prematurely, he did not arrive until three months after his father (also named Isaac) had died. Newton was a sickly child who was raised mostly by his grandmother after his mother, Hannah Ayscough, remarried and left to go live with her new husband, the Reverend Barnabas Smith.[2]

When Reverend Smith died in 1653, Newton's mother returned home. Not long afterward, Newton was sent to the King's School in nearby Grantham, where he lived and studied for several years. When he was approaching adulthood his mother called him back to Woolsthorpe, where she set the teenage Newton to work as a farmer. Farming did not suit his temperament or abilities and, fortunately for him and for history, his former master at the King's School persuaded Newton's mother that young Isaac should continue his education. After excelling in his remaining years at the school, Newton enrolled at Trinity College of the University of Cambridge in 1661.[3]

Although Newton's father had been head of an English manor, and although his mother had inherited considerable wealth from her second husband, he received little financial support from his family when he enrolled at Trinity College. To help pay his way he served as a subsizar, a servant to a wealthier

---

[i] According to the Gregorian calendar used in Catholic countries he was born on January 4, 1643.

Figure 9.1 Portrait of Isaac Newton from Voltaire's *Elemens de la Philosophie de Neuton* (1738). Image courtesy History of Science Collections, University of Oklahoma Libraries.

student or a Fellow of the College. Newton must have carried out his duties as subsizar, and his formal coursework as a student, well enough to remain at Trinity, but he soon grew interested in things that were not part of the official curriculum.[4]

Newton began to supplement his assigned readings with additional study in the areas of mathematics and natural philosophy. By the time Newton attended Trinity College the Copernican system was widely accepted in England. The acceptance of the Copernican theory by English mathematicians and astronomers can be traced back to Thomas Digges' early translation of, and advocacy for, the theory. The theory gained wider acceptance within educated English society because of a few works of popular fiction, such as Francis Godwin's *The Man in the Moone*, published in 1638, which followed the exploits of Domingo Gonsales who is carried to the Moon by wild swans. In the same year John Wilkins published a popular scientific work entitled *The Discovery of a World in the Moone*. Both works presented a Copernican view of the world, with particular emphasis on the similarities between the Earth and the Moon.[5] At Trinity, Newton studied astronomy from Thomas Streete's *Astronomia Carolina*, which presented planetary motions from a Copernican–Keplerian perspective.[6] From Streete, Newton would have learned about Kepler's elliptical orbits and harmonic law, although not his area law.[i]

Newton also studied Descartes' *La Géométrie* and *Principia philosophiae*. He was heavily influenced by Descartes but did not fully adopt the Cartesian *plenum*, the view that all space is filled with some form of matter. Newton instead leaned toward Gassendi's *atomism*, which viewed space as mostly empty with tiny atoms of matter moving through the void. Newton embraced one of the fundamental elements of the "mechanical philosophy" promoted by both French philosophers: that the forces that act on bodies are the result of collisions with (sometimes invisible) particles. However, Newton was uncertain about the mechanical explanation of gravity because he realized that whatever caused an object to fall had to act on the entire body of the object, not just its outer surface as one might expect colliding particles to do.[7]

In 1664 he was awarded a scholarship that allowed him to drop his role as subsizar and devote himself to his extracurricular studies, mostly in mathematics.[8] Then in August of 1665 Cambridge University closed because of a plague outbreak. Newton returned home to Lincolnshire, but since his intellectual pursuits were already disconnected from the Trinity curriculum he had no trouble in continuing his studies. He had already made important steps toward what would be his greatest mathematical achievement: the development of the calculus. He later recalled the plague year as the time when he finally figured out the "direct method of fluxions" and started on the "inverse method of fluxions" (what we would now call differential and integral calculus, respectively).[9]

Newton also spent time during the plague year thinking about motion. According to his later recollection, during this time he determined (independently of Huygens) that the centrifugal force on an object in uniform circular motion must be proportional to the diameter of the circle and inversely proportional to the square of the period of circular motion.[10] Later he used this result to show that the centrifugal force on a body sitting on the rotating Earth was far smaller than the weight that held the body to the Earth. The fact that objects rested firmly on Earth's surface was no argument against a rotating Earth.[11]

Newton also turned his attention to the Moon. What was the centrifugal force of the Moon's motion? Using the information he had available, including an inaccurate value for Earth's diameter, Newton compared the centrifugal force of the Moon to the weight the Moon would have if it was at Earth's surface.

---

[i] In place of Kepler's area law, Streete used an equant approximation based on theories proposed by Ismael Boulliau and Seth Ward.

He found that the Moon's weight at the surface would be more than 4000 times as great as its centrifugal force in orbit. Newton's centrifugal force calculations are described in Appendix A.17.

Newton then applied his new understanding of centrifugal force to celestial motions. By combining his centrifugal force law with Kepler's Third Law of Planetary Motion, treating the planetary orbits as circular, Newton found that the centrifugal forces of the planets were inversely proportional to the square of their distances from the Sun (see Appendix A.18 for details). Along the lines advocated by Borelli, Newton viewed the orbital motion of the planets as a balance between the centrifugal force pushing outward and another force that must push the planet inward toward the center. There had to be some other force pushing the planets toward the Sun to hold them in their orbits, a force that might result from particles of ether in a Cartesian vortex colliding with the planetary bodies. As the planet moved in its orbit the centrifugal force would alternately become greater or less than the force pushing the planet toward the Sun, so the planet would move in and out to generate an elliptical path.[12]

Newton's conclusion that the centrifugal forces on the planets followed an inverse-square law, combined with his result that the Moon's weight at Earth's surface was 4000 times as great as the centrifugal force on the Moon in orbit, may have been suggestive. The distance of the Moon from Earth's center is about 60 times Earth's radius. If the Moon was subject to an inverse-square force, then the force on the Moon in orbit would be a factor of $60^2 = 3600$ smaller than the force it would experience at Earth's surface. The rough agreement between this factor and the factor of over 4000 calculated by Newton could have suggested that the force that keeps the Moon in orbit is connected to the force that gives objects their weight on the Earth, but the difference was enough that Newton did not pursue the idea further.

Around this time Newton also began investigating light. By passing a beam of sunlight through glass prisms Newton was able to show that sunlight was composed of many different colors of light. The path of each color of light was bent by the prism at a different angle. Newton defined a quantity he called "refrangibility" that provided a quantitative measure of the bending of light by a prism. The refrangibility of a given color was fixed, but he found that violet light was more refrangible than red light. This discovery helped to explain the problem of chromatic aberration in lenses: the different refrangibilities meant that different colors were focused to different points. Newton conducted other optical experiments that endangered his own vision, including staring at the Sun in order to observe the "afterimage" produced by his eye and sticking a blunt needle between his eyeball and eye socket. Although it took some time to recover from these experiments he managed not to blind himself.[13]

After the plague abated, Newton returned to Cambridge where he was soon elected a Fellow of Trinity College. He continued his optical studies, and in late 1668 he built the first ever reflecting telescope, using a mirror to focus light and thus avoid the problems of chromatic aberration in lenses.[14] Newton also continued his work on mathematics with the encouragement of Isaac Barrow, the first Lucasian Professor of Mathematics at Cambridge, among others. In 1669, when Barrow left his position to pursue an administrative career, he helped Newton become the second Lucasian Professor.[15]

Up to that point Newton was mostly unknown to those outside Cambridge, with the exception of a few mathematicians to whom his work had been circulated. That changed in 1671 when Newton sent a reflecting telescope to be demonstrated at a meeting of the Royal Society of London, which had been founded in 1660 as one of the first scientific societies in the world. Soon Newton was elected a member of the Society and he published a paper detailing his optical researches in the Society's *Philosophical Transactions*.[16]

Newton's ideas about the nature of light were strongly criticized by the Society's Curator of Experiments, Robert Hooke. Hooke had his own theory of light, which suggested that colored light was a modified form of white light, in direct contradiction to Newton's belief (now known to be correct) that the color of light is fixed and white light is just a combination of different colors of light. Newton was incredibly sensitive to criticism and Hooke's commentary was enough to make him withdraw completely from public discussion of natural philosophy.[17]

In fact, Newton seems to have nearly given up natural philosophy altogether. Instead he focused his energies on secretive studies of alchemy and theology. At some point he privately rejected the orthodox Trinitarian theology of the Anglican church, instead adopting Arian views that denied the divinity of Christ. This heretical belief posed a serious problem for Newton. Cambridge Fellows were expected to become ordained in the Anglican church, which Newton could not do in good conscience. In 1675, while Newton was preparing to leave Cambridge, Barrow convinced King Charles II to exempt the holder of the Lucasian Chair from that requirement. Newton could remain at Cambridge and continue his solitary studies, but although a run-in with Robert Hooke had led Newton to seek seclusion, another encounter with Hooke would set him on the path to worldwide fame.[18]

9.1.2   Robert Hooke

From about 1656 to 1662 Robert Hooke had served as an assistant to Robert Boyle, the chemist and natural philosopher who had become famous for his experiments with air pumps. In fact, Boyle's first air pump was built by Hooke. Hooke was an early participant in the Royal Society and in 1662 he was appointed the Society's Curator of Experiments. As Curator he was responsible for performing experimental demonstrations, of his own devising or based on the work of others, at Society meetings.[19]

Hooke had wide-ranging interests. He is now best known for his early studies with the microscope and his work with the architect Christopher Wren to rebuild London after the Great Fire of 1666. He was also very interested in the study of motion. His experiments with conical pendulums and balls rolling on conical surfaces led him to conclude that circular motion consisted of a straight line motion tangent to the circle together with a motion toward the center of the circle. In his 1666 lecture entitled "Planetary Movements as a Mechanical Problem" Hooke suggested that the motion of the planets could likewise be understood as a straight-line inertial motion combined with an attraction toward the Sun.[20]

Also in 1666, Hooke conducted experiments to see if the weight of an object diminished when it was placed deep in a well. The idea that weight might be reduced inside the Earth went back at least to Francis Bacon, who stated in 1628 that "it is very probable, that the Motion of Gravity worketh weakly, both far from the Earth, and also within the Earth; The former, because the Appetite of Union of Dense Bodies with the Earth, in respect of the distance, is more dull; The latter, because the Body hath in part attained his nature, when it is some depth in the Earth."[21] While Bacon thought in terms of an object attaining its natural place, Hooke thought in terms of attractive, perhaps magnetic, forces, suggesting that "if all parts of the terrestrial globe be magnetical, then *a body* at a considerable depth, *below the surface of the earth, should lose somewhat of its gravitation, or endeavour downwards,* by the attraction of the parts of the earth placed above it."[22] In other words, the downward gravitational attraction of the parts of Earth below the object would be partially cancelled by the upward attraction of those parts that are above the object, thus reducing the object's weight.

Occasionally Hooke even tried his hand at observational astronomy. In 1671 he set out to demonstrate the orbital motion of Earth by measuring the annual parallax of the star Gamma Draconis, which passed almost directly overhead for observers in London. Galileo, in his *Dialogo*, had suggested two methods for measuring the annual parallax of stars. One was to find two stars that appeared nearby on the sky, but which were at very different distances and would therefore display different parallax angles. Annual parallax would then show itself as an annual variation in the angular separation between the two stars. Galileo's second method was to carefully observe changes in the meridian transit altitude of a star.[23] It was this second method that Hooke employed.

Hooke cut a hole in the roof of his apartment to accommodate a zenith telescope, a telescope that pointed straight upward. He then set out to observe

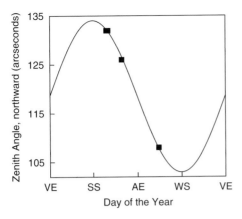

Figure 9.2 Plot of Hooke's 1671 measurements of the angle between the zenith and the transits of Gamma Draconis. The curve shows the predicted pattern of variation for this star from annual parallax, with the total parallax angle adjusted to fit the data.

the angle between the zenith and the point where Gamma Draconis crossed the meridian. Unfortunately, he had a terrible time getting his telescope to work properly and that, combined with illness, led him to make only four transit observations. Even so, Hooke's observations suggested that the transit altitude for Gamma Draconis had shifted about 24″ to the south between early July and late October.

Hooke published his results in 1674 under the title *An Attempt To prove the Motion of the Earth from Observations*. Although Hooke did not explicitly point it out, his observations fit perfectly with the expected pattern for the annual parallax of Gamma Draconis (see Figure 9.2). What he did say was "'Tis manifest then by the observations of July the Sixth and Ninth: and that of the One and twentieth of October, that there is a sensible parallax of the Earth's Orb to the fixt Star in the head of Draco, and consequently a confirmation of the Copernican System against the Ptolomaick and Tichonick."[24] The parallax curve shown in Figure 9.2 gives a parallax angle of $\theta = 16″$, implying that the star lay at a distance of nearly 13,000 AU. However, Hooke's four observations made using an unreliable telescope were not sufficient to convince most of his peers that he had truly measured the annual parallax of a star.

Although most of Hooke's *Attempt* focused on his parallax measurements, it did include some other interesting material. Hooke had made his October transit observation during the afternoon. This observation may have been the first of a star during bright daylight, and Hooke noted that the star's angular diameter was much smaller than when it was observed at night. Hooke pointed out that this tiny angular diameter, less than one second of arc, showed that the argument against the Copernican system based on star sizes (by Tycho and Riccioli) was based on incorrect angular diameters for the stars.[25]

Hooke ended his *Attempt* by briefly describing

> a System of the World different in many particulars from any yet known, answering in all things to the common Rules of Mechanical Motions: This depends upon three Suppositions. First, That all Coelestial Bodies whatsoever have an attraction or a gravitating power towards their own Centers, whereby they attract not only their own parts, and keep them from flying from them, as we may observe the Earth to do, but that they do also attract all the other Coelestial Bodies that are within the sphere of their activity; and consequently that not only the Sun and Moon have an influence upon the body and motion of the Earth, and the Earth upon them, but that ☿ also ♀, ♂, ♃, and ♄ by their attractive powers, have a considerable influence upon its motion as in the same manner the corresponding attractive power of the Earth hath a considerable influence upon every one of their motions also. The second Supposition is this, That all bodies whatsoever that are put into a direct and simple motion, will so continue to move forward in a straight line, till they are by some other effectual powers deflected and bent into a Motion, describing a Circle, Ellipsis, or some other more compounded Curve Line. The third Supposition is, That these attractive powers are so much the more powerful in operating, by how much the nearer the body wrought upon is to their own Centers. Now what these several degrees are I have not yet experimentally verified; but it is a notion which if fully prosecuted as it ought to be, will mightily assist the Astronomer to reduce all the Coelestial Motions to a certain rule, which I doubt will never be done true without it.[26]

This statement from the *Attempt* was the first published suggestion that gravity, here conceived as a force that attracts bodies to the Earth, also acts to attract celestial bodies to each other.[27] Hooke's new notion of gravity was not quite a universal force: he thought that some substances, such as the matter composing comets (which he believed travelled in nearly straight lines through the solar system), might have a reduced gravitational attraction or perhaps none at all. Even so, Hooke was tantalizingly close to the set of ideas that would form the basis of a new universal physics.[28]

Hooke's problem was that he was not sure how to work out the mathematical details. As he says in the *Attempt*, all that was needed was someone who could carry out the necessary observations and calculations that Hooke was, he claimed, too busy to do himself. He closed by saying, "But this I durst promise the

Undertaker, that he will find all the great Motions of the World to be influenced by this Principle, and that the true understanding thereof will be the true perfection of Astronomy."[29] Meanwhile, Hooke's unknown "Undertaker" had turned his back on natural philosophy to study alchemy and theology in Cambridge.

9.2   Letters between rivals

As the 1670s drew to a close, both Hooke and Newton had gained important insights into planetary motion and its connection to gravity, but both men were still confused on some important points. Newton, when he had thought about planetary motions at all, had thought in terms of a balance of centrifugal and centripetal (center-seeking) forces. Then he had turned his back on astronomical problems completely. Hooke accepted Kepler's erroneous "distance law" (that the speed of a planet is inversely proportional to its distance from the Sun) and believed that comets moved in straight lines. Furthermore, although Hooke was a capable mathematician he did not know how to calculate the details of planetary motions.[i] Each had some pieces of the puzzle but lacked others. Each could have helped the other, but their dispute over the theory of colors had not left them on speaking terms.

Then, in 1677 Hooke was appointed Secretary of the Royal Society. One of his duties was to correspond with the Society's members in order to keep them active in Society business. Although Hooke himself had driven Newton away from the Society, he hoped to lure him back. Hooke wrote to Newton on November 24, 1679, asking him to contribute something for discussion at a Society meeting.[31] He also asked what Newton thought of his idea of "compounding the celestiall motions of the planetts of a direct motion by the tangent and an attractive motion towards the centrall body."[32]

Newton replied on November 28 to say that he was no longer interested in matters of natural philosophy, so much so that he had not even heard of Hooke's ideas about planetary motion. But Newton also congratulated Hooke on his successful parallax measurement, which suggests that he was familiar with Hooke's *Attempt*, in which those ideas were clearly presented. In any case, Newton suggested that, now that Hooke had proved the orbital motion of Earth, there might be a way to prove Earth's rotational motion as well. Newton stated that an object dropped from a high tower should fall not to the point on Earth's surface directly below, but slightly to the east of that point.[33] Newton reasoned that when the

---

[i] Hooke did make a serious, but ultimately unsuccessful, attempt to work out the mathematical details of producing planetary orbits from straight line inertial motion and a central attractive force using a graphical method.[30]

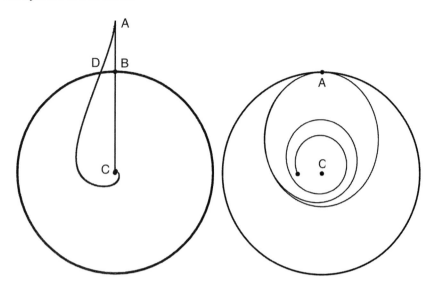

Figure 9.3 Computer simulation reproductions of the sketches from Newton's letter of November 28, 1679 (left) and Hooke's letter of December 9, 1679 (right). The left (Newton) diagram shows the view from an observer on the rotating Earth and treats the Earth as a homogeneous sphere, while the right (Hooke) diagram shows the view from an (inertial) observer floating in space and treats the Earth as a point mass. Both diagrams exaggerate the rate of Earth's rotation and include a slight resistance to motion within the Earth that is proportional to the object's speed. The Hooke diagram also shows an elliptical path assuming no resistance.

object was on the tower it was moving faster than the point on Earth's surface directly below, because it must complete a (slightly) larger circular motion in the same period of time. Thus, as the object fell it would outrun the point below it and land slightly ahead (eastward).[i]

Newton included a rough sketch showing the path of the falling object (Figure 9.3, left). Although it was not relevant to his suggestion, Newton showed the path continuing inside the Earth and eventually spiralling in toward the Earth's center. Although Newton did not spell out all of his assumptions in drawing his diagram, we know that he was assuming the tower was at Earth's equator and he drew the sketch from the perspective of an observer on the rotating Earth (so that the path starts off straight down but then appears to deflect toward the east). A path similar to that drawn by Newton can be obtained if we also assume that the force on the object, once it is inside the

---

[i] This eastward deflection is now viewed as a manifestation of the Coriolis effect.

Earth, is proportional to the distance from the center and that there is also some resistive force slowing the object's motion.

Hooke replied on December 9. He agreed with Newton that falling bodies should experience an eastward deflection due to Earth's rotation if the object was on the equator. At the northern latitude of London he thought that there must also be a southward deflection. However, he disagreed that the object would spiral into Earth's center. Instead, he claimed that if the "gravitation to the former Center remained as before" the path would "resemble An Elleipse ..."[34] Hooke illustrated his idea with a sketch of his own (Figure 9.3, right). His sketch showed the elliptical path, as well as an ellipto-spiral path that would be followed if the object was subject to resistance inside Earth. It is important to note that Hooke's sketch was drawn from the perspective of an observer floating in space, not rotating with the Earth. The initial motion of the object is eastward (in the direction of Earth's rotation). That motion is then gradually deflected, by an attractive force toward the center of Earth, into an elliptical path, analogous to Hooke's suggestions about the motion of the planets around the Sun.

Like Newton, Hooke was not very clear about the assumptions that went into his sketch. Newton thought he meant that the force on the object would have a constant magnitude (or strength) while the object was inside the Earth. He realized that the path produced by such a force would not be an ellipse and he replied to Hooke on December 13 to point out the error.[35]

On January 6, 1680, Hooke wrote to Newton to correct the misunderstanding. He stated that "my supposition is that the attraction always is in duplicate proportion to the distance from the center reciprocall, and consequently that the velocity will be in a subduplicate (proportion) to the attraction, and consequently as Kepler supposes reciprocall to the distance." By "duplicate proportion to the distance from the center reciprocall" Hooke meant exactly the inverse-square law that Newton had glimpsed back in the plague year of 1666. Hooke thought that was the correct law for points outside the Earth, but he realized that it probably did not apply inside the Earth. Instead, he thought "that the more the body approaches the center the lesse will it be urged by the attraction, possibly somewhat like the gravitation on a pendulum." In other words, Hooke thought that inside the Earth the gravitational force would be proportional to the distance from the center.[36]

Hooke was not just interested in theoretical speculation. He had set out to test Newton's idea of the eastward deflection of falling bodies. In his letter of January 6 he reported:

> I have (with as much care that I could) made 3 tryalls of the experiment
> of the falling body, in every of which the ball fell towards the south-east

of the perpendicular, and that very considerably, the least being above a quarter of an inch, but because they were not all the same I know not which was true.[37]

In another letter, of January 17, Hooke wrote

I can now assure you that by two tryalls since made in two severall places within doors it succeeded Also. Soe that I am perswaded the Experiment is very certaine, and that It will prove a Demonstration of the Diurnall motion of the earth as you have very happily intimated.

It seemed the matter of Earth's rotation was settled to Hooke's satisfaction, but as with his parallax measurements his peers would find his results unconvincing (for good reason, as we will see). Hooke's January 17 letter closed by handing the problem of the path of a body subject to an inverse-square force over to Newton. "I doubt not but that by your excellent method you will easily find what that Curve must be, and its proprietys, and suggest a physicall Reason of this proportion." Although Newton never replied to that letter, we know that he set out to do just what Hooke suggested.[38]

Newton and Hooke ceased to communicate with each other after this series of letters. Hooke gave up on getting anything from Newton and moved on to other pursuits (but not his promised System of the World). Newton, however, kept thinking about planetary motions and gravity. Hooke's letters had set his thinking on a clearer path. He set aside centrifugal forces and Cartesian vortices and started thinking of orbital motion as a combination of inertial motion and an attractive **centripetal** force (a center-seeking force that points inward to the center of attraction). His decision to embrace attractive forces may have been encouraged by his alchemical studies, which had led him to the idea that fundamental particles must have certain attractions that gave rise to chemical reactions.[39]

Newton didn't just start thinking about orbital motion differently, he also started calculating differently. He developed a computational method for finding the path followed by a body moving under the influence of a force toward a fixed center. He was able to prove that centripetal forces produce motion that follows Kepler's area law: the line from the fixed center to the moving body sweeps out equal areas in equal times. Newton had ignored the area law in his previous work, even though most English astronomers were familiar with the law and considered it accurate, though difficult to use in practice. Newton's proof that central forces reproduce the area law was an important step in the development of his physics.

Newton also used geometrical methods to examine the force needed to produce elliptical motion with one of the foci playing the role of the fixed center of force. He found that an inverse-square force was required,[40] but he told no one of his discovery.

In November 1680 a comet appeared in the skies, only to disappear behind the Sun in December. Another comet appeared a few days later, on the opposite side of the Sun and moving in a different direction. John Flamsteed, who in 1675 had been appointed as the first Astronomer Royal and given charge of the new Royal Greenwich Observatory, proposed that these two comets were in fact the same object. He argued that the comet had passed close to the Sun and that the Sun had exerted some influence that had changed the comet's direction of motion.

Newton, still thinking that comets moved in straight lines, raised objections to Flamsteed's theory, but in 1682, when a new comet appeared, Newton made a thorough study of comet motion and decided that comets did not move in straight lines. Their paths curved, with the greatest curvature corresponding to their point of closest approach to the Sun. It seemed that the one type of celestial object that Hooke had exempted from gravitational forces might not be exempted after all.[41] All of the pieces were in place for Newton to develop a new celestial physics, but still he refused to make any of his work public.

Newton's brilliant new approach to celestial motion might have remained hidden forever if it were not for a casual conversation that took place in January 1684. Hooke, Christopher Wren, and Edmund Halley, a young astronomer best known at the time for his catalog of stars in the southern hemisphere, chatted about whether an inverse-square attraction to the Sun could explain the motions of the planets. Hooke claimed that he had already proved it. Halley admitted that he had tried, but failed. Wren offered a prize, a book valued at 40 shillings, to whoever could supply a convincing proof in the next two months. Those two months went by and Wren never had to pay up.[42]

Although the prize was lost, Halley did not give up on the quest. When he was in Cambridge later in 1684 he paid a visit to Newton to ask him what path a planet would follow if it were subject to an inverse-square force toward the Sun. Newton replied right away that it would be an ellipse, saying that he had calculated it several years ago. He claimed he was unable to find his proof during Halley's visit, but promised to redo it and send it to Halley. In November Halley received the result, a short work now known as *De motu corporum in gyrum* (On the motion of bodies in orbit) in which Newton showed that all three of Kepler's laws followed from an inverse-square attraction to the Sun.[43]

Halley was amazed and asked Newton for permission to publish *De motu*. Newton, always fearful of criticism, wanted to work out more of the details before publishing anything. From the summer of 1684 until the spring of 1686 Newton did almost nothing but work on these details, frequently forgetting even to eat. Of the very few letters he wrote during this time, most of them were to ask Flamsteed for his comet observations, which Newton needed in order to bring comets into his new theory of celestial dynamics. He even gave up alchemy and theology for the time being.[44]

As Newton closed in on completing his manuscript, Halley convinced the Royal Society to order the printing of the book, although Halley had to cover the printing costs himself. Not every member of the Society was thrilled to hear of Newton's work, though. Halley wrote to Newton to inform him that Hooke had accused Newton of stealing his ideas. It is not hard to understand Hooke's anger. He had spelled out the fundamental ideas (inertial motion plus an inverse-square attraction to the Sun) in his published works and in personal letters to Newton. What Hooke lacked was a mathematical demonstration that these ideas reproduced Kepler's laws.

Once Newton learned of Hooke's charges he threatened to suppress the last portion of the book. He also revised his manuscript to eliminate those few places where he had acknowledged Hooke's contributions. Halley supplied Newton with a constant stream of praise and made light of Hooke's priority claim. That was enough to calm Newton and convince him to proceed with the full work as planned, although he made substantial changes to the final part. By July 1687 the printing of Newton's *Philosophiae Naturalis Principia Mathematica* (Mathematical Principles of Natural Philosophy), the greatest scientific book ever written, was complete.[45]

9.3   The *Principia*: Books I and II

Newton's *Principia* consists of some introductory material followed by three Books.[46] The introductory material lays out a set of mathematical rules that govern the motion of bodies subject to forces, and Book I shows how these rules can be used to determine the motion of a body subject to different centripetal forces, or else determine the force on a body that moves in a specified way. Book I is hypothetical in character: Newton did not claim in Book I that such motions and forces actually exist in the world, he simply worked out the mathematical details associated with those motions or forces. In that sense Book I is a little bit like the medieval investigations of accelerated motion.

In Book II Newton extended his analysis to bodies that are subject to resistance, particularly to bodies moving through a fluid medium. It is in Book III that Newton applied his grand theoretical structure to concrete problems. In Book III he showed that the mathematical results of Book I could explain an incredibly wide variety of phenomena that were observed in the real world.

Newton's *Principia* is of such great importance in the history of astronomy and physics that it is worth going through it (or at least through Books I and III) step by step. The introductory material starts with a short preface followed by a series of definitions.[47] Newton defined "quantity of matter" as a measure of matter given by multiplying the volume of an object by the object's density. Later Newton referred to this quantity as the **mass** of an object, and today physicists would call it the inertial mass. Newton noted that mass was proportional to weight, but he was careful not to identify it with weight, because he knew that the weight of an object could change from place to place.[i] In fact, Newton's definition of mass, or quantity of matter, is not entirely clear since we now think of density as mass divided by volume, which would make Newton's definition circular. However, in Newton's day the relative densities of several substances were reasonably well known, so it was not inappropriate for him to make use of that concept. What Newton was searching for here was a property of an object that could not be increased or decreased (without adding or removing matter). As we will see, Newton used mass as a measure of an object's resistance to *changes* in motion.

Newton then defined an object's "quantity of motion" to be the product of its mass and its velocity (what physicists now call "momentum"). Note that "quantity of motion," like velocity, is a *vector* quantity. That means it has both a magnitude, or amount, and a direction.[ii] For example, the speed of an object is the magnitude of the object's velocity, but to fully specify the velocity you must state both the speed and the direction of the object's motion.

Newton proceeded to define three different kinds of "force." The first of these he called the "inherent force of matter" or the "force of inertia." He defined this force as the power of a body to resist changes to its motion. This resistance is proportional to the mass of the body, so as noted above mass measures resistance to change in motion. We no longer think of inertia as a force, but for Newton a force was something that helped determine an object's motion. As we will see, the inertia of an object plays an important role in determining how the object will move, so Newton treated it as a force.

Newton's other two forces were "impressed force" and "centripetal force." Impressed force is anything that acts on a body to change the body's motion. Centripetal force is a particular type of impressed force that acts always to push the body toward a particular fixed point. Note that the definition of centripetal force implies that a force has a direction as well as a magnitude, which means that forces are also vectors just like velocity or quantity of motion.[48]

---

[i] Experiments with pendulums had shown that gravity was weaker in some locations on Earth than in others.

[ii] Newton's definitions did not make it clear that quantity of motion, and other vector quantities, had a direction as well as a magnitude. However, the vector nature of these quantities was clear from the way Newton used them in the *Principia*.

After presenting these and other definitions, Newton included a scholium (or explanatory note). He attempted to provide philosophical definitions of absolute time and space. He distinguished between absolute motion (the change of an object's location within absolute space) and relative motion (the change of an object's location relative to other objects). He claimed that the only way to tell the difference between absolute and relative motion was to consider the forces that produce the motions. The absolute motion of a body could only be changed by impressed forces acting on that body, while the relative motion of a body could be changed without any forces acting on the body.

Philosophically, Newton's claims about absolute time and space are hard to justify and they don't seem to play an important role in the rest of the *Principia*. Note, however, that the relative motions are the same in both the Copernican and Tychonic theories. Newton's scholium can be seen as an attempt to convince the reader that it is only through an understanding of the forces acting on bodies that we can tell which of these theories is true. The rest of the *Principia* would supply that understanding of forces.

The scholium is followed by three Axioms or Laws of Motion.[49] These three laws form the fundamental basis of Newton's physics, and all three can still be found (perhaps slightly reworded) in introductory physics textbooks used today. The **First Law** states, "Every body perseveres in its state of being at rest or of moving uniformly straight forward, except insofar as it is compelled to change its state by forces impressed."[50] This law just restates Descartes' idea of inertial motion. It serves to distinguish motions that are "natural" and require no explanation (either rest or straight line motion at constant speed) from motions that must be explained in terms of impressed forces (any motion that is not straight and/or not at constant speed).

The **Second Law** states, "A change in motion is proportional to the motive force impressed and takes place along the straight line in which that force is impressed."[51] So when a force acts on an object the change in that object's quantity of motion will be greater when the force is stronger. That change will take place in the direction that the force is acting, so a downward force will add a downward component to an object's motion, while a force pointing eastward would add an eastward component. Newton's Second Law is often restated as: the total force on an object is equal to the mass of the object times the object's acceleration. This restatement is logically equivalent to Newton's version provided we assume that the mass of the object remains constant.

Newton explained how to use the Second Law in Corollary 1 of his Axioms: "A body acted on by [two] forces jointly describes the diagonal of a parallelogram in the same time in which it would describe the sides if the forces were acting separately."[52] Consider the case of a projectile fired at an angle above

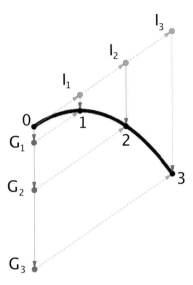

Figure 9.4 The motion of a projectile according to Newton. The projectile's motion is composed of its inertial motion and a downward motion with constant acceleration. The composition of these two motions forms a parabolic path.

horizontal, shown in Figure 9.4.[i] Newton treated gravity as an impressed force constantly pushing the projectile downward. The motion of the projectile is then a combination of inertial motion and the motion produced by this downward force.

The projectile is fired at time 0. Let's look at its inertial motion first. During the first interval of time the projectile's inertial motion would carry it in a straight line to the point I_1 in Figure 9.4. In the next (equal) interval of time its inertial motion would carry it the same distance (since it is moving at constant speed) in the same direction (since it is moving in a straight line), so it ends up at I_2. After three equal intervals it will be at I_3.

Now let's consider the motion produced by the force of gravity. From Galileo's work on falling bodies (Section 8.3) we know that the distance an object falls is proportional to the square of its time of fall. During the first interval the object falls from O to G_1. After two time intervals (twice as much total time) the object should have fallen four times as far, so at the end of the second interval it reaches the point G_2 (where the distance from O to G_2 is four times the distance from O to G_1). By the end of the third interval it should have fallen nine times as far as it fell in the first interval (because $3^2 = 9$), so it will be at G_3.[ii]

[i] Newton treated this case in a scholium that follows Corollary 6 of the Axioms.
[ii] Note that we could find the same results by applying Galileo's rule that in successive equal time intervals the object falls over distances that follow the pattern of odd numbers:

The true motion of the projectile, then, is given by the combination of the inertial and gravitational motions. To find the position of the projectile at the end of the first interval we just form a line from the starting point to I_1 and another from the starting point to G_1. We form a paralellogram from these two lines as shown in the figure, and the true location of the projectile is along the diagonal of the parallelogram at point 1. After the second interval the projectile is at 2 and after the third it is at 3. The sequence of points generated in this way forms a parabola, in agreement with Galileo's claim in the *Dialogo*.

Newton's **Third Law** states, "To any action there is always an opposite and equal reaction; in other words, the actions of two bodies upon each other are always equal and opposite in direction."[53] By action, Newton meant force. What he was saying is that forces always come from *interactions* between two objects. Object A exerts a force on object B, but at the same time object B exerts a force, of equal magnitude but in the opposite direction, on object A. Part of this law stems from Kepler's crucial insight that physical forces must come from physical objects, not from empty points in space, but Newton's new insight was that when one object is pushing on another, the second object pushes back on the first.

For example, consider a book sitting on a table. We know the table must be pushing up on the book with a force equal to the book's weight, otherwise the book would fall. Newton's Third Law tells us that the book must also push down on the table with the same amount of force. That example is not too surprising, but others are much less intuitive. Newton treated gravity as a force of attraction toward the Earth, so the Earth exerts an attractive force that pulls the book downward. That's what we call the book's weight. According to Newton's Third Law the book must also exert a force on the Earth, pulling the Earth upward with a force equal to the book's weight. This example seems counterintuitive: if we drop a book the book falls toward the Earth, the Earth does not rise up toward the book! But keep in mind that although the objects exert equal *forces* on each other, that does not mean that the *effects* of those forces are the same. The weight of the book will cause the book to accelerate downward at a rate of 9.8 meters per second per second, in modern units. If the book has a mass of 1 kilogram, then its "quantity of motion" (momentum) will change at a rate of 9.8 kilogram meters per second per second.[i] The Earth will experience a force of equal magnitude,

---

1, 3, 5, ... The distance from G_1 to G_2 is three times the distance from O to G_1, and likewise the distance from G_2 to G_3 is five times the distance from O to G_1.

[i] The unit "kilogram meter per second per second" is now known as the "Newton" and it is the standard metric unit for the magnitude of force. Note that Newton only claimed that the change in motion was *proportional* to the force, but here we are assuming that the change in motion is *equal* to the force. Both force and change in motion (or change in momentum) are vector quantities, so if they are equal they must have the same magnitude and the same direction.

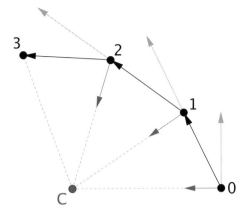

Figure 9.5 The motion of a body subject to impulsive, central forces. At each numbered point the body is subject to an instantaneous force directed toward the center C. The motion of the body between the numbered points is a combination of inertial motion and the motion due to the impulsive force.

but because it has a *much* greater mass it will experience almost no acceleration (or change in velocity).

So much for the introductory material of the *Principia*. Book I then begins with an explanation of Newton's "method of first and ultimate ratios," a mathematical technique that allowed Newton to deal with curved geometrical shapes using approximations based on shapes constructed from straight lines. His approximations involved the mathematical concepts of limits and sums of infinite series, which form the basis for Newton's "fluxions" (what we now call calculus), but Newton didn't present his method of fluxions in the *Principia*.[54]

The rest of Book I is an examination of the motion of bodies under various forces. In Proposition 1 Newton showed that a centripetal force always produces motion that follows Kepler's area law.[55] His method is illustrated in Figure 9.5. This figure illustrates the motion of a body subject to impulsive centripetal forces that act on the body at equal time intervals. An impulsive force is just a force that acts instantly to change the object's motion, rather than acting continuously over time (like gravity). The forces are centripetal because they always point toward the fixed point C.

At point 0 the object is already moving upward (toward the top of the page). During the first time interval its inertial motion would carry it along the path shown by the light gray arrow, but at the start of this time interval the impulsive force pushes the body toward C. The motion resulting from this force is shown by the dark gray arrow. The combination of the inertial motion and the motion from the impulsive force causes the body to move along the diagonal of the parallelogram formed from the two gray arrows.[i] This combined motion is shown as a black arrow that takes the body to point 1. At that point the process repeats: the motion during the next interval will be a combination of the inertial motion (which will be in the same direction and with the same speed as the overall motion during the first interval) and the motion due to the impulsive force (possibly with a different magnitude) acting at point 1 to push the body toward C. These two motions combine, according to the parallelogram law, to take the body to point 2, and so on.

Using this construction and some basic geometry, Newton showed that the areas of triangles C01, C12, and C23 are equal. Therefore, the line from the force center C to the body sweeps out equal areas during each of these equal time intervals. His result only applied to impulsive forces, but Newton pointed out that the actual duration of the time intervals does not come into play in showing that the areas are equal: all that matters is that the time intervals are equal. If we shrink the time intervals down toward zero, then the points 0, 1, 2, ... form a continuous curve and the force acts on the body continuously rather than at discrete times, but the line from C to the object must still sweep out equal areas in equal times. Newton also showed, in Proposition 2, that any body that moves such that the line between the body and a fixed point sweeps out equal areas in equal times must be subject to a force that always points toward the fixed point (this is the converse of Proposition 1).

Newton then presented a general method for finding the force on a body following a given curved path with a specified force center.[56] He applied this technique to a wide variety of curves, but the most important ones for astronomical purposes are ellipses, parabolas, and hyperbolas (known as conic sections because they can be generated by intersecting a flat plane with a double cone). Newton showed that if a body follows one of these conic section paths, with a force pointing always toward the focus of the curve, then that force must vary inversely as the square of the distance from the focus. Newton also proved that for a body moving in such an elliptical path the square of its period will be proportional to the cube of the major axis of the ellipse (which is Kepler's third, or harmonic, law).

Note that Newton did not quite address Halley's question from 1684. Halley wanted to know what path an object would follow if subject to an inverse-square attraction. Newton replied that it was an ellipse, but what Newton proved in the *Principia* was the converse of this statement: he showed that if an object moves in an ellipse with a force toward the focus, then that force must be an inverse-square force. The truth of a statement does not imply the truth of its converse.

---

[i] This result follows from Corollary 1 to the Laws of Motion.

For example, it is true that a square is necessarily a rectangle, but not true that a rectangle is necessarily a square. Newton claimed that an inverse-square force toward a fixed center must produce a conic section orbit, but in the first edition of the *Principia* he didn't offer a real proof. In later editions he sketched out a proof but without providing any real details.[57]

Much of the rest of Book I consists of the mathematics needed to analyze conic section orbits, including how to determine the specific orbit followed by an object given just a few pieces of information (such as the location at a few different times). Newton then proceeded to consider the motion of two bodies that attract each other by a force that acts along the line between the two bodies. He showed that this problem can be reduced to one in which each of the bodies moves as though it is subject to the same force pointing toward a special fixed point, known as the "center of gravity,"[i] which lies along the line between the two bodies and closer to the body with the larger mass.[58]

Newton claimed that two bodies that attract each other by an inverse-square force must move around their common center of gravity in conic section orbits. For example, the two bodies might move in elliptical orbits with the center of gravity at one focus. Although Newton didn't state it until later in the *Principia*, the implication was clear: if a planet is attracted to the Sun by an inverse-square force it would move in an ellipse around the center of gravity, *and so would the Sun*. The great mass of the Sun might put the center of gravity much closer to the Sun than the planet, but the Sun would still have an elliptical motion around this point. Perhaps the Sun is not stationary after all. In a similar manner, the Moon does not orbit around the Earth but instead around the center of gravity of the Earth–Moon system, and the Earth must move around this same point.[ii] When Newton tried to extend his analysis to systems with three or more bodies he found that he could not work out a general solution. However, in the *Principia* he does present approximate solutions for certain special cases.[59]

Throughout most of Book I Newton treated all bodies as though they were mathematical points with no size, but toward the end of Book I he extended his examination of inverse-square attractions to spherical bodies. He assumed

---

[i] Note that center of gravity is a suggestive term. Newton first defines the term in Corollary 4 of the laws, but in Book I he uses the term in a mathematical way without connecting it to the concept of gravity. Only in Book III would it become clear why this term is called center of *gravity*. Today physicists would use the term "center of mass" instead.

[ii] The center of gravity of the Earth–Moon system lies inside the body of the Earth, about 3000 miles from the Earth's center, so the Earth's "orbit" around this center is more like a wobble. However, the center of gravity of the Sun–Jupiter system lies just outside the body of the Sun.

that each little bit of matter in the body attracts every other little bit of matter by a force that is inversely proportional to the distance between the bits. From that assumption and some sophisticated mathematics Newton showed that a particle placed inside a homogeneous (uniform density) sphere will experience an attraction toward the center of the sphere that is directly proportional to the distance from the center. A particle placed outside such a sphere will experience an attraction toward the center that is inversely proportional to the square of the distance from the center. Likewise, two spherical bodies will attract each other by a force that is inversely proportional to the square of the distance between their centers.[60] These results allowed Newton to take an important step from the fictitious point particles of his earlier analysis toward physical objects such as planets. He showed that real, solid spheres will behave just like those point particles as long as we assume that every bit of matter attracts every other bit by an inverse-square force.

In Book I Newton considered a few different types of forces, but he spent the vast majority of his effort examining inverse-square attractive forces. He showed that motions that fit with Kepler's Laws of Planetary Motion can be explained by inverse-square attractions, whether we treat the planets as point particles attracted toward a fixed center located at the Sun or as spherical bodies attracted to the spherical body of a Sun that is not fixed in place. However, in Book I Newton did not make the connection between his mathematical results and the actual motions of planets in our solar system. He simply showed what motions result when bodies move according to certain mathematical rules.

In Book II Newton examined the motion of bodies subject to resistive forces, including projectiles moving through a resisting fluid (like air). He examined pendulum motion and waves traveling through an elastic medium (like sound waves through air). Historically, the most important part of Book II may be the final section in which Newton considered the circular motion of fluids.[61] He applied his results to see if Descartes' celestial vortices could reproduce Kepler's Laws of Planetary Motion. He found that Kepler's Laws, which were shown in Book I to be entirely consistent with inverse-square attractive forces, were completely inconsistent with the motion of vortices.[62] Thus, Descartes' explanation of planetary motions was not valid.

9.4   The *Principia*: Book III

In Book I Newton worked out the mathematical details of the motion of bodies subject to inverse-square forces, but without claiming that any such motions or forces exist in the real world. In Book III he showed how his results correspond to the actual observed motions of bodies in the heavens and on the

Earth. Books I and II are almost exclusively about mathematics. Book III, which is titled *The System of the World*, is about natural philosophy.

Before he started examining the connections between his mathematics and the real world, Newton laid out four rules for the study of natural philosophy, which will be discussed in Section 9.5. Then Newton presented several observed phenomena related to celestial motions. Specifically, he stated that the moons orbiting Jupiter, the moons orbiting Saturn, and the planets orbiting the Sun all conform to Kepler's second and third laws (the area and harmonic laws). He also suggested, but without supplying much evidence, that the Moon orbiting around the Earth follows Kepler's area law.[63]

The remainder of Book III is devoted to showing how an incredible variety of natural phenomena can be explained (or predicted) by inverse-square attractive forces. Newton pointed out that the motions of the planets, as well as those of the moons orbiting the Earth, Jupiter, and Saturn, can be explained by an inverse-square attraction to the body they orbit. Then Newton made the crucial connection: the inverse-square force that attracts the Moon toward the Earth and keeps the Moon in orbit is the same force that makes objects near the surface of Earth fall toward the Earth's center. In other words, this inverse-square force is what had always been called gravity.[64]

To demonstrate this connection Newton revisited his 1666 comparison of the force on the Moon in its orbit and the force it would experience due to gravity at the Earth's surface. This time he had better data, including an improved value for Earth's diameter found by Jean Picard in 1670. Newton showed that the force of gravity that the Moon would experience at Earth's surface was 3600 (or 60^2) times greater than the force that keeps the Moon in orbit. Since the radius of the Moon's orbit is very nearly 60 times the radius of the Earth, this result is exactly what would be expected if gravity is an inverse-square force.

Originally Newton had composed a version of Book III that was less mathematical in style and was intended for a broad audience. That version was published in 1728, shortly after Newton's death, as *A Treatise of the System of the World*. In that version Newton provided a different, less mathematical explanation of the connection between gravity and the Moon's orbit, which was illustrated with the drawing in Figure 9.6. The figure shows a cannon placed (at V) on top of a very high mountain on Earth. The cannon fires a ball horizontally. If the ball is fired at a relatively low speed then it will land not far from the base of the mountain (at D). If it is fired at a greater speed then it will travel farther along Earth's surface before it lands (at E, F, or even G). If the cannonball is fired fast enough, said Newton, it will go into orbit around the Earth and return to the point V from which it was launched. In other words, the motion of a cannonball subject to gravity is no different than the motion of the Moon in its orbit. Gravity

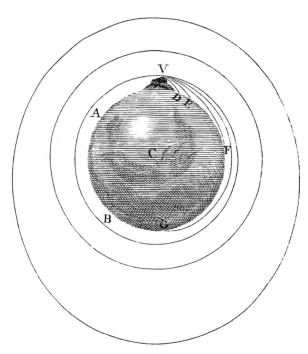

Figure 9.6 A cannonball launched horizontally from a high mountaintop lands farther from the base of the mountain the faster it is launched. If it is launched fast enough it can orbit the Earth, just as the Moon does. Its orbital motion is a combination of inertial motion and gravitational attraction toward the Earth's center. Image courtesy History of Science Collections, University of Oklahoma Libraries.

pulls the Moon toward the Earth's center, and the combination of gravity and the Moon's inertia sets the Moon orbiting around the Earth.[65]

From that point on in the *Principia* Newton no longer wrote about general "inverse-square attractions" but instead wrote about gravity. Gravitational attractions exist between the Sun and planets, between Jupiter and its moons, between the Earth and its Moon, and even between the Earth and a falling cannonball. He generalized this result to claim that *all* bodies are attracted to each other by an inverse-square gravitational force. But what else, other than the distance between the bodies, determines the strength of these "universal" gravitational forces?

Newton claimed that the gravitational force on any object is proportional to that object's quantity of matter (mass).[i][66] His main evidence for this assertion was from a series of experiments that he performed on pendulum motion. He used two identical round wooden boxes, filled with different substances, as

---

[i] In modern terms we would say that an object's gravitational mass is proportional to its inertial mass.

pendulum bobs. He filled the boxes with equal weights of different materials and hung them from strings of equal length. When he set the pendulums in motion he found that they kept perfect time with each other. If the gravitational force on the pendulum bob was not proportional to the bob's mass then otherwise identical pendulums composed of different substances might swing with different periods. Since Newton never found this to be the case, no matter what substances he used, he concluded that the gravitational attraction must be proportional to the mass. Furthermore, if the gravitational force on the entire object is proportional to the total mass of the object, then the force on a tiny part of that object must be proportional to the mass of that tiny part. Thus, every piece of an object is subject to gravitational attraction in proportion to the mass of that piece.

If object A exerts a gravitational force on object B, then that force is proportional to the mass of object B. But by Newton's Third Law of Motion we know that object B also exerts an equal and opposite force on object A. By the logic given above, that force must be proportional to the mass of object A. Therefore, the gravitational forces between two objects are proportional to the masses of the two objects multiplied together, and inversely proportional to the square of the distance between the objects.

Newton then showed that universal gravitation could do more than just explain the motion of the planets, moons, and falling bodies, it could also be used in conjunction with Kepler's harmonic law to compare the masses of any body that was orbited by some other body. Newton compared the masses of the four bodies in the solar system that were known to have orbiting companions (the Sun, Earth, Jupiter, and Saturn).[67] He found that the mass of the Sun was 169,282 times that of Earth, 3021 times that of Saturn, and 1067 times that of Jupiter.[i] The details of these calculations are shown in Appendix A.19.

From these masses, and determinations of the size of each object (based on their relative distances in the Copernican–Keplerian system and their observed apparent diameters), Newton determined their relative densities. He found that the Sun and Jupiter had very similar densities, while Saturn was a bit less dense and the Earth was about four times as dense. This was the first time that anyone had been able to say something certain about the material composition of the planets,[ii] and it can be viewed as the beginning of the field now known as planetary science.

---

[i] The numbers for Jupiter and Saturn are reasonably accurate but the number for Earth is too small by about a factor of two. The reason for this error is that Newton did not have an accurate value for the Earth–Sun distance. He thought that distance was about 79 million miles when in fact it is about 93 million miles. That distance was not accurately determined until the 1760s (see Section 10.3.1).

[ii] Although Kepler made an attempt in his *Epitome*.

Next Newton applied universal gravitation to the shape of a rotating body such as the Earth. Consider a chunk of Earth located on the surface at the equator. This chunk is moving in a big circle as a result of Earth's rotation, and thus some of the gravitational force pulling that chunk toward Earth's center is used up just to keep the chunk from flying off along the tangent to the circle. The remaining part of the gravitational force is left to pull the chunk in toward the center. On the other hand, a similar chunk located at one of the poles is not moving due to Earth's rotation, so the full gravitational force is available to pull that chunk toward the center. As a result, the Earth should be squished inward along the poles and bulge outward along the equator, a bit like a tangerine. This shape is known as an oblate spheroid, and Newton argued that all rotating planets should be oblate spheroids. In particular, he claimed that the diameter of the Earth along the equator should be about 17 miles greater than the diameter from pole to pole, a slight difference considering the average radius of the Earth is about 4000 miles.[68]

Newton then demolished Galileo's theory of the tides, showing that the tides arise not from the combination of Earth's orbital and rotational motion, as Galileo thought, but from the gravitational attraction of Earth's sea water to the Moon and (to a lesser extent) the Sun. The water on the side of the Earth closest to the Moon is attracted to the Moon somewhat more than the Earth, as a whole, is. The water on the side farthest from the Moon is attracted slightly less. The result is that the water bulges away from Earth's center on the side toward the Moon, and also on the side opposite the Moon. As the Earth rotates, any given point on the surface will pass through one of these bulges about every 12 hours, giving rise to high tides at those times.[i] The Sun produces a similar, but smaller effect, so that the greatest high tides (known as spring tides) occur when the Sun and Moon are in the same or opposite directions from Earth (i.e. at new or full moon).[69]

Newton also attempted to explain the details of the Moon's motion using his theory of universal gravitation.[70] In some ways this was the least successful part of the Principia, and the problem of the Moon's motion was the subject of continued work, much of it by French physicists, for more than a century after Newton's death. More successful was Newton's theory of the precession of the equinoxes. He showed that the gravitational attraction of the Sun and Moon acting on the oblate shape of the Earth resulted in a slow wobble of the Earth's rotational axis. While his theory had a few minor flaws, he was able to calculate a rate of precession (50″ per year) that was in good agreement with both contemporary estimates and modern measurements.[71]

---

[i] The timing is slightly modified by the Moon's orbital motion so that high tides are really about 12.4 hours apart, on average.

Newton devoted the last portion of Book III to comets.[72] He claimed that comets followed conic section orbits with the Sun at a focus. Some comets might follow parabolic or hyperbolic paths, in which case they would never return, but others might orbit the Sun along highly elongated elliptical paths. These comets were only visible when they were in the portion of their elliptical orbit that was closest to the Sun. Since one end of an elongated ellipse can be well approximated as a parabola, Newton was able to treat cometary orbits as parabolas and determine their paths from observation. He used the observations of Flamsteed and others to investigate the comet of 1680 but couldn't quite get the theory to work out. However, in the second edition of the *Principia* he included a new computation by Halley, based on Newton's physics, showing that the comet seen in November 1680 heading toward the Sun was indeed the same object as the comet seen in December heading away from the Sun, just as Flamsteed had suspected. Halley even went on to determine that the comet seen in 1682 was likely the same one that had been seen in 1607 and 1531. Since it seemed that this comet had a period of about 76 years, he predicted that it would return in 1758.[73]

The first edition of the *Principia* ended with this material on comets. The book was a difficult one and few readers could follow all of Newton's arguments. Nevertheless, the incredible variety of phenomena that Newton could explain using only three laws of motion and a universal gravitational force was truly astounding. In England he was celebrated as a genius and granted almost god-like status. English scientists began to think of themselves as carrying on the Newtonian tradition, even though their work was often more empirical and less mathematical than Newton's was.[74]

On the Continent, Newton's work was not as well received. In part this was because Descartes' philosophy was dominant and Newton's approach seemed to contradict the generally accepted understanding of how natural philosophy was to be done. Descartes had insisted that all phenomena should be explained in terms of mechanical causes – particles running into other particles and thereby changing their motion. Newton proposed attractive forces acting over great distances with no mechanism for transmitting the force from one body to another. Many Continental natural philosophers viewed attractive forces as "occult qualities" and felt that explaining Nature in such terms represented a huge step backward for science.[75] Newton's notion of universal gravitation was particularly opposed by the Swiss mathematician Johann Bernoulli and the German mathematician Gottfriend Wilhelm Leibniz.[i]

[i] Newton and Leibniz were involved in a priority dispute over the invention of the calculus. The evidence seems to indicate that the two men developed the calculus independently, although Newton's work preceded that of Leibniz. There is no doubt that Leibniz was the first to publish his work on the calculus, and furthermore his notation became widely adopted by Continental mathematicians.

When Newton published the second edition of his *Principia* in 1713 he added to the end a General Scholium in which he attempted to address the accusation that he had provided no valid explanation for what gravity really was. Newton admitted that he had not fully explained the nature of gravitation:

> I have not as yet been able to deduce from phenomena the reasons for these properties of gravity, and I do not feign hypotheses. For whatever is not deduced from the phenomena must be called a hypothesis; and hypotheses, whether metaphysical or physical, or based on occult qualities, or mechanical, have no place in experimental philosophy. In this experimental philosophy, propositions are deduced from the phenomena and are made general by induction. ...And it is enough that gravity really exists and acts according to the laws that we have set forth and is sufficient to explain all the motions of the heavenly bodies and of our sea.[76]

Rather than making up plausible hypotheses about how Nature must operate and trying to explain natural phenomena according to those hypotheses, as Descartes had done, Newton claimed that he instead derived mathematical laws from the phenomena. As long as the phenomena fit the mathematical laws, then the laws must be accepted as real and no more explanation should be expected from the natural philosopher.

In his General Scholium Newton also expressed his belief that, while universal gravitation could explain the current motions of bodies in the solar system, it could not possibly explain the fact that the planets all orbit in nearly the same plane and in the same direction. "This most elegant system of the sun, planets, and comets could not have arisen without the design and dominion of an intelligent and powerful being."[77] Indeed, Newton's private letters reveal that he saw God's active hand at work throughout the universe. Perhaps even gravity itself was a continual action of God that sustained the solar system.[78]

In spite of the initial criticism of Newton's work on the Continent, the overwhelming power of his theories and mathematical results could not be ignored. Eventually Continental authors embraced Newton's work. One author who led the way was Francois-Marie Arouet, better known as Voltaire.[79] In 1738 Voltaire published *Éléments de la philosophie de Newton* (Elements of the philosophy of Newton), a popular account of Newton's *Opticks* and *Principia*. Voltaire was greatly assisted in this work by his mathematically gifted companion Émilie du Châtelet (see Figure 9.7). In 1749 du Châtelet completed a translation into French of the *Principia*, to which she added her own commentary and some new ideas including the important concept of conservation

Figure 9.7 Frontispiece of Voltaire's *Éléments de la philosophie de Newton* (1738).
A light shines from Newton in the heavens and is reflected from a mirror held by
a muse representing Émilie du Châtelet, who translated and explained Newton's
work to Voltaire. The light finally illuminates the manuscript that Voltaire is
writing. Image courtesy History of Science Collections, University of Oklahoma
Libraries.

of energy. Once Newton's ideas became accepted on the Continent, French
and German mathematicians began to extend and refine his mathematical
physics.

After the triumph of Newtonian physics, the victory of the Copernican–
Keplerian theory was assured. Although Descartes had made the Copernican view palatable to most European astronomers, after Newton it became
impossible to argue that the Sun orbited the Earth. Whether or not Newton
had provided a sufficient explanation for gravitational "attractions," it was clear
that his physics could at least accurately describe the motions of the heavens.
In a Newtonian universe, it made sense for the small Earth to circle around
the massive Sun. The idea of the Sun orbiting the Earth, as in the Ptolemaic or
Tychonic systems, might work geometrically but it made no sense physically.
As Kepler had anticipated, an understanding of physical causes was needed to

decide between these geometrically equivalent theories. It was Newton who supplied that understanding.

After the publication of the *Principia* Newton was justly famous. He was twice elected to Parliament and was appointed President of the Royal Society. In 1696 he was given the position of warden of the Royal Mint, and later became Master of the Mint. In these roles he helped to reform English currency and worked ruthlessly to stamp out counterfeiters. In 1705 he was knighted by Queen Anne. According to the English calendar, Sir Isaac Newton died on 20 March 1726.[i]

9.5   Reflections on science

At the beginning of Book III of the *Principia* Newton presented four "rules for the study of natural philosophy." These rules were carefully stated so that Newton could argue from his mathematical results and from observed phenomena to the conclusion that all bodies attract each other through a universal gravitational force. It was very important to Newton that his idea of universal gravitation was derived from observed phenomena and not "feigned" (or invented) as a hypothesis. Here we will examine each of Newton's rules and how he used them to argue for the truth of a universal gravitational force.

We will start by looking at Newton's first two rules.

> Rule 1: No more causes of natural things should be admitted than are both true and sufficient to explain their phenomena.
>
> Rule 2: Therefore, the causes assigned to natural effects of the same kind must be, so far as possible, the same.[80]

These rules are both aspects of what is sometimes called the law of parsimony, or Ockham's razor after William of Ockham, the 14th-century English scholar who promoted the idea. The basic idea is that our explanations should be as simple as possible to explain the observed phenomena. If we can explain something in terms of a single cause, then we should not use an explanation that involves many different causes. If we can explain lots of different things with the same cause, then we should not propose a separate cause for each of those things.

Newton used these rules to great effect in arguing for a single, universal, inverse-square gravitational force. Newton showed mathematically that a single inverse-square attractive force could account for the observed motions of the planets (or of Jupiter's moons, etc.). This inverse-square force was both true

---

[i] According to the Gregorian calendar Newton died on March 31, 1727. The difference in the year arises because the English calendar began the new year on March 25 rather than January 1.

(it fit the phenomena) and sufficient (nothing else was needed to explain the phenomena). In contrast, Descartes' vortices were not true (they were not consistent with the phenomena) unless something else was added. By Rule 1 it was best to accept a single, simple inverse-square force that could explain all of the planetary motions.

Newton also used this rule to connect the force that attracts a falling apple to the Earth with the force that keeps the Moon in its orbit. It was possible that these could be two separate forces, one of which only acts on objects near Earth's surface and another which acts on celestial bodies outside Earth's atmosphere. But a single inverse-square attraction to the Earth is sufficient to explain both the falling apple and the orbiting Moon, and by Rules 1 and 2 we should not introduce two different forces when one force is sufficient.

In addition, Newton used Rule 2 to argue that the same cause was responsible for falling bodies on Earth, the orbit of the Moon, the orbits of the moons of Jupiter and Saturn, and the orbits of the planets around the Sun. Each of these phenomena could be explained in terms of an inverse-square attractive force, so they were "natural effects of the same kind." By Rule 2 they should be assigned the same cause. Since the force responsible for making bodies fall to Earth had always been called gravity, Newton argued that gravitational forces were responsible for all of the celestial motions as well.

Newton's next two rules justify his claim that gravity is a *universal* force.

> Rule 3: Those qualities of bodies that cannot be intended and remitted [i.e., qualities that cannot be increased and diminished] and that belong to all bodies on which experiments can be made should be taken as qualities of all bodies universally.
> Rule 4: In experimental philosophy, propositions gathered from phenomena by induction should be considered either exactly or very nearly true notwithstanding any contrary hypotheses, until yet other phenomena make such propositions either more exact or liable to exceptions.[81]

Rule 3 says that if we find that all of the objects we can test have a certain property, and that property cannot be increased or reduced (intended or remitted), then we should assume that *all* objects have that same property. Newton showed that falling objects and the Moon gravitate toward Earth, the moons of Jupiter gravitate toward Jupiter, and the planets gravitate toward the Sun, all with a force proportional to the mass of the object (which cannot be increased or diminished). Therefore, by Rule 3 Newton was justified in claiming that *all* objects attract each other through a gravitational force that is proportional to the masses of the objects.

Rule 4 states that a theory derived from observational data, and that fits all of the known data, should be taken as true until it is contradicted by some new observation. Such a theory should not be dismissed just because it conflicts with some "hypothesis" (Newton used that word to mean an unfounded assumption or an invented theory that was not derived from phenomena). Newton claimed that he derived the theory of universal gravitation from observations and experiments, in conjunction with the first three rules for natural philosophy. By Rule 4 the theory of universal gravitation should be considered true until it is contradicted by observational or experimental evidence. It should not be discarded simply because it doesn't fit the preconceived notions of the "mechanical philosophy" that was advocated by Descartes and his followers.

One important thing to note about Rule 4 is that it includes a suggestion of what to do if new observations *do* contradict a previously well-established theory. Newton stated that such a contradiction should lead to a "more exact" theory or else it should make the theory "liable to exceptions." If a theory has worked in many different circumstances but is then found to fail in some new circumstance, it need not be entirely cast aside. Instead, it can be modified or perfected so that it works in the new case as well as the old, or it can be given a limited scope so that the theory is only held to work in certain situations but not in others.

It is not only new data that can lead to correcting or limiting a theory. In fact, Newton's theory of universal gravitation both corrected and limited the previously successful theories of Kepler and Galileo. Kepler's three laws of planetary motion were very successful, even if most astronomers after Kepler viewed them as nothing more than accurate approximations. However, according to Newton's physics, Kepler's laws were not exactly true. A single planet orbiting the Sun would move in an elliptical orbit with the Sun–planet line sweeping equal areas in equal times, but the Sun would not actually reside at the focus of that ellipse. Rather, the center of gravity of the Sun–planet system would lie at the focus. Once we include other planets, which exert gravitational forces on the original planet, then the orbit of the planet would no longer be a perfect ellipse and the Sun–planet line would not sweep equal areas in equal times (because the total force on the planet would no longer point exactly toward the Sun). Newton's physics showed how to correct Kepler's laws and even showed that in certain situations, such as three bodies of similar mass lying at comparable distances from each other, Kepler's laws broke down completely. Thus, it was Newton's theory, not empirical data, that led to the correction and limitation of Kepler's laws.[82]

In a similar fashion, Galileo's theories that all bodies accelerate toward the Earth with constant acceleration and that projectiles follow a parabolic path are inconsistent with Newtonian physics. According to Newtonian physics, the force on an object near Earth was inversely proportional to the square of its distance from Earth's center. In practice, that distance changed relatively little as the object fell. If the object was dropped from a 100-foot high tower then its distance from Earth's center would change by 100 feet out of a total of nearly 4000 miles. The force, and acceleration, experienced by the object would be approximately, but not exactly, constant. Likewise, according to Newtonian physics the path of a projectile was actually a portion of an ellipse with the center of Earth at one focus,[i] but it was a highly elongated ellipse and the part followed by the projectile could be closely approximated as a parabola.

Newton's theory of universal gravitation helped to explain why Kepler's laws and Galileo's theory of falling worked so well, but also showed that they were not strictly correct. This important characteristic of universal gravitation is something that we have come to expect from any successful new theory. The new theory must explain all of the old data that were explained by the previous theory, and thus it must be able to show why the previous theory was at least approximately correct, a process sometimes called "theory reduction." At the same time, the new theory must give us something new. We must be convinced that the new theory is "more exact" than the old theory.

One final thing to note about Newton's physics is its mathematical character. Newton did not simply claim that all bodies attract each other through a gravitational force. He specified a mathematical form for that force (proportional to the masses of both bodies and inversely proportional to the square of the distance between them) and through his laws of motion he gave a mathematical method for determining the observable effects of those forces. Because the observable effects fit with his mathematical theory, Newton could get away without specifying a mechanism for gravity. In effect, the mathematics made universal gravitation seem real even though Newton could offer no plausible *cause* for gravity.

In that sense, Newton was using a hypothetico-deductive method much like Galileo. He proposed mathematical models and showed that real motions fit with his models. These models could then be used to explain and predict many things that were already known: the orbits of planets, moons, and comets, the precession of the equinoxes, the fall of bodies near Earth's surface, the motion of projectiles, and the tides. Newton's physics also predicted something new: the nonspherical shape of the Earth. Newtonian physics could also be used to determine quantities such as the mass and density of a distant planet that were

---

[i] Even that elliptical path is only correct if the Earth is a perfectly homogeneous sphere. Since it is not, the path of the projectile will not be a perfect conic section of any kind.

previously unknowable. In short, Newton's physics did everything that one might hope for ... except provide a mechanical cause for gravity.

Newton worried about causes. While Galileo had dismissed the whole notion of causes, Newton only claimed that he had not yet found the cause of gravity. His successors, though, were content with having a mathematical model that worked, and soon grew less fearful of the idea of real attractions. They stopped asking about the cause of gravity and took gravitational attractions as explanations in themselves. It was not until the 20th century and the development of Einstein's general relativity that an underlying cause for gravity was found.

In the meantime, Newtonian physics flourished. If the Copernican idea of Earth's motion was not completely accepted before Newton, it was hard to deny it after Newton. Newton showed that the only point in our system that could possibly be taken as a stationary center was the center of gravity of the Sun and planets, and that point lay inside, or at least very close to, the Sun. The tables had been turned on Aristotle: now a moving Earth made much more sense, in the context of Newtonian physics, than a stationary Earth.

Even so, at the beginning of the 18th century there was no direct observational evidence for the Earth's orbital motion or its rotation. Astronomers knew what kind of evidence to look for: annual parallax of the stars, the oblate shape of the Earth, deflections of falling bodies, but none of these effects had been detected (in spite of Hooke's claims to the contrary). Furthermore, Newton had shown that "universal gravitation" operated throughout the solar system, but was that enough to make it "universal?" Did gravity act even among the stars?[i]

---

[i] Newton himself had some doubts about gravitational interactions between stars, because he believed that the stars must fall together if they attracted each other. He eventually decided that they did attract each other, but that the forces mostly cancelled out and God intervened to prevent the residual forces from causing the stars to collapse together.[83]

10

Confirming Copernicus: evidence for Earth's motions

10.1 Evidence for Earth's orbit

10.1.1 *The aberration of starlight*

In the quest for direct evidence of the Earth's orbit around the Sun, the gold standard was always the detection of annual stellar parallax. While few astronomers were convinced that Hooke had successfully detected parallax in 1671, some sought to make their own detection. One notable attempt was made by John Flamsteed, who reported a successful detection of parallax in the star Polaris in 1699. Flamsteed believed he had observed an annual parallax of about 15″ in the North Star, which put that star at a distance of about 14,000 AU, similar to the parallax of Gamma Draconis reported by Hooke.[1]

Stellar distances had important implications for Newtonian physics. Newton claimed that all objects exerted gravitational forces on all other objects, so presumably stars might exert gravitational forces on the planets of our solar system. If some stars were as close as Flamsteed and Hooke claimed, then they might disrupt the motion of the planets and destabilize the solar system. But no such effects were seen. Could it be that Newton's "universal" gravity did not operate among the stars?

Shortly after Flamsteed published his work, doubts were raised about his results. The Danish astronomer Ole Rømer and the French astronomer Jacques Cassini both pointed out that Flamsteed's measurements did not fit the expected pattern for the parallax of Polaris.[2] To really claim a successful parallax measurement, an astronomer could not just present a few observations showing that a star had apparently shifted its position on the celestial sphere. A convincing parallax measurement would require a continuous series of measurements over the course of (at least) a year that fit the expected pattern.

Parallax shifts the apparent position of a star toward the apparent position of the Sun on the celestial sphere. The exact pattern depends on the star's location on the celestial sphere, but for the star Gamma Draconis, which lies near the ecliptic pole, the pattern is fairly simple. Parallax should shift that star northward on the summer solstice, westward on the autumnal equinox, southward on the winter solstice, and eastward on the vernal equinox. The north–south component of this pattern can be seen in Figure 9.2. Hooke's data did seem to fit the expected parallax pattern for Gamma Draconis, but his results were considered unreliable.

In December 1725 Samuel Molyneux and James Bradley set out to replicate Hooke's work using a more reliable telescope. They installed a 24-foot long zenith telescope, which Bradley believed to be accurate to 1″, at Molyneux's mansion at Kew, England. They soon noticed that Gamma Draconis seemed to be moving southward. The star continued to shift farther south until March, when it was 20″ south of its December position. Then the star began moving northward, eventually reaching its December position in June and continuing on until it was 20″ north of its December position in September. This was not at all what was expected. The star seemed to be going through an annual variation, but the pattern was shifted by three months from the expected parallax pattern.[3]

Initially Molyneux and Bradley thought they had discovered an annual wobble of the Earth's rotational axis, akin to the wobble that causes the precession of the equinoxes but with a much smaller angle and a much shorter period. That idea could explain what they were seeing in Gamma Draconis, but when they observed another star (35 Camelopardolis) they found that it too displayed annual motions but the motions were inconsistent with the idea of a wobbling axis.

After Molyneux was called to serve as Lord of the Admiralty, Bradley continued the zenith observations using a new telescope designed by the instrument maker George Graham, installed at the home of Bradley's aunt. The telescope was shorter (12 feet long) and had a wider field of view so that it could be used to observe more stars. Bradley tested the telescope thoroughly until he was confident it was accurate to half an arcsecond. He measured transits of many stars and found that they all exhibited an annual variation that was shifted from the parallax pattern by three months, although the exact pattern depended on the star's location on the celestial sphere.

Eventually Bradley hit on an idea that could explain his data.[4] The idea can be traced back to Bradley's earlier observations of the moons of Jupiter. Bradley had carefully observed eclipses of Jupiter's moons in hopes of using these eclipses as time markers for determining longitude at sea. In the course of his observations he confirmed an effect that had previously been noticed by Ole Rømer: the timing of the eclipses was affected by the relative motion of Earth and Jupiter. Specifically, when Jupiter and Earth were getting closer together the time between two consecutive eclipses of a particular moon were shorter than average. When Jupiter and Earth were getting farther apart, the time was longer than average.

Rømer had proposed that the changes in eclipse timing could be explained if light traveled at a finite speed. When Jupiter and Earth were approaching each other, the light from the second eclipse had less distance to travel in order to reach Earth than the light from the first eclipse. As a result, the light from the second eclipse would arrive sooner than expected and the time between observed eclipses would be shortened. When Jupiter and Earth were moving away from each other the opposite would happen: the light from the second eclipse would have farther to go and observations of that eclipse would be delayed. Rømer had even estimated the speed at which light travels: fast enough that light took only 11 minutes to go from the Sun to the Earth.

Bradley's observations of the moons of Jupiter had convinced him that Rømer was right, and he realized that the finite speed of light could also help to explain the apparent motions he was seeing in the stars.[5] Bradley's idea is illustrated in Figure 10.1. Let's assume that we wish to observe a star that is directly overhead at s. If we point the telescope straight upward we won't see the star because, as the starlight is traveling through the telescope tube, the tube moves (due to the motion of Earth) and therefore the starlight will hit the side of the tube rather than reaching the eyepiece at the bottom. In order to see the star we must tilt the telescope in the direction of Earth's motion (toward the right in Figure 10.1) so that the starlight reaches the front of the moving telescope at point p_1 when the telescope is in position T_1, and travels down the tube to reach the eyepiece at point p_3 when the telescope is in position T_3.

The combination of Earth's orbital motion and the finite speed of light results in an apparent shift in the star's position, which came to be called "the aberration of starlight." The shift is always in the direction of Earth's orbital motion. Since the Earth moves in a roughly circular orbit around the Sun, the direction of its motion changes over the course of a year. In fact, the Earth is always moving toward the point on the celestial sphere that is roughly 90° from the apparent position of the Sun. Another way to say this is that the Earth is always moving toward the apparent location that the Sun had three months before. Thus, the pattern of motion resulting from aberration will mimic the pattern of parallax but with a three month shift.

The angle at which the telescope is tilted in Figure 10.1 depends on the ratio of the speed of the Earth in its orbit to the speed of light. Details are given in Appendix A.20, but because Earth's orbital motion is much slower than the speed of light this angle is very small. Bradley found that his data for Gamma Draconis

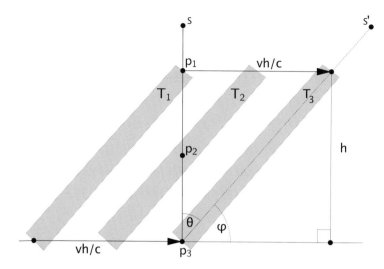

Figure 10.1 The aberration of starlight. Light travels straight downward from the star at s, passing through the points p_1, p_2, and p_3. The telescope, shown at three different times, moves toward the right due to Earth's orbital motion. In order for the starlight to pass down the middle of the telescope tube the telescope must be tilted in the direction of Earth's motion. The apparent position of the star is s′. The aberration angle θ is greatly exaggerated in this figure. Adapted from Timberlake (2013) with the permission of the American Association of Physics Teachers.

fit the expected pattern for aberration with a maximum tilt of 20″ (north or south), as shown in Figure 10.2. An aberration tilt of 20″ indicates that light must travel over 10,000 times as fast as the Earth. That means light must take a little more than 8 minutes to travel from Sun to Earth, a bit faster than Rømer had thought.

Bradley found that this aberration effect, with the speed of light quoted above, could explain all of his observations of various stars. Furthermore, he showed that it could also explain Flamsteed's observations of Polaris. Independently of Bradley, the Bologna astronomer Eustachio Manfredi made similar observations which showed annual movements of the stars Sirius and Arcturus.[6] Although Manfredi did not offer an interpretation for his observations, they were found to be entirely consistent with Bradley's aberration theory.[i] The only data that

---

[i] Manfredi observed east–west variations in right ascension, while Bradley had observed north–south variations in declination, so their observations were really complementary. But Manfredi, working in Italy where Copernicanism was still forbidden, never accepted Bradley's interpretation and was not convinced that the effect could be seen in all stars. He described the apparent motions he had detected as "aberrations." Bradley adopted this term and the name stuck.

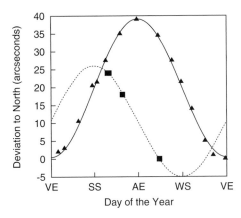

Figure 10.2 Bradley's data for Gamma Draconis (triangles) and the prediction of the aberration theory (solid curve). Hooke's data (squares) and the parallax prediction (dotted curve) are shown for comparison.

contradicted Bradley's aberration theory were those of Hooke. Hooke's data fit the pattern for parallax, not aberration. More importantly, they directly conflicted with Bradley's more precise observations of the same star. Hooke, apparently, had been deceived by the inaccuracy of his telescope into seeing what he wanted to see.

Bradley's observations were so systematic and precise, and his explanation so thoroughly convincing, that the aberration of starlight soon gained widespread acceptance, at least in Protestant countries where the Copernican model was generally acknowledged as correct. Aberration provided the first direct evidence for the Earth's orbital motion while simultaneously providing evidence for the finite speed of light.[i] It did not, however, tell astronomers anything about the distances to the stars. Aberration depends only on the speeds of Earth and light and a star's location relative to the ecliptic poles. It does not depend on a star's distance. To measure the distance to a star, parallax was still the only reliable way.

10.1.2   *Parallax at last*

Once Bradley understood the explanation for the apparent motions of Gamma Draconis and other stars, he was able to "reduce" (or correct) his data by subtracting the predicted shift due to aberration. The result was a set of

---

[i] Technically the Earth's rotation also produces an aberration effect, but with a period of one day rather than one year. However, the speed of Earth's orbital motion is a factor of more than 70 greater than the speed of Earth's rotational motion at the equator. For points away from the equator the rotational motion is even slower. Therefore, the rotational aberration is negligibly small compared to the orbital aberration.

positions the star would have had if aberration didn't exist. The hope was that these reduced observations might reveal a small annual parallax that was hiding underneath the larger aberration effect, but Bradley could find no trace of parallax in his reduced data. He was convinced that his observations and his corrections for aberration were accurate enough that he would have detected a parallax as small as $1''$.[7] No wonder, then, that parallax had not been successfully detected before. If stellar parallaxes were all smaller than $1''$ then there was no way they could have been seen before aberration was discovered and understood.

Bradley's reduced data did reveal one new effect, though. He found that the apparent positions of stars seemed to vary with a period of about 18.6 years, the same period as the precession of the Moon's nodes.[i] This time he found that he *could* explain the observed motions using a wobble of Earth's rotational axis. This new wobble came to be known as **nutation**. Even after correcting for nutation Bradley could find no trace of parallax.[8]

The stars that Bradley had observed were not particularly bright. It was possible that brighter stars might be closer and therefore have a larger parallax. In addition, there might be ways of looking for parallax that could avoid the problems caused by aberration, nutation, precession of the equinoxes, and the refraction of light by Earth's atmosphere (see Section 6.5). One such method had been proposed by Galileo in his *Dialogo*: observe two stars that were close together on the sky, but at very different distances. The closer star will have a larger parallax than the more distant one, and therefore the angular separation between the stars will change over the course of a year.[9] Since the two stars are close together on the sky they will be affected by aberration, nutation, precession, and refraction in almost exactly the same way, so any observed change in the angle between the stars must be due to differences in their parallax.[10]

In the 1780s, William Herschel, a professional musician who had emigrated from Germany to England as a child and later became an avid astronomer, compiled an enormous list of so-called "double stars." Double stars were stars that appeared as a single star to the eye, but were revealed to be two (or more) stars when viewed through a telescope. Herschel found these double stars in part to demonstrate the power of the large reflecting telescopes that he and his sister Caroline had built in their workshop, but he also hoped to use double stars to detect parallax. He found some promising candidates in which one star of the double was much brighter than the other, suggesting that the stars were at very different distances, but at that time Herschel did not make the kind of systematic measurements required to reveal parallax.[11]

---

[i] The precession of the Moon's nodes is responsible for the small difference between the draconic and sidereal months discussed in Section 3.2.

Later, in 1802, Herschel observed some of these double stars again. His new observations revealed changes in some of the double stars – but it was not just the angle between the stars that had changed. In some cases the orientations of the stars relative to each other were different.[12] His data suggested that these double stars were actually part of what is now called a binary star system: a pair of stars that orbit around each other. These observations destroyed Herschel's hopes for finding parallax because two stars in a binary system must be about the same distance from Earth and thus will not show any differential parallax. His observations did, however, reveal two important facts. Because the stars seemed to move in elliptical orbits around each other, the observations suggested that stars are indeed subject to Newtonian gravity. Because the stars differed in brightness, but not in their distance from Earth, these observations also indicated that stars do not all have the same intrinsic brightness (or luminosity).

The double star method seemed hopeless, but other astronomers pursued the more conventional approach of measuring tiny changes in the positions of stars on the celestial sphere. One such astronomer was Giuseppe Piazzi, who worked at the Palermo Observatory in Italy. Piazzi's efforts did reveal some motions among the stars, but these were "proper motions," slow movements of stars along the celestial sphere that had first been discovered by Edmund Halley.[13] These motions were steady and progressive, not cyclical like annual parallax. However, stars with large proper motions were assumed to be relatively close to Earth and thus might serve as good candidates for future parallax measurements.[i]

It seemed as though there would never be a successful measurement of annual parallax. Then, suddenly, there were three. The first successful measurement of parallax was made by Friedrich Bessel (Figure 10.3) at Königsberg Observatory in Prussia.[ii] Bessel made extensive measurements of the "flying star" 61 Cygni, a star with a large proper motion measured by Piazzi. After reducing his observations Bessel found that his data fit the expected pattern for parallax with a parallax angle of $0''.3136$, equivalent to a distance of about 660,000 AU.[14]

Bessel published his results in late 1838, and in January 1839 another parallax measurement was made public. Thomas Henderson, a Scottish astronomer working at the Cape of Good Hope in South Africa, had measured the parallax of the bright southern star Alpha Centauri. He found a parallax of about $1''$, which put Alpha Centauri about 200,000 AU away.[15] Henderson was fortunate in his

---

[i] Piazzi's careful measurements of stellar positions also led to his discovery of Ceres, the first asteroid to be identified. Like Pluto, Ceres was initially considered to be a planet but is now classified as a dwarf planet.
[ii] Königsberg is now the city of Kaliningrad, Russia.

Figure 10.3 Portrait of Friedrich Bessel. Image courtesy History of Science Collections, University of Oklahoma Libraries.

choice of star: Alpha Centauri turned out to be the closest star to us other than the Sun.[i]

Later that year, Wilhelm Struve at Dorpat Observatory (later called Tartu Observatory, in Estonia) published a parallax measurement for the bright star Vega (Alpha Lyrae). Struve found a parallax of $0''.261$, corresponding to a distance of nearly 800,000 AU. Struve had actually presented a preliminary result for the parallax of Vega in 1837, over a year before Bessel's announcement for 61 Cygni.[16] However, Struve's value ($0''.125$ or nearly 1.7 million AU) was based on just a few observations and it differed so greatly from his later result that most astronomers, including Struve himself, considered Bessel to be the first to successfully and accurately measure stellar parallax. We now know that Struve's first value was actually more accurate than his second.[17]

These three parallax measurements not only gave conclusive, direct evidence for Earth's orbital motion, they also indicated that the nearest stars lay hundreds of thousands of Astronomical Units from our solar system. Copernicus was right: the stars really are so far away that the size of Earth's orbit was like a point in comparison. Furthermore, although the stars do participate in universal

---

[i] Alpha Centauri is actually a triple star system. Henderson was aware that it was a binary, but the third (and closest) star, Proxima Centauri, was not discovered until 1915.

gravitation, they are all far enough away that we need not worry about their gravitational forces disrupting our solar system.

10.2   Evidence for Earth's rotation

10.2.1   *The figure of the Earth*

In the *Principia* Newton had predicted that the Earth must have an oblate shape, with its diameter between the poles smaller than the diameter of the equator, on account of its rotation. In favor of this prediction he presented evidence that Jupiter had an oblate shape, with its north–south diameter (roughly aligned with Jupiter's axis of rotation) smaller than its east–west diameter, as seen in telescope observations. He also presented evidence based on the motion of pendulums at different locations on Earth, which showed that the effect of gravity was weaker at the equator, as predicted by Newtonian physics for a rotating Earth.[18]

However, in 1718 Jacques Cassini announced that his measurements of the length of a degree of latitude on Earth showed that the Earth had a prolate shape. Figure 10.4 illustrates the difference between a true sphere, an oblate spheroid, and a prolate spheroid. The length of a degree of latitude on Earth is determined by the distance you must travel, north or south, in order for the direction of the line perpendicular to Earth's surface to change by 1°. The flatter the Earth is, the farther you have to go in order to produce this 1° change. An oblate spheroid is flatter near the poles, so a degree of longitude will be longer there. A prolate spheroid is more curved near the poles, so a degree of longitude there will be shorter. Cassini's measurements indicated that the degree of longitude was shorter to the north, which suggested that the Earth was prolate.[19]

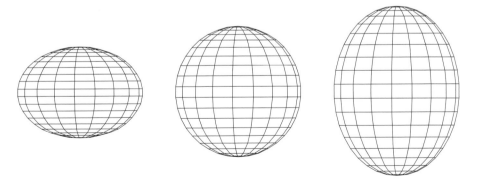

Figure 10.4 Possible shapes for the Earth: oblate spheroid (left), sphere (center), or prolate spheroid (right).

A prolate Earth directly contradicted the predictions of Newton's physics. In 1734 Johann Bernoulli presented a purported proof, based on Cartesian vortex theory, that the Earth should have a prolate shape, and he cited Cassini's measurements as evidence in favor of this prediction. The shape of the Earth became a centerpiece of the contest between Newtonian and Cartesian physics.

Pierre-Louis Maupertuis was one of the few French scientists at the time who had fully adopted Newton's physics. He had made a name for himself by publishing a book on the shapes of stars in which he treated the stars as rotating fluid bodies subject to Newtonian gravity. Maupertuis was perfectly positioned to become an advocate for the Newtonian side of the debate over the Earth's shape. He felt that Cassini's results were inconclusive because the measurements were all made within France, with only a small change in latitude between locations. Maupertuis proposed expeditions to measure the length of a degree of latitude near the equator and near the north pole in an effort to settle the matter.[20]

In 1736 French expeditions were sent to what is now Ecuador and to Lapland (in Finland). Maupertuis led the expedition to Lapland himself. Maupertuis returned to France in 1737, but the South American expedition encountered incredible difficulties and no member of the expedition made it back to France until the mid 1740s. Maupertuis was left to compare his Lapland measurements with those made by Cassini in Paris. He found that the Lapland degree was longer than the Paris degree, as expected for an oblate Earth. Cassini criticized Maupertuis' methods and even his instruments, but later measurements (including a new measurement of the Paris degree by Cassini's son, as well as the South American measurements) confirmed that the Earth was oblate. The difference between the Earth's radius at the equator and at the poles is about 13 miles, somewhat less than the 17 miles predicted by Newton.[21]

Although the Earth's rotation was no longer in doubt by the 1730s, the demonstration of the Earth's oblate shape can be viewed as the first direct evidence that the Earth spins. Historically, it was more important for demonstrating the correctness of Newtonian physics, which soon gained widespread acceptance in France and throughout continental Europe.

10.2.2   Falling bodies

Before he predicted the oblate shape of the Earth in the *Principia*, Newton had suggested another way to demonstrate the rotation of the Earth. In his 1679 letter to Robert Hooke he had proposed that the rotation of the Earth would cause an object dropped at the equator to fall slightly toward the east because the object would start off moving eastward with a slightly greater speed than the surface of the Earth. In subsequent letters he and Hooke agreed that for locations north of the equator, such as London or Cambridge, the object would fall slightly toward the southeast (see Section 9.2).

The prediction of the southeastern deflection is correct if "down" is considered to be the direction toward the Earth's center, because the body will move in a plane formed by the direction of its initial eastward velocity relative to Earth's center (resulting from Earth's rotation) and the direction of the gravitational force on the object (toward Earth's center). The intersection between that plane and the Earth's surface does not lie along a line of constant latitude. Rather, the intersection curve just grazes the object's starting latitude and otherwise passes over lower latitude lines.

In practice, though, down is usually taken as the direction of a plumb line (a heavy object, or "bob," hung on a string that is allowed to come to rest). The rotation of the Earth actually alters the direction of a plumb line because the bob on the string is moving in a circle around Earth's axis. Therefore, the string must not only cancel out the gravitational force on the bob, it must also supply the centripetal force needed to keep the bob moving in this circle. As a result, a plumb line hung at a northern latitude will point in a direction that is slightly south of the Earth's center. This fact is connected with the oblate shape of the Earth: the direction of the plumb line is perpendicular to the Earth's surface, and in a spheroid a line perpendicular to the surface does not necessarily pass through the center. In the northern hemisphere the perpendicular line will pass south of the center if Earth is oblate, north of the center if Earth is prolate (see Figure 10.4).

An object dropped in the northern hemisphere will fall slightly to the south as measured relative to the direction toward Earth's center, but because a plumb line will hang slightly south of that direction, the object will end up falling almost exactly along the plumb line. Therefore, the southward deflection will be nearly undetectable if measurements are made relative to a plumb line, as they are in practice. Hooke should not have seen any southward deflection, and the fact that he did calls all of his measurements into question. In any case, Hooke never published his experiments on the deflection of falling bodies.

After Hooke, many more attempts were made to measure the deflection of falling bodies from the vertical (as defined by a plumb line). Giovanni Battista Gulielmini dropped balls down the Asinelli tower in Bologna (where Riccioli had performed his experiments on falling bodies) and found a consistent deflection to the southeast, but there were problems with his experimental procedure. Johann Benzenberg conducted the experiment in the tower of St. Michael's Church in Hamburg and he, too, found a southeastern deflection. In 1831, Ferdinand Reich dropped 106 balls down a mineshaft in Freiburg and once more found southeastern deflections. In most of these cases the eastward deflection agreed moderately well with the theoretical prediction based on a rotating Earth and Newton's physics, but the southward deflection was unexplained.[22]

In 1902 Edwin Hall tried to resolve the perplexing question of the southward deflection by dropping 948 balls from the tower of the Jefferson Physical Laboratory at Harvard University. He found an eastward deflection of 0.15 cm, which agreed fairly well with the predicted value of 0.18 cm. He also found that, on average, the balls were deflected southward, but the amount was only 0.005 cm, which was well within the margins of error for the experiment.[23] About the same time a similar experiment was performed in France by Camille Flammarion and his results indicated a small *northward* deflection.[24] Taken as a whole, these experiments on falling bodies demonstrated the eastward deflection that was expected on a rotating Earth. The southward deflection found in most of the experiments remains unexplained. Southward deflections are not impossible, particularly if the density of the Earth is not uniform, but they are almost certainly too small to have been found in these experiments.

10.2.3   *The Foucault pendulum*

Probably the best, and certainly the most famous, demonstration of the Earth's rotation was carried out by the French physicist Léon Foucault in 1851.[25] Foucault was an expert instrument designer and had noticed something odd when he was doing some machining work for a new instrument. He found that a vibrating steel rod maintained its plane of vibration even as the rod rotated on a lathe. That observation suggested to Foucault that the plane of oscillation of a pendulum might remain constant even as the Earth rotates underneath it. If so, observers on Earth would see the plane of the pendulum's oscillation precess (or slowly rotate) in the opposite direction from Earth's rotation.[26]

The easiest way to think about the motion of a pendulum on a rotating Earth is to consider a pendulum hung at Earth's north pole. The pivot point of the pendulum lies on Earth's rotational axis and therefore it does not move as the Earth rotates. The pendulum, if released from rest, would simply swing back and forth in a plane that is fixed relative to the distant stars. Meanwhile, the Earth is rotating counterclockwise as viewed from above Earth's surface. Observers standing nearby will rotate with the Earth and as a result the plane of the pendulum's oscillation will appear to slowly rotate, or precess, clockwise relative to the Earth's surface. The plane of oscillation would precess 360° in one sidereal day. At lower latitudes the situation is a little more complicated, but the same general effect is observed, with the period of precession increasing as you move away from the pole and becoming infinite (no precession) at the equator.[i]

---

[i] The precession of the Foucault pendulum is now thought of as a manifestation of the Coriolis effect mentioned in Section 8.4.

Figure 10.5 The Foucault pendulum at the Tellus Museum in Cartersville, Georgia, USA (photo credit: Michael Bailey). The pendulum bob is seen on the right. The pendulum's plane of oscillation precesses due to Earth's rotation and, as a result, the pendulum sequentially knocks over the dominos arranged in a circle around the center point.

Because the precession is so slow, you have to use a pendulum whose oscillations won't die out quickly if you want to see this effect. It is best to use a very long pendulum (which will oscillate slowly) with a very heavy pendulum bob (which won't be greatly affected by air resistance). In his first demonstration, for the Académie des Sciences at the Paris Observatory, Foucault used an 11-meter long pendulum. He followed that up with a public demonstration using a 67-meter pendulum with a 28-kg bob suspended from the ceiling of the Panthéon in Paris. The plane of the pendulum's oscillation precessed clockwise so that it turned 360° in about 32 hours. The rotation of the Earth was demonstrated for all to see.

Foucault's experiment quickly became famous and Foucault pendulums were installed in academic buildings and museums throughout Europe and the USA (see Figure 10.5). A replica of Foucault's 67-meter pendulum still hangs in the Panthéon today.[27]

10.3   Loose ends

10.3.1   Measuring the Astronomical Unit

The theories of Copernicus, Kepler, and Newton allowed ever more refined determinations of the distances between the Sun and planets in terms of the fundamental Astronomical Unit (AU), but when Newton wrote the *Principia* only rough estimates were available for the value of the AU in more conventional units. Tycho had believed the diurnal solar parallax to be 3′, which would mean an Earth–Sun distance of about 7.3 million km. Kepler thought that the solar parallax must be smaller than this (and thus the distance must be larger). In the second edition of the *Principia* Newton had used an estimated value of 10″.5 for the solar parallax (meaning an AU of 125 million km).[28] All of these values were only rough estimates.

It was not necessary to actually measure the Sun's diurnal parallax in order to find the Earth–Sun distance. The distance between Earth and *any* planet was known in AU, so a parallax measurement for any planet would suffice to determine the value of the AU in conventional units. We have seen that Tycho attempted, but failed, to measure the diurnal parallax of Mars. Giovanni Dominico Cassini attempted the same measurement in 1672. He measured the position of Mars from Paris while his assistant, Jean Richer, measured it from Cayenne on the coast of South America. Their result, which put the AU at 87 million miles, was widely accepted even though the uncertainties of their measurement made it untrustworthy.[29]

Edmund Halley was unconvinced by Cassini's measurement. He thought a better determination of the solar parallax could be obtained using a method proposed by James Gregory in 1663. Gregory had suggested using a transit of Venus or Mercury to measure parallax. Recall that a transit occurs when Mercury or Venus passes across the face of the Sun as seen from Earth (see Figure 10.6). The conditions required for a transit are much like those for an eclipse (see Section 3.2): in order for a transit to occur the planet must be at inferior conjunction and also near a node (one of the points where the planet's orbit crosses the ecliptic plane). Transits of Venus are rare. They occur in pairs with eight years between the two transits in a pair, but with over one hundred years between consecutive

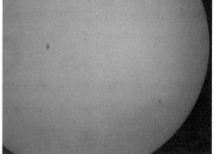

Figure 10.6 Photographs of the 2012 Venus transit (left, photo by Kalen Maloney) and the 2016 Mercury transit (right, photo by Todd Timberlake) taken from Berry College. Venus is obvious in the left photo. Mercury is the tiny dot toward the lower right of the Sun in the right photo.

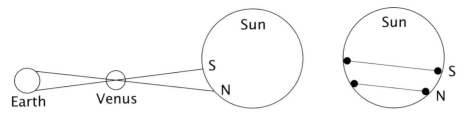

Figure 10.7 Using a transit of Venus to measure parallax. Two observers on Earth, one at a northern latitude (N) and one at a southern latitude (S), observe Venus pass across the Sun as shown on the left. The observers will see Venus cross the Sun along paths that are slightly shifted due to parallax, as shown on the right. The diagram is not to scale.

pairs. Transits of Mercury are much more common and occur about once every seven years on average.[30]

Transits can be used to measure parallax because observers at different locations will see the transiting planet pass across the Sun along slightly different paths, as shown in Figure 10.7. Careful measurements of the timings of first contact (when the planet first appears to touch the edge of the Sun), second contact (when the planet first appears completely within the Sun's disk), and so on can reveal the angular shift between these different paths. If the distance between the observers on Earth is known, that angular shift can be used to determine the distance to the planet.[i]

In 1677, while he was on the island of Saint Helena mapping the stars of the southern hemisphere, Halley made careful observations of a transit of Mercury. When he returned to England he discovered that only one astronomer had made similar observations from Europe so it was impossible to get a good result for the solar parallax. Halley was convinced that a transit of Venus, easier to observe than a transit of Mercury, could yield a reliable value for the solar parallax and the AU. He proposed an effort to observe the next Venus transit from a wide variety of locations on the Earth.[31]

The next transit of Venus was not until 1761 and Halley did not live to see it, but astronomers worked to carry out his plan. Observations were made by several astronomers in Europe and also by Charles Mason and Jeremiah Dixon at the Cape of Good Hope in South Africa. The observations were compiled and analyzed by James Short, who published a value for the solar parallax of 8″.5

---

[i] The parallax of the Sun must also be taken into account in these measurements. For example, during a transit of Venus the Sun is approximately four times as far away as Venus is, so its parallax will be one-fourth that of Venus. The observed parallax shift for Venus, relative to the Sun, will be about three-fourths of the true parallax shift.

on the day of the transit. On that day the Earth was in a part of its orbit that was somewhat farther than its average distance to the Sun and Short computed that the mean solar parallax would be 8".7 (making the AU about 151 million km). Unfortunately, the data were not entirely consistent, mostly as the result of inaccurate longitude values for some of the observing locations.[32]

In some ways the 1761 transit served as a trial run. Astronomers learned many lessons from their experiences in 1761 and applied their new knowledge to observing the next transit of Venus in 1769. Observations were made from all over the world: Tahiti, India, Russia, Europe, South Africa, Canada, and Baja California. The results were compiled by Thomas Hornsby, who published a mean solar parallax of 8".78, corresponding to an AU of 150 million km, very close to the modern accepted value.[33] However, the observations were plagued by the so-called "black drop effect," in which the black disk of Venus is distorted into a teardrop shape around the time of second and third contact. This effect made it hard for astronomers to determine the exact times of these contacts, and other astronomers found values for the solar parallax that differed slightly from Hornsby's result.

The exact value for the AU was refined by later transit observations and other methods, but after the 1769 Venus transit astronomers had an accurate understanding of the distance scale for the solar system. When Thomas Henderson found that Alpha Centauri had a parallax of about 1", astronomers knew that the corresponding distance of 200,000 AU was equivalent to about 30 million million km. The stars, as it turned out, were just as outrageously far away as Copernicus said they might be.

10.3.2   Star size and diffraction

If stars were as tremendously far away as Copernicus thought, then they ought to appear to us as pinpoints of light with no discernable size. Why, then, did astronomers consistently measure the apparent diameters of stars to be several seconds, or even minutes, of arc depending on the instrument used? The lunar occultations observed by Crabtree and Horrocks, as well as similar occultations later observed by Halley, showed that the true apparent diameters of stars must be much less than 1", but why, then, did the diameters seem so much larger when the stars were viewed with the naked eye or through a telescope?

William Herschel investigated the apparent diameters of stars and found that they depended on the brightness of the stars but also on the optical system used to view the star. In general, brighter stars had larger apparent diameters, but increasing the aperture or magnification of the telescope reduced the apparent diameter as measured in arcseconds. In 1828, William's son John noted that at high magnification some stars appear to be surrounded by faint rings of light. He suggested that the apparent disks of stars, with their surrounding rings, might be explained by the wave theory of light that had been developed by Thomas Young in 1804 and extended by Augustin Fresnel.[34]

In 1835 George Airy wrote a detailed analysis of the formation of a star's image in a telescope based on this wave theory. He showed that the wave theory, specifically an effect now known as diffraction, could account for the observed features of these images. Airy found that the most important factor in determining the apparent size of a star's telescopic image was the aperture of the viewing instrument: telescopes with larger openings produced smaller star images. The next most important factor was brightness, with brighter stars appearing somewhat larger when viewed through the same telescope. What did not matter at all in determining the appearance of the stars was the star's true physical size.[35]

In fact, the first measurement of the true physical diameter of a star was not made until 1920, when Albert Michelson and Francis Pease determined the diameter of the star Betelgeuse (Alpha Orionis) using an interferometer, a device that uses the interference of light waves to carry out precise measurements. They found a true apparent diameter of $0''.047$, while they estimated that the star's annual parallax was about $0''.018$.[36] Those results indicated that Betelgeuse had a physical diameter of about 2.6 AU, larger than the diameter of Earth's orbit! Tycho Brahe had rejected the idea that stars lay at vast distances because in that case his measurements of their apparent diameters suggested that they were vastly larger than the Sun. Now we know that Tycho was fooled by diffraction, and stars really are outrageously far away – but some stars really are vastly larger than our Sun.

10.3.3   *Testing Newtonian physics*

Newton's *Principia* explained an incredible variety of known phenomena, but it also provided a framework for predicting new things. We have already seen that Newton successfully predicted the oblate shape of the Earth. Newton's physics also predicted that comets were, in principle, just like planets orbiting the Sun. The orbit of a comet could be a hyperbola or a parabola, but it could also be an ellipse like the orbits of the planets. Cometary orbits might be greatly tilted relative to the ecliptic plane, and comets could even orbit clockwise while the planets orbited counterclockwise, but even so it was possible to calculate the orbit of a comet using Newton's methods. If the comet's orbit was elliptical, it would someday return to the inner part of the solar system where it might become visible again.

Newton had applied this technique to the comet of 1680, but Edmund Halley extended this work to other comets and even improved on Newton's methods.

After computing the orbits of several comets Halley noticed that his results for the comets of 1531, 1607, and 1682 were all very similar. He thought these three comets might all be the same object, which orbited the Sun with a period of 75 or 76 years. The only problem was that the time between the cometary perihelions of 1531 and 1607 was slightly different from the time between the perihelions of 1607 and 1682. If these were all the same object, then something had caused the period of the comet to change. Halley suspected that the gravitational attraction of Jupiter and Saturn might be to blame.

As he continued his work Halley realized that a comet seen in 1456 also had a similar orbit. He did his best to determine the orbit of the 1682 comet and account for the alterations caused by Jupiter and Saturn, which were likely to be significant since the comet had passed relatively close to Jupiter in 1681. Halley made the bold prediction that this comet would return to become visible near the end of 1758.[37]

Halley died more than a decade before the predicted return of his comet, but other astronomers remembered his prediction and even improved upon it. The French astronomer Alexis Clairaut, assisted by Jérôme Lalande and Nicole-Reine Lepaute, recalculated the comet's orbit to account for the perturbing effects of Jupiter and Saturn over three full periods of motion. They predicted the comet would reach perihelion in April 1759, with an estimated error of one month.

As it turned out, the comet was first seen not by professional astronomers but by a German farmer (and amateur astronomer) George Palitzsch in December 1758. The comet reached perihelion in March 1759, about one month earlier than predicted by Clairaut. The return of "Halley's comet" was viewed as a triumph for Newtonian physics in spite of this small error.[38]

Clairaut speculated that some of the error in the predicted perihelion date for Halley's comet might be due to the gravitational pull of unknown objects beyond Saturn's orbit. This suggestion proved prophetic when, in 1781, William Herschel noticed what he thought was a comet while hunting the skies for double stars. Further observations showed that this object moved almost exactly along the ecliptic and lacked the tail that was characteristic of most comets. Herschel realized that he had discovered the first new planet in recorded history. He named it the "Georgian Star" after King George III, who then appointed Herschel to a permanent position as King's Astronomer, but the name of the planet was later changed to Uranus.[39] Kepler's vision of six planets encased within the five Platonic solids crumbled to dust: there was no sixth Platonic solid to place between Saturn and Uranus.

After Herschel announced the discovery of his new planet, astronomers worked to determine the exact orbit of Uranus from observations. They were able to show that Uranus had actually been seen before 1781, but it was thought to be just a small star. However, they could not seem to make the old observations and the new observations mesh into a single elliptical orbit. Furthermore, the variations in the orbit could not be explained by the perturbing effects of Jupiter and Saturn.

Two possibilities were considered: either Newton's law of universal gravitation acted differently at great distances, or there was some *other* planet out there perturbing the orbit of Uranus. Most astronomers put their faith in Newtonian gravitation and thought that there must be an eighth planet beyond Uranus.[40]

In principle, astronomers could use the differences between the observed motions of Uranus and those predicted by Newtonian physics (including the effects of the Sun, Jupiter, and Saturn) to determine the mass and location of the unknown planet. But in practice such a calculation required new mathematical techniques that had never been used before.

In England, John Couch Adams worked out the orbit of the unknown planet in 1843 by making use of some simplifying assumptions. He assumed that the new planet had a circular orbit and that it followed the Titius–Bode law, an arithmetical rule that seemed to fit the relative orbital sizes of the known planets.[i] Adams computed a more refined orbit, along with predicted future locations for the planet, in September 1845. He communicated his results to some prominent British astronomers but did not publish his work.[41]

Meanwhile, in France, Urbain Le Verrier was working on his own computation for the orbit of the unknown planet. He published a predicted position for the new planet in June 1846. Although he didn't know about Adams' prediction, both astronomers had predicted nearly the same location. Le Verrier published a more complete orbit and a more precise predicted location in September of that year.

Le Verrier sent his results to the astronomers at the Berlin Observatory and encouraged them to look for the new planet. There, Johann Galle and Heinrich d'Arrest scanned the skies near Le Verrier's predicted position and compared what they saw to available maps of the stars. They soon found a "star" that was not on any of their maps. The planet that came to be known as Neptune was discovered about $1°$ away from Le Verrier's predicted position.[42] Meanwhile, astronomers in Britain had started to look for the new planet using the predictions of both Adams and Le Verrier, but their method was different. They observed the same patch of sky on different dates, carefully charting the stars

---

[i] The Titius–Bode law states that, in units of tenths of an AU, the semi-major axes of the planets are given by $a = 4 + x$ where $x = 0, 3, 6, 12, 24, \ldots$ The value of x doubles each time after the first step. This rule fits well with the orbits for Mercury, Venus, Earth, Mars, Ceres, Jupiter, Saturn, and Uranus, with errors less than 5% in each case.

they saw in hopes of finding one that moved relative to the others. They actually saw Neptune more than once, but did not recognize that it was not just a star.[43] As with Halley's comet, a prediction first made in England was made even better in France and confirmed in Germany.

In hindsight, we now know that the orbits calculated by both Adams and Le Verrier were not very accurate. Both had assumed the correctness of the Titius–Bode law, but the orbit of Neptune does not fit that "law" very well so the sizes of their orbits were incorrect. However, the predictions of both men happened to coincide with the true position of Neptune around the time of its discovery. In any case, the discovery of Neptune near its predicted position was viewed as yet another vindication of Newtonian physics.

Following his success in predicting Neptune's location, Le Verrier undertook a detailed examination of the orbit of Mercury. As with the orbit of Uranus, he found that Newtonian gravity could not account for the detailed motions of the planet even if he included the effects of other planets such as Venus and Earth. Specifically, Le Verrier found that the position of Mercury's perihelion point precessed around the Sun at a rate of $565''$ per year, while the gravitational attractions of other planets could only account for $527''$ per year. He found that he could fix the problem by increasing his value for the mass of Venus, but that led to perturbations of Earth's orbit that were inconsistent with observations of the Sun.[44]

In 1859 Le Verrier suggested that the extra precession of Mercury's perihelion might be explained by an unknown planet orbiting very close to the Sun. What had worked to explain the odd motions of Uranus might also work for Mercury. But any planet that was large enough to perturb Mercury's orbit as needed would also be large enough to be seen, especially during a solar eclipse or transiting across the Sun's face. Le Verrier was well aware of this problem and suggested that instead of a single planet there might be a collection of smaller planetoids orbiting tightly around the Sun, reminiscent of Scheiner's proposed explanation of sunspots.

Not long after Le Verrier's suggestion was made public, an amateur astronomer by the name of Edmond Lescarbault observed the transit of an unknown object across the Sun.[i] Le Verrier was informed and he announced the discovery. The name "Vulcan" was proposed for the new planet. However, further observations could find no trace of Vulcan and calculations showed that Vulcan could not explain the perihelion precession of Mercury even if it was really there.[45]

---

[i] In fact, several observations of objects within Mercury's orbit were reported, but most turned out to be obvious errors or fabrications.

Astronomers proposed other solutions to the problem of Mercury's perihelion precession. Perhaps the Sun had a nonspherical shape that altered its gravitational force on Mercury? Perhaps dust scattered through the ecliptic plane could explain the extra precession? None of these proposals worked out. The most radical proposal was the one suggested by Asaph Hall in 1894. Hall proposed that gravitational forces might not conform exactly to Newton's inverse square law, but instead might fall off as the reciprocal of the distance raised to the power $2 + \delta$, where δ was some very small number. This slight modification of the inverse square law could explain the motion of Mercury, but its effect on Venus would be small enough that it might not have been noticed. The effect on planets more distant from the Sun would be negligible. However, this idea was later ruled out by Ernest Brown based on a study of the Moon's motion.[46]

The extra precession of Mercury's perihelion was finally explained by the general theory of relativity, a new theory of gravity proposed by Albert Einstein in 1915.[47] In most cases the predictions of general relativity are almost perfectly approximated by Newton's universal gravitation. However, there can be noticeable differences between the two for motions near very massive objects such as that of Mercury near the massive Sun. In the case of Uranus, deviations from the Newtonian predictions led to the discovery of a new planet and the vindication of Newtonian physics. The deviations of Mercury, on the other hand, led to the overthrow of Newtonian gravity and its replacement with an even better theory. But that's another story.

10.4   Reviewing the revolution

10.4.1   Why did it succeed?

The success of the Copernican Revolution was not just the result of brilliant insights by a handful of geniuses. While it is true that there were a few individuals who played critically important roles in advancing the heliocentric theory, the success of the whole enterprise depended upon the hard work, intelligence, and insight of numerous astronomers, mathematicians, and natural philosophers. But there were also a few attitudes and methods that helped the revolution to succeed.

One key to the success of the Copernican Revolution was the realist attitude of men such as Nicholas Copernicus, Tycho Brahe, Johannes Kepler, Galileo Galilei, and Isaac Newton. These men were convinced that they could discover at least a few facets of the true nature of the world. They were not content with generating empirical theories that just fit the data. They wanted theories that revealed the truth. In searching for truth they employed criteria that went well beyond empirical adequacy: explanatory power, harmony and beauty, simplicity and elegance, and consistency with other accepted theories. They also made use of assumptions about how the world works to guide their investigation. For example, both Kepler and Newton insisted on theories that involved physical interactions between material bodies.

Because these scientists used a variety of criteria to judge theories, they often found that a theory could be successful according to some criteria but poor according to others. They had to make choices about which criteria were most important. For example, many of these men were willing to accept contradictions with Aristotelian physics in order to embrace the explanatory power and harmony of the Copernican model. But they did not simply accept the contradictions with Aristotle. They also worked to resolve contradictions between scientific theories by modifying one theory or the other. From these efforts, driven by the desire to uncover the true nature of the world, these scientists created a unified physics and astronomy that could explain and predict an incredible variety of phenomena with a few simple, elegant laws.

Of course, these astronomers did insist that their theories had to match with past and future observations. However, they recognized that empirical tests of theories are only as good as the data used to carry out the test. In some cases, differences between theory and observation could be attributed to errors in the observations. An approximate theory that matched low-precision observations might be good enough for some people, but for a scientist trying to uncover the truth about Nature such a theory was inadequate. The desire to know the truth led to a push for better instruments and more precise observations that could indicate whether or not a theory was really correct. Sometimes these new data increased confidence in an accepted theory, but in some cases they revealed flaws in the theory that led astronomers to improve the theory.

In order to provide the most stringent possible test of a theory, both the theory and the observations used to test it had to be quantitative. Mathematics thus played a crucial role in the development of astronomy. A mathematical theory that supplied detailed, quantitative predictions that were found to agree with observations was worthy of much greater confidence than a qualitative theory that could only explain general characteristics of natural phenomena. When a mathematical rule is shown to fit with observations in great detail we are inclined to accept the theory as "real" even if the mechanism responsible for the mathematical rule can't be specified (as was the case for Newtonian gravity).

One final characteristic that helped the Copernican Revolution succeed, and which is really critical to any progress in science, was the open but skeptical minds of the scientists involved. Progress in science can only occur if scientists are willing to explore new ideas, develop new models, and design new technologies in an effort to better match the data, better explain phenomena, and better fit with other theories. Scientific progress relies on the willingness of scientists to set aside established theories to look for something better. At the same time, though, scientists cannot accept new innovations too easily. New ideas, models, and instruments must be subjected to scrutiny or else we may be too easily led to abandon a good theory to embrace a fantasy.

In practice, the perfect balance of innovation and skepticism is hard to maintain in any single individual. Some scientists may be particularly good at generating new ideas, but may also tend to get carried away by their own ideas without exercising proper restraint. Others may be excessively conservative, refusing to abandon an established theory or use a new type of instrument because they are overly skeptical of any innovation. All types of individuals can make important contributions to science, though. Science progresses when the *community* of scientists maintains a balance of innovation and skepticism. That balance keeps the scientific community searching for new and better answers without too easily throwing away what has worked in the past.

10.4.2   Was it really a revolution?

One could argue that the Copernican Revolution was neither Copernican, nor a revolution. It is certainly true that Copernicus was not solely responsible for the change from a geocentric to heliocentric worldview. He initiated that change, but he did not see it through to its end. However, his work convinced at least a few astronomers that the advantages of a heliocentric model were worth considering in spite of the many problems that such a model created. The work of these astronomers to resolve those problems led us to a fundamentally new understanding of the world and our place within it.

But was the Copernican Revolution really a revolution? There is, of course, the double association of the word revolution with the work of Copernicus. His book "on the revolutions" not only described the turning of the celestial spheres, it also initiated a dramatic change in astronomy and cosmology. In fact, one might think that the word "revolution" came to mean a radical change *because* of its association with Copernicus' book, but that is not so. The word was associated with dramatic change as early as 1400, more than a century before Copernicus' work was published.[i]

Now we think of a revolution, in a social or political context, as a sudden and radical overthrow of an existing system. In that sense, the word "revolution" might not seem applicable to the change from a geocentric worldview to

---

[i] For example, Chaucer uses the word in this way (in Middle English) in his *Romaunt of the Rose*, probably written in the late 14th century.

a heliocentric one. The change was certainly not sudden. Copernicus published his great work in 1543. It is much harder to date when the heliocentric system became widely accepted, but a reasonable choice might be with the publication of Descartes' *Principia philosophiae* in 1644. The "revolution" was not complete until Newton published his *Principia* in 1687 and Newtonian physics became generally accepted about 50 years later. One can argue that the Copernican "revolution" took over 200 years from start to finish. Such a slow change might not seem very revolutionary.

On the other hand, the advent of heliocentrism does seem revolutionary. The change from the Aristotelian–Ptolemaic worldview to the Copernican–Newtonian one was a genuinely radical change. It included not only a fundamental change in astronomy and physics, but also a crucial change in humanity's understanding of its place in the cosmos. The heliocentric cosmos held new possibilities, such as the possibility that other planets harbored living creatures similar to those found on the Earth. At the same time it created new mysteries, such as whether the universe was finite or infinite.

If we compare the pre-Copernican worldview to the post-Newtonian one, the change seems radical, but on a finer scale we can see that the "revolution" consisted of many smaller changes. After all, Newtonian physics could explain why the Ptolemaic model served as a good approximation for describing the observed motions of the planets, Sun, and Moon. The universe of Newton could be viewed as the outcome of a process of refining and adjusting the theories of the ancient Greeks. There is a continuity to the story of the Copernican Revolution, and even those who opposed the new heliocentric theory had important roles to play in that story. In the end, perhaps it doesn't matter very much whether this important part of human history should be called the "Copernican Revolution" or not. What is important is the story itself and the human desire to understand the world that carried the story forward.

10.5   Reflections on science: why did it take so long?

Why did it take two centuries for the Copernican Revolution to be completed? Attempting to answer this question gives us an opportunity to reflect on the way scientific theories are judged and either accepted or discarded.

The process of replacing an existing theory with another theory is rarely sudden and dramatic. This process is usually slow and incremental. There are very good reasons why this should be so. A theory that has gained widespread acceptance must serve an important purpose (or several purposes). Such a useful theory is not likely to be quickly discarded, even if it is found to be flawed. After all, a flawed theory is better than no theory at all. In order for an old theory to be discarded, there must be something better to take its place.

Was Copernicus' theory better than the theories of the ancient Greeks? Not obviously. It is often the case that new scientific theories have flaws of their own. It is not until these theories are refined and improved that they become worthy of replacing an older, successful theory.

Another important factor that slows scientific change is that scientific theories are not judged in isolation. Theories aren't just tested against data, they are also judged in relation to other theories. Scientific theories are interconnected, so replacing a theory in one area often creates conflict with theories in other areas of science. The Copernican model conflicted with Aristotelian physics, and it was only after Galileo and Newton offered a better physics that the Copernican theory could gain widespread acceptance.

Furthermore, the data that we want our theories to explain and predict are not static. New observations are made, and new instruments are developed that give us access to whole new realms of data. Our scientific theories are constantly tested anew as more data become available. Sometimes these new observations are made, and new instruments are developed, to answer specific questions that arise from our scientific theories. Copernicus tested his theory against data that were essentially the same as available to Ptolemy, but Galileo turned his telescope to the skies in hopes of finding support for his Copernican ideas, and as a result he introduced entirely new phenomena into astronomy.

Scientific change is also slowed because scientists are imperfect. Science is a human endeavor and human beings are fallible. Sometimes they make mistakes, sometimes they use bad judgement, sometimes they are guided by motives that are less than pure, and sometimes they are genuinely doing the best they can but they are led astray because they don't have the knowledge to properly interpret their observations. Even so, it is the curiosity of individual people, their desire to know the truth about the world, that drives science onward and leads to genuine progress in our knowledge.

Because science consists not of theories in isolation, but of a connected network of knowledge, every new advance generates new questions. Errors, whether in measurement, mathematics, logic, or interpretation, may lead to erroneous theories, but eventually these theories will come into conflict with other areas of knowledge. When these problems arise, scientists work to resolve them and often (though not always) find the original errors and correct them or discard them. It takes time for all of the moving parts of this interconnected network of knowledge to come into alignment, and even then the alignment is never perfect. Errors and unanswered questions always remain. Science, though, does progress. We really do know more about the workings of Nature today than was known in the 2nd century BC, thanks to the tremendous effort of a large number of people over a long period of time. The progress of science may be a bit like the apparent motion of the planets – highly irregular and occasionally retrograde – but over sufficiently long times science does advance and improve.

Toward what, then, is science progressing? Is it getting closer to the TRUTH, in some absolute sense? We can't really know. The methodology of science is not like that of mathematics or syllogistic logic. We cannot achieve absolute certainty in our knowledge of the world. The results of science are always tentative and approximate, but our scientific knowledge has evolved and adapted over the years to better fit with our experience, even as our experience has been extended into regions undreamed of in the past: the subatomic world, the realm of the galaxies, etc.

These new realms of experience can lead to fundamentally new scientific theories but those new theories must still fit with those parts of the old theories that worked well. In that sense we can trust much (most?) of the scientific knowledge we have today. The science of the future may be quite different from what we know today, but it will have to conform with, and even explain, the incredible successes of our current theories.

The more mature and well-tested a theory, the less likely it is to be radically overturned in the way that Ptolemaic astronomy was. We can be confident that many of today's best scientific theories are at least good approximations to the even better theories that will come tomorrow. We now know that the Earth rotates on its axis and orbits the Sun ... and that knowledge will never be overturned by some future theory.

Appendix

Mathematical details

A.1   Angular measure

The measurement of angles is very important to both ancient and modern astronomy. Before we look at the use of angles in astronomy, however, it is important to be clear about what exactly angles are and to introduce some terminology.

An **angle** is a figure formed by two line segments diverging from a common point. Two angles are shown in Figure A.1. On the left, line segments $\overline{vu}$ and $\overline{vw}$ converge at the point v; this point is called the **vertex** of the angle. On the right, line segments $\overline{yx}$ and $\overline{yz}$ converge at the vertex y. The angles themselves may be denoted by ∠uvw and ∠xyz; these are read as "angle u-v-w" and "angle x-y-z." In this nomenclature the vertex is always placed between the endpoints.

All angles are basically alike, but some diverge more dramatically than others. In Figure A.1, ∠xyz is "more spread out" than ∠uvw. We would like to somehow quantify this "spread-outness," and here is where the measurement of angles comes into play. We decide, arbitrarily, that if two line segments are perpendicular, then they are separated by **90 degrees of arc**. A single degree of arc is therefore 1/90 of this so-called **right angle**. If two line segments go out in opposite directions from the vertex, then those segments are separated by 180 degrees. Degrees are indicated by the symbol °.

Just as distances between points are rarely equal to an exact number of meters, angles between lines are rarely equal to an exact number of degrees. In the case of distances, we break up meters into 100 centimeters and each of these into 10 millimeters, etc. In the case of angles, we play a similar game. Each degree is broken down into 60 equal (very small) angles; each of these angles is said to have a value of a single **minute of arc**. See Figure A.2. Further, each minute of

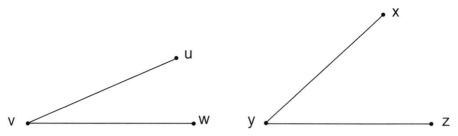

Figure A.1 Two angles. Angle uvw is smaller than angle xyz.

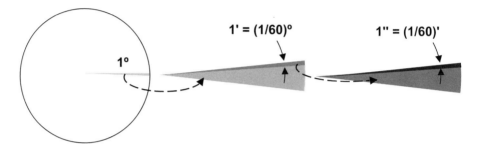

Figure A.2 A single degree of arc may be broken down into 60 minutes of arc; each minute of arc may be further broken down into 60 seconds of arc.

arc is comprised of 60 (tiny) angles; each of these has a value of a single **second of arc**. Therefore there are 3600 seconds of arc in one degree. Minutes of arc are indicated by a single prime (′) and seconds of arc are indicated by a double prime (″). We may then say that an angle has a value of so many degrees, so many minutes, and so many seconds. As an example, $12° \ 34' \ 56''$ indicates an angle of a little more than $12°.5$ (since $0°.5 = 30'$) and very slightly less than $12° \ 35'$ (since $60'' = 1'$).

Why is this important for astronomy? The stars, Sun, Moon, and planets may be thought of as images projected onto the inside surface of an enormous hollow sphere centered on the Earth. This so-called **celestial sphere** is discussed in detail in Chapter 2. One question that comes up again and again is: How far apart do these objects appear to be? Since we don't know the "distance" to the celestial sphere, the only way to answer these questions is by using angular measurements.

As a simple example, consider Figure A.3. The stars Altair and Deneb are shown on the celestial sphere. Imagine a line between your head and Altair, and another between your head and Deneb. Your head is the vertex and the angle between these two lines is $39° \ 13' \ 12''$. Now, it is possible to play this same game with your head and *any* two points on the celestial sphere, whether or not there's anything "there." For example, imagine a point on one side of the Moon's disk,

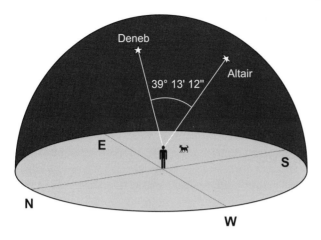

Figure A.3 Angular measurements can be made between any two points on the celestial sphere. Your head is always the vertex of the angle.

and the point on the other side of the Moon's disk. Since these are two different points, the angle between them is not zero. The angle between these points is, on the average, about $31'\ 6''$. (Interestingly, the angular size of the Sun is very close to this number as well.) Angular measurements can be used to define coordinates on the sky, or on the celestial sphere, as discussed in Chapter 2.

A.2   Apparent diameter

As is familiar to everyone, the farther away an object is the smaller it appears. The apparent size of an object can be characterized by its *apparent diameter*, or *angular diameter*. The apparent diameter of an object is the angle between the line of sight to one side of the object and the line of sight to the opposite side of the object. For simplicity we will only consider the apparent diameter of spherical objects. We will explore the relationship between the sphere's physical diameter (the actual diameter of the sphere in distance units), the distance of the sphere from an observer, and the apparent diameter of the sphere as seen by that observer.

Figure A.4 illustrates the geometry involved in determining the apparent diameter of a sphere. The observer is at O and three spheres are shown, with diameters D_1, D_2, and D_3. Note that spheres 1 and 2 lie at the same distance from the observer ($R_2 = R_1$), but because sphere 2 is larger ($D_2 > D_1$) it has a larger apparent diameter ($\phi_2 > \phi_1$). On the other hand, spheres 1 and 3 have the same apparent diameter ($\phi_3 = \phi_1$), but sphere 3 is farther away ($R_3 > R_1$) and thus has a correspondingly larger physical diameter ($D_3 > D_1$).

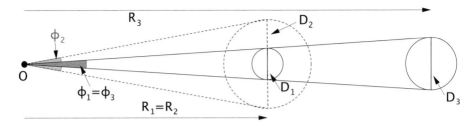

Figure A.4 The apparent diameter of a sphere. The observer is at O. The figure illustrates the apparent diameter (ϕ) of three different spheres with physical diameters indicated by D at distances indicated by R.

We can derive an approximate equation relating physical diameter (D), distance (R), and apparent diameter (ϕ) by treating the diameter of the sphere as though it is a small section, or arc, of a circle centered on O. The actual circles are indicated by dotted lines in Figure A.4. The length of a circular arc is proportional to the angle as measured from the center of the circle. We know that the circumference of the circle is $2\pi R$, and that the full circumference occupies an angle of 360°. Because the length and angle are proportional we must have

$$\frac{\phi}{D} = \frac{360°}{2\pi R}. \tag{A.1}$$

We can solve Equation A.1 for ϕ to get

$$\phi = \frac{180°D}{\pi R}. \tag{A.2}$$

For example, a sphere with a physical diameter of 10 cm at a distance of 20 m will have an apparent diameter of

$$\phi = \frac{180°(0.1 \text{ m})}{\pi (20 \text{ m})} = 0°.286, \tag{A.3}$$

or about 17′.

Alternately, we could solve Equation A.1 for D to get

$$D = \frac{\pi R\phi}{180°}. \tag{A.4}$$

We know the Moon has an apparent diameter of about 0°.5. Hipparchus found that the Moon lies about 240,000 miles away (see Section 3.3.2). From this information we find that the physical diameter of the Moon must be

$$D = \frac{\pi(240,000 \text{ mi})(0°.5)}{180°} = 2094 \text{ mi}, \qquad (A.5)$$

or roughly one-fourth the diameter of Earth as determined by Eratosthenes.

A.3 Trigonometry

The triangle shown in Figure A.5 has interior angles a, b, and c; the last is a right angle. This figure is therefore called a **right triangle**. Such triangles are of great importance to astronomy. The sides of the triangle are labeled A, B, and C; these labels are also used to indicate the lengths of the sides. Notice that a opens toward A, b opens toward B, and c opens toward C. We then say that A is the side *opposite* to a; similarly for the other sides and angles. The side opposite to the right angle is always the longest and is called the **hypotenuse**. The two shorter sides are called **legs**. We say that A is the leg *adjacent* to b and that B is the leg adjacent to a.

Pythagoras of Samos (6th century BC) found a simple relationship among the lengths of the sides of a right triangle. Referring to Figure A.5, he found that

$$A^2 + B^2 = C^2. \qquad (A.6)$$

This relationship, called the **Pythagorean theorem**, is true for all right triangles regardless of the size of the triangle or the values of the angles a and b. In the triangle above, let the length of A be 3 units and the length of B be 1.75 units. We can find C from the Pythagorean theorem:

$$C^2 = 3^2 + 1.75^2 = 12.0625, \qquad (A.7)$$

and

$$C = \sqrt{12.0625} \approx 3.47 \text{ units.} \qquad (A.8)$$

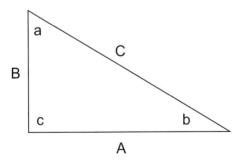

Figure A.5 A right triangle with right angle c and hypotenuse C.

Also of great importance are the three principal trigonometric functions **sine**, **cosine**, and **tangent**. These are functions of angles, which means that when one computes the sine, cosine, or tangent, the "input" is an angle. The functions themselves should be thought of as ratios of lengths of sides. Definitions and examples from Figure A.5 follow.

- The **sine** of an angle is the ratio of the length of the opposite leg to that of the hypotenuse. Therefore

$$\sin a = \frac{A}{C} = \frac{3}{3.47} \approx 0.864. \tag{A.9}$$

- The **cosine** of an angle is the ratio of the length of the adjacent leg to that of the hypotenuse. Therefore

$$\cos a = \frac{B}{C} = \frac{1.75}{3.47} \approx 0.504. \tag{A.10}$$

- The **tangent** of an angle is the ratio of the length of the opposite leg to that of the adjacent leg. Therefore

$$\tan a = \frac{A}{B} = \frac{3}{1.75} \approx 1.714. \tag{A.11}$$

The same game can be played for angle b. In these examples, the angles are the "input" and the ratios A/C (the sine), B/C (the cosine), and A/B (the tangent) are the "output." Can we also determine the angles themselves, given these ratios? Yes, using the **inverse trigonometric functions**. The only one we will need to use is the **inverse tangent**, denoted $\tan^{-1}$. For the inverse tangent function, the ratios are the "input" and the angles are the "output." Therefore, if the tangent is to be thought of as a ratio, the inverse tangent is to be thought of as an angle. The inverse tangent function can be calculated using any scientific calculator. Thus, the inverse tangent of a ratio (number) is the angle whose tangent is equal to that ratio (number). In the case of angle b, the input to the inverse tangent function would be the ratio $B/A = 1.75/3 \approx 0.5833$:

$$b = \tan^{-1}(0.5833) \approx 30°.26. \tag{A.12}$$

So we find that the angle b is 30°.26.

Now that we know the value of b we can determine the angle a using the fact that the sum of the interior angles for any triangle (not just right triangles) is equal to $180°$. That is, $a + b + c = 180°$. Therefore, since we know that $b = 30°.26$ and $c = 90°$, we can write

$$a = 180° - 30°.26 - 90° = 59°.74. \tag{A.13}$$

Note that $\sin a = \sin(59°.74) \approx 0.864$, in agreement with our result above. (The values of the trigonometric functions for any angle can be found using any modern scientific calculator.)

A.4   Finding the Sun's altitude from shadows

Using the definition of the tangent function given in Appendix A.3 and the diagram in Figure 2.5 we see that

$$\tan\theta = \frac{h}{\ell}, \tag{A.14}$$

where θ is the Sun's altitude, h is the height of the gnomon, and ℓ is the length of the gnomon's shadow. We can determine the length of a gnomon's shadow from its height and the Sun's altitude by solving Equation A.14 for ℓ:

$$\ell = \frac{h}{\tan\theta}. \tag{A.15}$$

If the Sun's altitude is $32°$, then a 1.2-meter gnomon will cast a shadow that is $(1.2 \text{ m})/\tan(32°) \approx 1.92$ m long. Note that $\tan(32°) \approx 0.625$ can be evaluated on any scientific calculator.

We can also rewrite Equation A.14 using the inverse tangent function in order to solve for the Sun's altitude:

$$\theta = \tan^{-1}\left(\frac{h}{\ell}\right). \tag{A.16}$$

So if a 1.2-meter gnomon casts a 0.5-meter long shadow, then the Sun's altitude is $\theta = \tan^{-1}(1.2/0.5) \approx 67°.4$. Again, the inverse tangent function can be evaluated on most scientific calculators.

A.5   Relative distances of Sun and Moon

Figure 3.3 shows the geometry of Aristarchus' method for determining the relative distances to the Sun and Moon.[1] The Moon is exactly at quarter phase as seen from Earth, so the angle at the Moon is a right angle. The angle θ is the angular separation of the centers of the Sun and Moon as seen from Earth. From trigonometry (see Appendix A.3) we know that

$$\cos \theta = \frac{d_{Moon}}{d_{Sun}}. \tag{A.17}$$

Taking the reciprocal[i] of both sides we find that

$$\frac{d_{Sun}}{d_{Moon}} = \frac{1}{\cos \theta}. \tag{A.18}$$

As discussed in Section 3.3.1, Aristarchus estimated $\theta = 87°$. Substituting this value into Equation A.18 we find $d_{Sun}/d_{Moon} = 1/\cos(87°) \approx 19.1$. The correct angle is $\theta = 89°\ 51' = 89°.85$, which gives $d_{Sun}/d_{Moon} = 1/\cos(89°.85) \approx 382$. So Aristarchus believed the Sun to be about 19 times as distant as the Moon, when in fact it is more than 380 times as distant.

A.6   Parallax and distance

The geometry of a parallax measurement is shown in Figure 3.4. Note that the point O_1, the center of the baseline, and the center of the Moon form a right triangle with the right angle at the center of the baseline. The angle at the Moon is θ, which we will refer to as the *parallax angle*. The parallax angle is half of the apparent shift of the Moon as we switch from location O_1 to O_2. Note that the distance d is measured from the center of the baseline to the center of the Moon (or whatever object we are observing). However, if the distance from the center of Earth to the center of our baseline is small compared to d, then d will closely approximate the distance from the center of Earth to the center of the Moon (or other astronomical object).

From trigonometry (see Appendix A.3) we know that

$$\tan \theta = \frac{b/2}{d} = \frac{b}{2d}. \tag{A.19}$$

If θ is a small angle (say, less than $1°$, which is the case for all astronomical parallax measurements), then

$$\tan \theta \approx \frac{\pi \theta}{180°}, \tag{A.20}$$

where θ is measured in degrees. Substituting Equation A.20 into Equation A.19 and solving for θ we find

$$\theta = \frac{90° \, b}{\pi \, d}. \tag{A.21}$$

---

[i] The reciprocal of a quantity is just one divided by that quantity.

The Moon is about 240,000 miles from Earth. If we were to observe the Moon from opposite sides of Earth (difficult to do in practice, since the Moon would be on the horizon for both observers!) we should find a parallax angle of

$$\theta = \frac{90°(8000 \text{ mi})}{\pi(240,000 \text{ mi})} = 0°.95. \tag{A.22}$$

Note that we have used 8000 miles for the diameter of Earth, based on Eratosthenes measurement (Section 2.3.2). The Moon's parallax is slightly less than $1°$, which means it shifts by about $2°$ (four times the Moon's angular diameter).

In practice, we usually measure a parallax angle and use it to determine the distance to the object. We can solve Equation A.21 for d to find

$$d = \frac{90° \ b}{\pi \ \theta}, \tag{A.23}$$

where θ is measured in degrees. We can use this equation to find the distance to the Sun. The parallax of the Sun, as measured from opposite sides of Earth, is about $9'' = 0°.00246$. So the distance to the Sun is

$$d = \frac{90° \ (8000 \text{ mi})}{\pi \ (0°.00246)} \approx 93,000,000 \text{ mi}. \tag{A.24}$$

Note that a parallax angle of $9''$ is far too small to measure with the naked eye. Such a small parallax angle is nearly impossible to observe for the Sun, which itself has an angular diameter of about $30' = 1800''$. The Sun's parallax was not accurately determined until the 18th century, and then only indirectly through measurements of the parallax of Venus during a transit (see Section 10.3.1).

A.7   Ptolemy: size of an inferior planet's epicycle

In Figure 4.4 there is a right triangle formed by the points e, f, and v. The side opposite the angle ϵ_{max} has length r_E and the side opposite the right angle (the hypotenuse) has length r_D. From the defintion of the sine function given in Section A.3 we know that

$$\frac{r_E}{r_D} = \sin \epsilon_{max}. \tag{A.25}$$

If we know the maximum elongation of the inferior planet, ϵ_{max}, then we can calculate the ratio r_E/r_D for that planet.

Using the greatest maximum elongation for Mercury from Table 3.1 we find $r_E/r_D = \sin(28°) \approx 0.47$. Using the greatest maximum elongation for Venus we find $r_E/r_D = \sin(47°) \approx 0.73$.

A.8   Ptolemy: size of a superior planet's epicycle

Figure 4.5 shows the geometry for opposition (left) and quadrature (right). In the diagram for quadrature we know that the angle ∠sem is a right angle because the planet is in quadrature, 90° from the Sun as seen from Earth. Likewise, we know that the angle ∠emf is a right angle because the line m̄f is parallel to the line s̄ē due to the synchronization of the superior planet's epicycle motion with the motion of the Sun.

Therefore, the triangle formed by e, f, and m is a right triangle. The side opposite the angle θ has length r_E and the side opposite the right angle (the hypotenuse) has length r_D. From the defintion of the sine function given in Section A.3 we know that

$$\frac{r_E}{r_D} = \sin\theta. \tag{A.26}$$

Unfortunately, we cannot directly measure θ because there is no visible object at f.

We can, however, calculate θ if we know the time from opposition to quadrature (T_q). At opposition (Figure 4.5, left) the angle ∠sem is 180°. Recall that the point f moves uniformly counterclockwise around the deferent, completing one full circle in the planet's tropical period (T_t). In the time T_q the point f moves through an angle $360° \, T_q/T_t$. In the same time, the Sun moves counterclockwise around its orbit through an angle $360° \, T_q/T_{ty}$, where $T_{ty} = 365.2422$ days is the period of the Sun's orbit, or one tropical year.

Putting these three angles together we can see that the angle ∠sef in the right diagram of Figure 4.5 must be equal to $180° + 360° \, T_q/T_t - 360° \, T_q/T_{ty}$. However, that angle is also equal to $90° + \theta$ as is clear from the diagram. Setting these two expressions equal to each other and solving for θ we find

$$\theta = 90° + \frac{360° \, T_q}{T_t} - \frac{360° \, T_q}{T_{ty}}. \tag{A.27}$$

Using the values of T_q and T_t for Mars from Table 3.1, as well as the length of the tropical year, we find that for Mars

$$\theta = 90° + \frac{360°\,(106 \text{ days})}{687 \text{ days}} - \frac{360°\,(106 \text{ days})}{365.2422 \text{ days}} \approx 41°.1. \tag{A.28}$$

So for Mars, $r_E/r_D = \sin(41°.1) \approx 0.657$. Following the same procedure for Jupiter we find $\theta \approx 11°.0$ and $r_E/r_D = \sin(11°) \approx 0.191$. For Saturn, $\theta \approx 7°.25$ and $r_E/r_D \approx 0.126$.

A.9   Copernicus: the orbital period of an inferior planet

In this section we will derive the formula for the orbital period of an inferior planet (T_{IP}) in the simplified Copernican system. The formula uses the Earth's orbital period, T_E, as well as the inferior planet's synodic period, T_s. Referring to Figure 5.8 we see that the Earth has gone through one full orbit (360°) plus an angle θ in the time T_s. The Earth goes through 360° in a time T_E. Since Earth moves at a uniform speed in the simplified Copernican theory, we know that the angles should be proportional to the times, so

$$\frac{360° + \theta}{T_s} = \frac{360°}{T_E}. \tag{A.29}$$

From Figure 5.8 we also know that the planet moves through exactly one full orbit (360°) more than Earth moved, so $720° + \theta$, in a time T_s. It moves through 360° in a time T_{IP}. Since the planet moves at a uniform speed in the simplified Copernican theory, we know that these angles are proportional to the times:

$$\frac{720° + \theta}{T_s} = \frac{360°}{T_{IP}}. \tag{A.30}$$

We can rewrite Equation A.30 as

$$\frac{360°}{T_s} + \frac{360° + \theta}{T_s} = \frac{360°}{T_{IP}}. \tag{A.31}$$

Now we can use Equation A.29 to substitute for $(360° + \theta)/T_s$ in Equation A.31 to get

$$\frac{360°}{T_s} + \frac{360°}{T_E} = \frac{360°}{T_{IP}}. \tag{A.32}$$

Finally, we can cancel the common factor of 360 degrees that appears in all three terms of Equation A.32 and rearrange to get our final result:

$$\frac{1}{T_{IP}} = \frac{1}{T_E} + \frac{1}{T_s} \rightarrow T_{IP} = \left(\frac{1}{T_E} + \frac{1}{T_s}\right)^{-1}. \tag{A.33}$$

Thus, if we know the planet's synodic period (T_s) and the Earth's orbital period (T_E) we can calculate the planet's orbital period (T_{IP}).

Note that Equation A.33 works even if the planet (and Earth) complete a different number of total orbits than is shown in Figure 5.8. Substituting the synodic period of Venus ($T_s \approx 584$ days) and the Earth's orbital period ($T_E \approx 365$ days) we find that the period of Venus is $T_{IP} \approx 225$ days. Likewise for Mercury ($T_s \approx 116$ days) we find a period of 88 days.

A.10   Copernicus: the orbital period of a superior planet

In this section we will derive the formula for the orbital period of a superior planet (T_{SP}) in the simplified Copernican system, using a procedure very similar to the one we used for an inferior planet. Again, the formula uses the Earth's orbital period, T_E, as well as the superior planet's synodic period, T_s. Referring to Figure 5.9 we see that the Earth has gone through an angle $720° + \theta$ in the time T_s. The Earth goes through $360°$ in a time T_E. Since Earth moves at a uniform speed in the simplified Copernican theory, we know that the angles should be proportional to the times, so

$$\frac{720° + \theta}{T_s} = \frac{360°}{T_E}. \tag{A.34}$$

From Figure 5.9 we also know that the planet moves through an angle $360° + \theta$ in a time T_s. It moves through $360°$ in a time T_{SP}. Since the planet moves at a uniform speed in the simplified Copernican theory, we know that these angles are proportional to the times:

$$\frac{360° + \theta}{T_s} = \frac{360°}{T_{SP}}. \tag{A.35}$$

We can rewrite the left side of Equation A.34 as

$$\frac{360°}{T_s} + \frac{360° + \theta}{T_s} = \frac{360°}{T_E}. \tag{A.36}$$

Now we can use Equation A.35 to substitute for $(360° + \theta)/T_s$ in Equation A.36 to get

$$\frac{360°}{T_s} + \frac{360°}{T_{SP}} = \frac{360°}{T_E}. \tag{A.37}$$

Finally, we can cancel the common factor of 360 degrees that appears in all three terms of Equation A.37 and rearrange to get our final result:

$$\frac{1}{T_{SP}} = \frac{1}{T_E} - \frac{1}{T_s} \rightarrow T_{SP} = \left(\frac{1}{T_E} - \frac{1}{T_s}\right)^{-1}. \tag{A.38}$$

Thus, if we know the planet's synodic period (T_s) and the Earth's orbital period (T_E) we can calculate the planet's orbital period (T_{SP}). Note that the only difference between the equation for inferior planets (A.33) and the equation for superior planets (A.38) is the sign of the final term. Equation A.38 will work even if the number of orbits completed by the Earth and the superior planet differs from what is shown in Figure 5.9.

Substituting the synodic period of Mars ($T_s \approx 780$ days) and the Earth's orbital period ($T_E \approx 365$ days) we find that the orbital period of Mars is $T_{SP} \approx 686$ days. Likewise, for Jupiter ($T_s \approx 399$ days) the orbital period is about 4300 days and for Saturn ($T_s \approx 378$ days) it is about 10,600 days.

A.11   Copernicus: the size of an inferior planet's orbit

The procedure for finding the size of an inferior planet's orbit in the simplified Copernican system is illustrated in Figure 5.10, where the angle ϵ_{max} is the planet's maximum elongation (the maximum angle between the planet and the Sun as seen from Earth). Once ϵ_{max} is measured, we can use trigonometry to find the Sun–planet distance in terms of the Sun–Earth distance. From trigonometry we know that the sine of an angle in a right triangle is equal to the ratio of the opposite side to the hypotenuse, so in Figure 5.10 the ratio r_p/r_e is equal to $\sin \epsilon_{max}$. For Venus, which has a maximum elongation $\epsilon_{max} = 47°$, the ratio is

$$\frac{r_p}{r_e} = \sin 47° \approx 0.73, \tag{A.39}$$

and for Mercury, with $\epsilon_{max} = 28°$ the ratio is

$$\frac{r_p}{r_e} = \sin 28° \approx 0.47. \tag{A.40}$$

What do these numbers tell us? They tell us that the Sun–Mercury distance is 47% of the Sun–Earth distance, and that the Sun–Venus distance is 73% of the Sun–Earth distance. To put this another way, the radius of Mercury's orbit is 0.47 Astronomical Units, where an Astronomical Unit (AU) is defined as the average Earth–Sun distance. The radius of Venus' orbit is 0.73 AU.

Note that we are assuming perfectly circular orbits, centered on the Sun. Copernicus didn't actually assume this. He placed the orbits of the planets somewhat off-center from the Sun, and also added small epicycles and other adjustments. As a result, the planets were not at fixed distances from the Sun. Since the distances calculated above made use of the *maximum* elongation, they actually correspond to the *maximum* distance of the planet from the Sun. For Venus this turns about to be very close to its average distance, but that is not the case for Mercury.

A.12   Copernicus: the size of a superior planet's orbit

To determine the size of a superior planet's orbit in the simplified Copernican system we must use our value for the period T_{SP} of the planet in question.

Appendix A.10 shows how to calculate this period from observational data. The geometry for determining the size of the planet's orbit is illustrated in Figure 5.11. We begin with the planet at opposition (at p_1 in the figure). We now wait not for the following opposition, but for that point in time when the planet and the Sun are 90° from one another as seen from Earth, known as quadrature. In Figure 5.11 the planet is at quadrature when the Earth is at e_2 and the planet is at p_2. The time that it takes for the planet to move from opposition to quadrature is denoted T_q. During this time the Earth has moved through the angle $\theta + \alpha$, while the planet has moved only through the angle θ. We want to know the fraction r_p/r_e, that ratio of the planet's orbital radius to Earth's orbital radius (1 AU). So how do we find this? From trigonometry we know that the cosine of an angle in a right triangle is the ratio of the adjacent side to the hypotenuse, so from Figure 5.11 we see that r_e/r_p, which is the reciprocal of the ratio we want, is equal to $\cos \alpha$. If we could determine the angle α we could figure out the ratio we want.

How do we find α? It is not an angle that we can measure directly. It is the angle between Earth and the superior planet as seen from the Sun – but we aren't on the Sun (thankfully). However, we can still figure out α. Let's start by looking at what we know about the superior planet's motion. We know that the planet goes through 360° in time T_{SP}, its orbital period. We also know that the planet moves through the angle θ in the time T_q. Since the planet moves at a uniform speed along this simplified Copernican orbit, we know that the angles should be proportional to the corresponding times:

$$\frac{\theta}{T_q} = \frac{360°}{T_{SP}}. \tag{A.41}$$

That doesn't tell us what α is, so we still have more work to do. Now we can use what we know about the Earth's motion. We know that the Earth passes through the angle $\theta + \alpha$ in the time T_q, and we know it passes through 360° in the time T_E, where $T_E = 365$ days is Earth's orbital period. Because the Earth moves at a uniform speed, the angles must be proportional to the times:

$$\frac{\theta + \alpha}{T_q} = \frac{360°}{T_E}. \tag{A.42}$$

We can expand the left side of this equation to get

$$\frac{\theta}{T_q} + \frac{\alpha}{T_q} = \frac{360°}{T_E}. \tag{A.43}$$

Now we can use Equation A.41 to replace θ/T_q in Equation A.43. The result is

$$\frac{360°}{T_{SP}} + \frac{\alpha}{T_q} = \frac{360°}{T_E}. \tag{A.44}$$

We can rearrange this equation to give

$$\frac{\alpha}{T_q} = \frac{360°}{T_E} - \frac{360°}{T_{SP}}. \tag{A.45}$$

Finally, we find α by factoring out the common factor of $360°$ on the right side of Equation A.45 and then multiplying both sides of the equation by T_q to get

$$\alpha = 360°T_q \left(\frac{1}{T_E} - \frac{1}{T_{SP}} \right). \tag{A.46}$$

Once we have used Equation A.46 to find α, we can calculate the reciprocal of the ratio we seek: $r_e/r_p = \cos\alpha$. To get the ratio we really want, we take the reciprocal of both sides: $r_p/r_e = 1/\cos\alpha$. It is important to note that in using Equation A.46 to find α we must use the same time units for T_E, T_{SP}, and T_q.

Now that we see how to do the calculation in principle, let's look at the specific case of Mars. The time T_q between opposition and quadrature for Mars is measured to be about 106 days, and we found the period of this planet to be 686 days, so the angle α is

$$\alpha = 360°(106 \text{ days}) \left(\frac{1}{365 \text{ days}} - \frac{1}{686 \text{ days}} \right) \approx 49°. \tag{A.47}$$

With this value of α we can find the size of Mars' orbit relative to Earth's:

$$\frac{r_p}{r_e} = \frac{1}{\cos\alpha} \approx \frac{1}{\cos 49°} \approx 1.52. \tag{A.48}$$

Thus, the radius of Mars' orbit is 1.52 AU. Similar calculations for Jupiter and Saturn indicate that the radius of Jupiter's orbit is about 5.3 AU while the radius of Saturn's orbit is about 7.9 AU. Note that these results are derived for a highly simplified version of Copernicus' planetary theory. However, the results are similar to those of Copernicus except for the radius of Saturn's orbit, which he found to be about 9.3 AU (in good agreement with modern values).

A.13   Kepler: the ellipse and area laws

Kepler's First Law of Planetary Motion states that planets move in elliptical orbits with the Sun at one focus of the ellipse. The geometry of an elliptical orbit is illustrated in Figure A.6. The Sun (S) lies at one focus of the ellipse, while the other focus (Q) serves as an approximate equant point. The center (C) of the

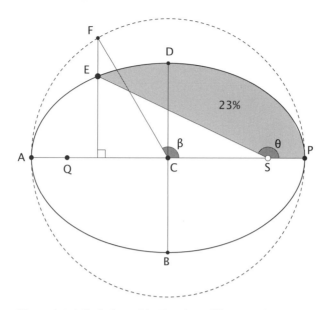

Figure A.6 A Keplerian orbit. The planet (E) moves along an elliptical path with the Sun (S) at one focus. The area swept out by the line segment $\bar{SE}$ is proportional to the elapsed time.

ellipse lies midway between the two foci. The point on the orbit closest to the Sun is the perihelion (P), while the point farthest from the Sun is the aphelion (A).

The major axis of the ellipse is the line segment $\bar{AP}$, which is the diameter of the long axis of the ellipse. Half of the major axis ($\bar{CA}$ or $\bar{CP}$) is known as the semi-major axis. Likewise, the diameter of the short axis ($\bar{BD}$) is known as the minor axis and half of the minor axis is called the semi-minor axis. The area of an ellipse is given by the product of its semi-major and semi-minor axes multiplied by π.

The shape of an ellipse (i.e. how elongated it is) can be characterized by a single parameter known as the eccentricity.[i] The eccentricity (e) of the ellipse shown in Figure A.6 is given by

$$e = \frac{\bar{CS}}{\bar{CP}} = \frac{d_A - d_P}{d_A + d_P},\tag{A.49}$$

where d_A and d_P are the distances of the planet from the Sun when the planet is at aphelion and perihelion, respectively. The eccentricity of the ellipse in Figure A.6 is $e \approx 0.73$, which you can check by making your own measurements.

---

[i] This eccentricity is related to, but different from, the eccentricity of an eccentric circular orbit as defined by Ptolemy. In this definition of eccentricity, all circles have an eccentricity of zero.

A circle has an eccentricity of zero, and the more elongated the ellipse, the closer its eccentricity will be to one. The actual planets of our solar system all have eccentricities relatively close to zero, so their orbits are nearly circular.

Kepler's Second Law of Planetary Motion states that the planet moves along its elliptical orbit in such a way that the area swept out by the line segment from the Sun to the planet is proportional to the elapsed time. In Figure A.6 the line from the Sun to the planet sweeps out the shaded area as the planet moves from perihelion (P) to point E. That shaded area is about 23% of the total area of the ellipse, so the time for the planet to move from P to E must be 23% of the planet's total orbital period (because in one full orbital period the Sun–planet line will sweep out the entire area within the ellipse). If, for example, this planet had an orbital period of 865 days, then the time for the planet to move from P to E would be 0.23 × 865 ≈ 199 days.

In practice it is difficult to compute the shaded area shown in Figure A.6. However, the area swept out by the Sun–planet line can easily be computed from the eccentricity of the ellipse and the area swept out by the segment $\overline{SF}$. Here F is a point that lies along an imaginary circle with the same center as the ellipse and with a radius equal to the semi-major axis of the ellipse. The point F is located so that the line $\overline{FE}$ intersects the major axis of the ellipse at a right angle, as shown in Figure A.6.

The goal of the astronomer is to find a relationship between the location of the planet E and the elapsed time. Astronomical calculations are generally done using angular quantities. The position of the planet can be determined from the **true anomaly** (the angle θ in Figure A.6) but it can also be calculated from the **eccentric anomaly** (the angle β shown in the figure) using some simple geometrical rules. The elapsed time is proportional to the shaded area in the figure, but it is convenient to convert this quantity into an angle known as the **mean anomaly** using the formula

$$\alpha = \frac{\text{Area}_E}{\text{Area}_T} \times 360°, \tag{A.50}$$

where Area_E is the area swept out by the Sun–planet line and Area_T is the total area of the ellipse.

In his *Astronomia Nova*, Kepler derived a relationship between the mean anomaly and the eccentric anomaly, which is now known as "Kepler's equation," and can be expressed in modern notation as[2]

$$\alpha = \beta - e\sin\beta. \tag{A.51}$$

If the eccentricity e of the ellipse is known, then Kepler's equation makes it fairly easy to compute the mean anomaly α given a value for the eccentric anomaly β.

Carrying out such a computation is essentially the same as asking "at what time will the planet reach a specified point on its orbit?" Unfortunately, this is not the question astronomers need to ask when they are compiling planetary tables. Instead, they want to know "where will the planet be on its orbit at a specified time?" So they must use Kepler's equation to find β given a value for α. That turns out to be very difficult to do. To solve Kepler's equation for β the astronomer must carry out repetitive calculations, each of which gives a slightly better approximation to the correct value. Only after many such repetitions will the astronomer have a satisfactory result. The difficulties of using Kepler's equation were one reason that the New Astronomy was not quick to catch on.[i]

Another reason was that Kepler's terminology was confusing for many astronomers. He introduced new terms and used traditional terms in new ways.[3] Some of his definitions seemed very strange. Look back at the defintion of the eccentric anomaly. Why should anyone bother with an angle that refers to an imaginary point that moves along an imaginary circle. Neither the circle nor the point F corresponds to anything physically real. The planet is actually at E and it moves along the ellipse. Kepler introduced the eccentric anomaly because it made his calculations easier, but it may have also made it harder for other astronomers to understand his new methods.

A.14   Kepler: the harmonic law

The modern version of **Kepler's Third Law of Planetary Motion** is that the square of a planet's period is proportional to the cube of the semi-major axis of the planet's elliptical orbit. If the planet's period is T and its semi-major axis is a, then Kepler's Third Law states that

$$T^2 = ka^3, \tag{A.52}$$

where k is some constant factor that applies to all objects orbiting a common body (e.g. all planets, comets, etc. that are orbiting the Sun).

We can make Kepler's Third Law easier to use by selecting convenient units. If we measure period in years, and semi-major axis in Astronomical Units, then for Earth both T and a have the numerical value 1. Substituting this result into Equation A.52 we see that in these units $k = 1$. So as long as we stick to these units, Kepler's Third Law just reduces to $T^2 = a^3$.

---

[i] Now astronomers can use digital computers to solve so-called "transcendental equations" such as Kepler's equation. Kepler, of course, did not have access to a digital computer and had to carry out these laborious calculations by hand.

Suppose we discover a new comet and find that it has a period of 75.3 years (as does Halley's Comet). Then we can use Kepler's Third Law to compute the semi-major axis of the comet's orbit:

$$a^3 = T^2 = (75.3)^2 \approx 5670, \tag{A.53}$$

and therefore

$$a = \sqrt[3]{5670} \approx 17.8 \text{ AU.} \tag{A.54}$$

Similarly, suppose we found a new planet that had a semi-major axis of 30.1 AU (as does Neptune). We can use Kepler's Third Law to compute the period of that planet's orbit:

$$T^2 = a^3 = (30.1)^3 \approx 27,271, \tag{A.55}$$

and therefore

$$T = \sqrt{27,271} \approx 165 \text{ years.} \tag{A.56}$$

In his *Epitome*, Kepler provided a physical explanation for his Third Law of Planetary Motion. He believed that the speed of an individual planet was governed by his (revised) distance law, and thus he thought that the force from the Sun that swept the planets around must have a strength that was inversely proportional to the distance from the Sun. Kepler also decided that the volume of each planet's body must be directly proportional to the planet's distance from the Sun, while the density of the planet must be inversely proportional to the square root of the planet's distance.[i] Kepler noted similarities between the densities of the planets and various substances, so that Saturn had the density of a diamond while Mercury had the density of quicksilver (i.e. mercury).[4]

Multiplying the density and volume laws Kepler found that the "quantity of matter" (akin to what we would now call the "mass") of a planet was directly proportional to the square root of its distance from the Sun. In a similar way, the "influence" of the solar force on a planet, which was found by multiplying the strength of the force by the volume of the planet that experiences the force,

---

[i] Kepler was convinced that either a planet's diameter, its surface area, or its volume must be proportional to its distance from the Sun, and he felt that volume fit best with observations of the angular sizes of planets. He gives arguments for why the density must be inversely proportional to the square root of the distance, but mostly it seems that he selects this law because it leads to the correct form for the Third Law.

ended up being the same on all planets. The speed of a planet was then found by dividing the influence by the "quantity of matter," so that the speed of a planet was inversely proportional to the square root of its distance. Finally, the period is found by dividing the circumference of the orbit (which is proportional to the Sun–planet distance) by the speed. Therefore, the period of a planet will be proportional to the three-halves power of the distance, or

$$T \propto a^{3/2}. \tag{A.57}$$

Squaring both sides we get back to the form of Kepler's Third Law given in Equation A.52. We now know, however, that Kepler's derivation of his harmonic law is completely invalid. Instead, the harmonic law is a consequence of Newtonian gravity.[5]

A.15   Galileo: measuring mountains on the Moon

Galileo used geometry to determine the height of a mountain on the Moon.[6] He noticed a bright point on the dark side of the Moon's terminator line. For simplicity let's assume that the Moon was at quarter phase so that the terminator line passed right through the center of the visible disk of the Moon. Galileo measured the distance from the terminator to the bright spot and found it was about 1/20 of the Moon's diameter. Because he knew the approximate distance and apparent diameter of the Moon, he knew that the Moon's physical diameter was about 2000 miles (see Section A.2). Therefore, the mountain peak must be about 100 miles from the terminator.

Galileo realized that the geometry of the situation must be like that shown in Figure A.7. The circle CAF, with center E, represents the body of the Moon. The line $\overline{GD}$ represents a ray of sunlight that just grazes the Moon's surface at the location of the terminator (C) and then strikes the peak of the mountain at D. The length of the segment $\overline{EC}$ represents the average radius of the Moon (what we might call the radius at "sea level" on Earth), which is 1000 miles (half of the diameter). The length of the segment $\overline{ED}$ is the distance from the center of the Moon to the top of the mountain peak.

The Pythagorean theorem states that the sum of the squares of the two shorter sides of a right triangle is equal to the square of the hypotenuse (see Appendix A.3). Applying the Pythagorean theorem to the triangle ECD, and using our previous results that the two shorter sides have lengths of 1000 and 100 miles, we find that the length of the longest side ($\overline{ED}$) is $\sqrt{1000^2 + 100^2} \approx 1005$ miles. Since we know that the length of $\overline{EA}$ is 1000 miles (the average radius of the Moon), we find that $\overline{AD}$, which represents the height of the mountain above the Moon's average surface level, must be about 5 miles.

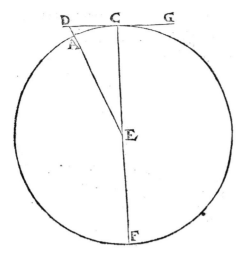

Figure A.7 Galileo's diagram for finding the height of a mountain on the Moon, from the *Sidereus Nuncius* (1610). Image courtesy History of Science Collections, University of Oklahoma Libraries.

Thus, using his observations and some geometry Galileo was able to show that the Moon has mountains that rise five miles above "sea level." For comparison, Mount Everest (of which Galileo was unaware) rises 5.5 miles above sea level. So even though the Moon is much smaller than the Earth, Galileo showed that it had mountains as tall as the tallest mountains on the Earth.

A.16 Galileo: falling bodies and projectiles

Galileo discovered that, in the absence of resistance from the air, bodies falling near the surface of the Earth fall with constant acceleration. Furthermore, the acceleration is the same for all objects. An object falling from rest with constant acceleration will have a speed that is proportional to the time of fall. In terms of ratios:

$$\frac{v_1}{v_2} = \frac{t_1}{t_2}, \tag{A.58}$$

where v is the speed, t is the elapsed time, and the indices label two different moments in time.

Galileo figured out that if the speed is proportional to the time, then the distance the object falls is proportional to the square of the time of fall, or

$$\frac{y_1}{y_2} = \frac{t_1^2}{t_2^2}, \tag{A.59}$$

where y is the distance of fall. This result can be derived from the mean speed theorem, because if the instantaneous speed is proportional to time then the average speed during the fall is also proportional to time (it is just half of the final speed) and the distance is the average speed multiplied by the time of fall. However, Galileo seems to have found this result from studying the relation between pendulum motion and falling bodies.[7]

We can combine Equations A.58 and A.59 to find a relationship between distance and speed. The result is that the speed of the object is proportional to the square root of the distance it has fallen, or

$$\frac{v_1}{v_2} = \frac{\sqrt{y_1}}{\sqrt{y_2}}. \tag{A.60}$$

We can use these proportionality relations to solve some practical problems. For example, suppose we know that an object falls 4.9 meters in the first second after it is dropped. How far will it have fallen two seconds after it was dropped. Using Equation A.59 we have

$$\frac{y_1}{4.9 \text{ m}} = \frac{(2 \text{ s})^2}{(1 \text{ s})^2} = 4, \tag{A.61}$$

so that $y_1 = 4 \times 4.9 = 19.6$ m. In a similar manner, if we know that the object has a speed of 9.8 m/s after one second of falling, then we can use Equation A.58 to find that it will have a speed of 19.6 m/s after two seconds of falling. We could have got the same result using Equation A.60 and our result that the object falls 19.6 m in two seconds.

Galileo also found that the horizontal motion of an object had a constant speed.[8] For motion with constant speed the distance traveled by the object is proportional to the elapsed time:

$$\frac{x_1}{x_2} = \frac{t_1}{t_2}, \tag{A.62}$$

where x is the horizontal distance traveled by the object.

The motion of a horizontally launched projectile is then a combination of horizontal motion at constant speed and vertical fall at constant acceleration. We can determine the relationship between the horizontal distance traveled (x) and the vertical distance fallen (y) by combining Equations A.62 and A.59 to find that

$$\frac{y_1}{y_2} = \frac{x_1^2}{x_2^2}. \tag{A.63}$$

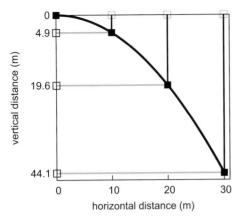

Figure A.8 Parabolic path of a projectile launched horizontally at 10 m/s.

In other words, for projectile motion the vertical fall is proportional to the square of the horizontal distance. That relationship indicates that the path of a projectile is a parabola.

As an example, consider the path of a projectile launched horizontally at 10 m/s. The path is shown in Figure A.8. The horizontal motion of the projectile is motion at constant speed, so the projectile moves 10 meters each second as shown by the open gray squares. The vertical motion of the projectile is a fall at constant acceleration. The projectile has fallen 4.9 meters after one second, 19.6 meters after two seconds, and 44.1 meters after three seconds, as shown by the open black squares. The combination of these two motions produces the parabolic path of the projectile as shown by the filled black squares and solid black curve in the figure.

A.17   Newton: gravity and the Moon

Early in his career Newton analyzed centrifugal forces[i] and used his results to compare the force on the orbiting Moon to the gravitational force on a body near the Earth's surface.[9] Suppose an object moves uniformly on the circle shown in Figure A.9. In some time t the object moves from A to G. If the object had moved in a straight line instead of a circle it would have moved to B and

---

[i] Today physicists would speak of the *centripetal* force required to keep an object moving in a circle, but in the 1660s Newton was still thinking of circular motion as a balance between an outward centrifugal force and an inward attractive force. In the *Principia* Newton discussed circular motion as resulting from inertial motion plus an inward centripetal force. The results shown in this appendix for Newton's centrifugal force would now be considered results for the centripetal force in uniform circular motion.

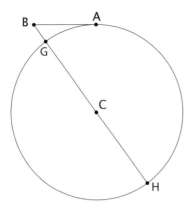

Figure A.9 Newton's diagram for determining centrifugal force, reproduced from one of his papers dating from the 1660s (with slightly different labels).

therefore its inertial motion would have carried it a distance GB outward from the circumference of the circle, so the centrifugal force acting over time t pushes the object outward a distance GB.

Galileo showed that objects starting from rest and subject to a constant force (like gravity) move in such a way that the distance they cover is proportional to the square of the elapsed time (see Appendix A.16). Suppose T represents the period of the object's movement around the circle and L represents the distance the object would be moved by the centrifugal force in the time T, then

$$\frac{L}{\text{GB}} = \frac{T^2}{t^2}.$$ (A.64)

Since the object moves at constant speed along the circle, the times are proportional to the distances traveled on the circle:

$$\frac{T}{t} = \frac{C}{\text{AG}},$$ (A.65)

where C is the circumference of the circle. Combining Equations A.64 and A.65 and solving for L we get

$$L = \frac{C^2}{\text{AG}^2} \times \text{GB}.$$ (A.66)

Newton knew, by Proposition 36 of Book III of Euclid's *Elements*, that in the case shown in Figure A.9,

$$\text{GB} = \frac{\text{AB}^2}{\text{BH}}.$$ (A.67)

But as t becomes very small BH becomes equal to the diameter D of the circle and AB becomes equal to AG. If we substitute these values into Equation A.67 and put the result for GB into Equation A.66 we find that

$$L = \frac{C^2}{D} = \frac{(\pi D)^2}{D} = \pi^2 D. \qquad \text{(A.68)}$$

So in the period of the circular motion the centrifugal force would push an object a distance π^2 times the diameter of the circle.

In some other time t (no longer assumed to be small), the centrifugal force would push the object a distance d, where

$$\frac{d}{\pi^2 D} = \frac{t^2}{T^2}, \qquad \text{(A.69)}$$

because we know the distance will be proportional to the square of the time. We can apply Equation A.69 to find how far the centrifugal force of Earth's daily rotation would push an object in one second. Newton knew that Earth's rotational period was one sidereal day, or about 86,000 seconds. He used a value of 3500 miles for Earth's radius, with each mile equal to 5000 feet of 12 inches each. Therefore, he assumed Earth's diameter was about 420 million inches. Inserting these values into Equation A.69 and solving for d we find that the centrifugal force of Earth's rotation would push an object, starting from rest, about 0.56 inches in one second.

From experiments with pendulums Newton had estimated that gravity at Earth's surface would push an object a distance of 196 inches in one second. Comparing this value with his result for the centrifugal force of Earth's daily rotation, Newton showed that gravity is about 350 times stronger than the centrifugal force. Thus, there was no need to worry about objects being flung off of the Earth by centrifugal forces. Gravity was far more than enough to keep objects on the surface of a rotating Earth.

Newton also rearranged Equation A.69 to get

$$\frac{d}{t^2} = \frac{\pi^2 D}{T^2}. \qquad \text{(A.70)}$$

For an object moving in any circle, then, the distance it would be moved by the centrifugal force in one second is proportional to the diameter of the circle and inversely proportional to the square of the period. We can use this relationship to compare centrifugal forces in two different circular motions. For two different forces acting on an object the distance moved by the object (starting from rest) in one second will be proportional to the force. So the ratio of the distances gives us the ratio of the forces:

$$\frac{F_1}{F_2} = \frac{d_1}{d_2} = \frac{D_1}{T_1^2} \times \frac{T_2^2}{D_2}, \tag{A.71}$$

where the last step makes use of Equation A.70. In other words, the centrifugal force is proportional to the diameter of the circle and inversely proportional to the square of the period:

$$F \propto \frac{D}{T^2}. \tag{A.72}$$

Newton applied Equation A.71 to compare the centrifugal force on the Moon in its orbit to the centrifugal force due to Earth's diurnal rotation. He knew that the diameter of the Moon's orbit was about 60 times the diameter of Earth. The period of the Moon's orbit is 27.3 days. Substituting these values into Equation A.71 he found that the centrifugal force due to Earth's rotation is about 12.5 times as great as the centrifugal force on the Moon. Combining this result with his earlier comparison to gravity he found that the force of gravity at Earth's surface was about $12.5 \times 350 = 4375$ times as great as the centrifugal force on the Moon.[i]

If gravity is an inverse-square law then one would expect the force of gravity at Earth's surface to be $60^2 = 3600$ times greater than the force on the Moon, since the Moon is 60 times as far away from the center as the Earth's surface. Newton's answer was not terribly far off, but it was not particularly close either. The problem was that he used an inaccurate value for Earth's radius. Around 1670 Jean Picard measured the distance that corresponds to one degree of latitude on Earth and found a value of about 68.6 miles, which corresponds to a radius of 3930 miles for Earth (where one mile is 5280 feet). Using this revised value in the calculations shown above we find that gravity on Earth's surface is about 295 times as great as the diurnal centrifugal force, and therefore about 3690 times as great as the centrifugal force on the Moon. This is much closer to the result expected from an inverse-square gravitational force. Newton presented the corrected calculation in the *Principia*.[11]

A.18   Newton: an inverse-square force on the planets

In Section A.17 we examined Newton's conclusion that the centrifugal force on an object moving uniformly in a circle is proportional to the diameter of the circle and inversely proportional to the square of the period (see Equation

---

[i] Newton wrote in his notebook that "the force of gravity is 4000 and more times greater than the endeavour of the Moon to recede from the centre of the Earth."[10]

A.72). Since the diameter of a circle is just double the radius, we can rewrite this proportionality in terms of the radius (R) and period:

$$F \propto \frac{R}{T^2}. \tag{A.73}$$

Newton combined this result with Kepler's Third Law to determine the behavior of the force that keeps the planets in orbit.[12]

Newton approximated the orbits of the planets as circles centered on the Sun. In that case, Kepler's Third Law states that the square of the orbital period is proportional to the cube of the orbital radius:

$$T^2 \propto R^3. \tag{A.74}$$

Substituting Equation A.74 into Equation A.73 we find that

$$F \propto \frac{R}{R^3} = \frac{1}{R^2}. \tag{A.75}$$

In other words, the centrifugal force on the planets is inversely proportional to the square of their distances from the Sun.[i]

A.19   Newton: universal gravitation

Newton concluded that any pair of bodies will exert gravitational forces on each other that are proportional to the mass of each body and also inversely proportional to the square of the distance between the bodies.[13] We can express this as

$$F \propto \frac{mM}{d^2}, \tag{A.76}$$

where F is the gravitational force, m and M are the masses of the two bodies, and d is the distance between them. If we wish to compare the gravitational forces in two different situations we can express this proportionality as a ratio of the two forces:

$$\frac{F_1}{F_2} = \frac{m_1 M_1}{d_1^2} \times \frac{d_2^2}{m_2 M_2}. \tag{A.77}$$

For example, suppose we want to know what happens if we double the mass of one object while leaving all other quantities the same. In that case $m_1 = 2m_2$ and we find that $F_1/F_2 = 2$. So doubling the mass of one object doubles the gravitational forces between the objects. Likewise, if we double the distance

---

[i] As noted in Appendix A.17, physicists today would describe this force as the centripetal force needed to keep a planet in a circular orbit, not as a centrifugal force.

between the objects ($d_1 = 2d_2$) we find that $F_1/F_2 = 1/4$, so the force is reduced by a factor of four.

Now suppose an object with mass m orbits in a circle around a central body with mass M_1 at a distance R_1 and experiences a gravitational force, or weight, F_1 toward that central body. Likewise, the same object orbiting a different central body with mass M_2 at distance R_2 experiences a gravitational force F_2. Equation A.77 (with $d = R$) and Equation A.71 (with $D = 2R$) give us two different ways to calculate the ratio of these forces. We must get the same answer in both cases, so it follows that

$$\frac{mM_1}{R_1^2} \times \frac{R_2^2}{mM_2} = \frac{R_1}{T_1^2} \times \frac{T_2^2}{R_2}. \tag{A.78}$$

We can rearrange this equation to find

$$\frac{M_1}{M_2} = \frac{R_1^3}{T_1^2} \times \frac{T_2^2}{R_2^3}, \tag{A.79}$$

where the common factor of m is cancelled.

Now consider Saturn's moon Titan, which orbits at a distance $R_1 = 0.00817$ AU from Saturn (as can be determined by angular measurements and the known Earth–Saturn distance) with a period of $T_1 = 15.9$ days. If Titan was instead orbiting the Sun at a distance of $R_2 = 1$ AU it would have the same period as Earth's orbit ($T_2 = 365$ days). Substituting these values into Equation A.79 we find that $M_ħ/M_⊙ \approx 0.000287$ or $M_⊙/M_ħ \approx 3500$. So the Sun is nearly 3500 times as massive as Saturn. A similar comparison using data for Earth's moon ($R_1 = 0.00257$ AU, $T_1 = 27.3$ days) shows that the Sun is about 330,000 times as massive as Earth. Using data from Jupiter's moon Callisto ($R_1 = 0.0126$ AU, $T_1 = 16.7$ days) we find that the Sun is 1050 times as massive as Jupiter.

In the previous paragraph we used modern values for all of the quantities. Newton had reasonably accurate data for most of these quantities, so the values given above match pretty well with the values in the *Principia* (see Section 9.4). The one value from the *Principia* that is seriously in error is the relative mass of the Earth. The reason for the error is that Newton did not have an accurate value for the radius of the Moon's orbit in Astronomical Units. He knew that the Moon's orbit had a radius of about 240,000 miles, but he thought that the Astronomical Unit was about 78 million miles when in fact it is about 93 million miles. This error led Newton to overestimate the Earth's mass relative to the Sun by about a factor of two.[14] The Astronomical Unit was not accurately measured until the late 18th century (see Section 10.3.1).

A.20   Bradley: aberration of starlight

Figure 10.1 shows the tilt of a telescope resulting from the aberration of starlight. Light first hits the telescope at p_1 and then travels vertically downward to the eyepiece at p_3. Let the distance from p_1 to p_3 be h. The amount of time that the light spends in the telescope will then be h/c, where c is the speed of light. During that time the telescope moves to the right a distance vh/c, where v is the speed of Earth in its orbit around the Sun.

Note that the angle θ in Figure 10.1 is part of a right triangle. The side opposite θ has length vh/c, while the side adjacent to θ has length h (the hypotenuse of the triangle is the same length as the telescope, but we don't need that for our analysis).

From trigonometry (see Appendix A.3) we know that

$$\tan \theta = \frac{vh/c}{h} = \frac{v}{c}. \tag{A.80}$$

Bradley found that the maximum aberration angle for a star that was located approximately at a right angle to Earth's direction of motion was about $20''$ (see Figure 10.2).[i] From this measurement we have

$$\frac{v}{c} = \tan (20'') \approx \tan (0°.00556) \approx 0.000097, \tag{A.81}$$

so the speed of Earth in its orbit is about 0.0097% of the speed of light.[ii]

We know that the Earth travels around the circumference of its orbit in one year (31,557,600 s). If we approximate Earth's orbit as a circle with a radius of 1 AU we find that its orbital speed is

$$v = \frac{2\pi R}{T} = \frac{2\pi (1 \text{ AU})}{31,557,600 \text{ s}} \approx 1.99 \times 10^{-7} \text{ AU/s}. \tag{A.82}$$

We can divide this result by the ratio v/c found above to get

$$c = \frac{v}{v/c} = \frac{1.99 \times 10^{-7} \text{ AU/s}}{0.000097} \approx 0.00205 \text{ AU/s}. \tag{A.83}$$

In other words, it takes light $1/0.00205 \approx 487$ seconds, or 8.1 minutes, to travel 1 AU from the Sun to the Earth (Bradley gave a value of 8.2 minutes).[16]

---

[i] Note that the "correct" north–south position of the star in Figure 10.2 is at the midpoint of the diagram near $20''$ on the vertical axis. Aberration causes the star's apparent position to move north and south from that point with a maximum deviation of $20''$.

[ii] Bradley expressed his result as $v/c = \sin\theta/\sin\phi = \sin\theta/\sin(90° - \theta)$ but that is mathematically equivalent to Equation A.80.[15]

Once the AU was determined to be about 150 million kilometers (see Section 10.3.1), this result could be used to find the speed of light in km/s:

$$c \approx 0.00205 \text{ AU/s} \times \frac{1.5 \times 10^8 \text{ km}}{1 \text{ AU}} \approx 307,500 \text{ km/s}, \tag{A.84}$$

which is in good agreement with the modern value of 299,792 km/s. In fact, once independent methods were devised for measuring the speed of light, precise measurement of the aberration angle could be used to determine a more precise value for the Astronomical Unit.

Notes

Chapter 1

1 For debunking of some of these Copernican myths, see Danielson (2001), Singham (2007), Graney (2012b), Danielson and Graney (2014).

Chapter 2

1 On the ancient Greek celestial sphere, see Evans (1998), 75–99. For a medieval textbook account of the celestial sphere, see Thorndike (1949), 118–132.

2 Hannah (2008), 110.

3 On ancient Greek models for solar motion, see Evans (1998), 53–58.

4 On the annual movement of the sunrise point along the horizon, see Hannah (2008), 5–9, and Evans (1998), 56–58.

5 On Ancient Greek use of shadows to track the Sun, see Evans (1998), 59–63, and Hannah (2008), 68–95.

6 Hannah (2008), 74–75, and Evans (1998), 95–96.

7 Hannah (2008), 125.

8 Evans (1998), 54–58.

9 Hannah (2008), 68–73.

10 On Greek parapegmata, see Hannah (2008), 49–59, and Evans (1998), 190–204.

11 On the Egyptian calendar, see Hannah (2008), 43–49, and Evans (1998), 175–182.

12 On the Julian calendar, see Evans (1998), 163–166.

13 On the Gregorian calendar reform, see Evans (1998), 166–171.

14 On the difference between the true local time and mean local time, sometimes called the "equation of time," see Evans (1998), 235–243.

15 On the signs of the zodiac, see Evans (1998), 39–40, and Hannah (2008), 15–18. For a medieval account, see Thorndike (1949), 124–125.

16 On ecliptic coordinates, see Evans (1998), 101.

17 On ancient Greek views of the Earth's shape, see Evans (1998), 47–53.

18 Aristotle (2009), 429–437.

19 On the "climes" of Earth, see Evans (1998), 91–95, and Thorndike (1949), 134–140.
20 Aristotle (2009), 437.
21 On Eratosthenes' determination of Earth's diameter, see Evans (1998), 63–65. See also Newton (1980) and Carman and Evans (2015).
22 On Ptolemy's estimate and its influence on Columbus, see Evans (1998), 65–66.
23 Evans (1998), 50–51.
24 On differences in local time and the creation of time zones, see Hannah (2008), 142–143.
25 On Hipparchus' discovery of the precession of the equinoxes, see Evans (1998), 245–259.
26 On Ptolemy's work on the precession of the equinoxes, see Evans (1998), 259–262.
27 On the controversy over Ptolemy, see Evans (1998), 265–274.
28 Evans (1998), 262.
29 On armillary spheres, see Evans (1998), 78–80.
30 Evans (1998), 320–321.

Chapter 3

1 Evans (1998), 311–312.
2 Evans (1998), 46.
3 On lunar calendars, see Hannah (2008), 18–26.
4 On lunisolar calendars, including the Metonic cycle, see Evans (1998), 182–190, and Hannah (2008), 31–42.
5 Hannah (2008), 34.
6 Evans (1998), 46.
7 Evans (1998), 316.
8 On Aristarchus' methods for determining the sizes and distances of Sun and Moon, see Evans (1998), 67–74.
9 On Hipparchus' estimate for the Moon's parallax, see Evans (1998), 72–73.
10 On Ptolemy's parallax estimate, see Evans (1998), 73.
11 On Babylonian planetary observations and their transmission to Greece, see Evans (1998), 297–298.
12 Evans (1998), 296.

Chapter 4

1 Goldstein and Bowen (1983), 333.
2 Aristotle (2009), 882–883.
3 See Bowen (2002) for evidence that Simplicius' account of the Eudoxan theory is not trustworthy.
4 The Eudoxan model is described in Evans (1998), 305–310, and in Linton (2004), 25–32.
5 Evans (1998), 308.
6 Evans (1998), 307.
7 This assumption is questioned in Goldstein (1997).
8 Schiaparelli's reconstruction is described in Evans (1998), 308–310, and in Linton (2004), 29–31.

9 Some alternative reconstructions are given in Yavetz (1998) and Mendell (1998), while Bowen (2002) argues for skepticism about any of these reconstructions.

10 Aristotle (2009), 882–883.

11 Evans (1998), 458n10.

12 Aristotle (2009), 883.

13 Evans (1998), 309.

14 Evans (1998), 310.

15 Evans (1998), 311–312.

16 See Carman (2015) and Bowen (2002) for evidence that Eudoxus and his contemporaries may have been unaware of changes in planetary magnitudes.

17 This account is found in Plato (1989), 311–312, and is discussed in Knorr (1990), 313–317.

18 Knorr (1990), 315.

19 The account can be found in Plato (1888), 107–137, and is discussed in Knorr (1990), 314–317.

20 Plato's account of the elements is given in Plato (1888), 187–209.

21 Aristotle (nd), Book II, 10.

22 Aristotle (nd), Book II, 12.

23 Aristotle (nd), Book II, 1.

24 Aristotle (nd), Book II, 4.

25 Aristotle (2009), 363–377.

26 Aristotle's physical theories are presented throughout his *Physica* and *De Caelo*. A characteristic and important passage is Book IV, Chapter 3 of *De Caelo* (Aristotle (2009), 459–461). A useful summary of Aristotelian physics and cosmology is given in Dijksterhuis (1950), 17–40, and in Kuhn (1957), 78–99.

27 Drabkin (1938), 76n37.

28 A detailed analysis of Aristotle's physics for violent motions is given in Drabkin (1938).

29 Drabkin (1938), 67–68.

30 Some competing theories are described in Kuhn (1957), 41–44.

31 Apollonius' model is described in Evans (1998), 337–342, and in Linton (2004), 45–49.

32 See Bowen (2002); Carman (2015) for discussions about whether the ancient Greeks were aware of variations in planetary magnitudes.

33 Evans (1998), 339–342.

34 Evans (1998), 342.

35 Evans (1998), 213.

36 Hipparchus' solar theory is discussed in Evans (1998), 210–226.

37 Evans (1998), 210.

38 Evans (1998), 220–226.

39 Evans (1998), 212.

40 Evans (1998), 217.

41 He does this in Book I, Ptolemy (1998), 35–47. See Taub (1993), 39–103, for a discussion of Ptolemy's arguments.

42 Ptolemy's solar theory is given in Book III, Ptolemy (1998), 131–172.

43 In Book IV, Ptolemy (1998), 173–216.

44 Ptolemy's planetary theories are laid out in Books IX-XIII, Ptolemy (1998), 419–647.

45 Ptolemy (1998), 442–448.

46 Evans (1998), 339.

47 See Evans (1984), 1082–1083, or Evans (1998), 339–342.

48 See Evans (1984), 1083–1085, or Evans (1998), 341–342.

49 Ptolemy describes this model in Ptolemy (1998), 443–444. See also Linton (2004), 75–81, and Brehme (1976).

50 See Evans (1984), 1085, or Evans (1998), 355–359, but see Rawlins (1987) for an alternative account of the development of the equant.

51 Evans (1998), 357.

52 Evans (1998), 358–359.

53 Ptolemy's latitude theory is presented in Book XIII of the *Almagest* (Ptolemy (1998), 597–647). A detailed analysis is given in Swerdlow (2005), 41–58.

54 Ptolemy (1998), 600.

55 Ptolemy (1998), 601.

56 Evans (1998), 102–103.

57 See Taub (1993), 129–133, for a discussion of Ptolemy's astrology.

58 See Evans (1998), 384–392, and Murschel (1995) for discussions of the *Planetary Hypotheses*. A partial translation is given in Goldstein (1967).

59 Ptolemy (1998), 419.

60 Ptolemy (1998), 419–420.

61 Goldstein (1967), 6–7.

62 See Murschel (1995).

63 Evans (1998), 391.

64 The figures in Table 4.1 can be found in Goldstein (1967), 7, as well as Evans (1998), 388.

65 See Goldstein (1967), 7, and Evans (1998), 388.

66 See Goldstein (1967), 8–9, and Evans (1998), 389.

67 Ptolemy first corrected the latitude theory of the *Almagest* in his so-called *Handy Tables*, and further refined his model in the *Planetary Hypotheses*. An analysis of these later models is given in Swerdlow (2005), 58–68.

68 Swerdlow (2005), 67.

69 A brief summary of the development of astronomy from Ptolemy through the Middle Ages is given in Evans (1998), 392–403.

70 On medieval encyclopedists, see Grant (1971), 7–9.

71 Detailed discussions of Medieval European astronomy and cosmology can be found in Grant (1996), Grant (1971), and Duhem (1987).

72 On the place of Aristotelian thought in the Medieval university, see Grant (1971), 20–35.

73 Thorndike (1949), 118–142.

74 Aiton (1987).

75 Evans (1998), 396.

76 Hetherington (2006), 77.

77 See Swerdlow (1972) and Dreyer (1953), 296–304.

78 On the source of the planet's light, see Grant (1996), 393–402, and Ariew (1987).

79 Carman (2015), 94, and Goldstein (1996b), 4.

80 For ben Gerson's measurements, see Goldstein (1996b), Goldstein (1996a), and Goldstein (2012).

81 On the problem of the order of the planets, see Grant (1996), 308–314, and Evans (1998), 347–351.

82 Goldstein (1967), 5–6.

83 Quoted from Aiton (1987), 23.

84 On the medieval impetus theory, see Grant (1964), Franklin (1976), Grant (1971), 48–52, and Dijksterhuis (1950), 179–185.

85 On condemnations of Aristotelian ideas by the medieval Church, see Grant (1996), 50–55, and Grant (1971), 24–29.

86 Grant (1971), 29–31.

87 Grant (1996), 642–647.

88 Dreyer (1953), 282–284.

89 Evans (1998), 274.

90 For an account of the theory of trepidation, see Evans (1998), 275–280.

91 Evans (1998), 279.

92 Evans (1998), 280.

93 Grant (1996), 315–318.

94 On this "empyrean heaven" see Grant (1996), 371–382.

95 Saliba (1995), 152–155.

96 Saliba (1995), 299–302.

97 The lunar theory of al-Shāṭir is described in Linton (2004), 105–106.

98 For more on how the voyages to the New World helped to originate the very notion of discovery, see Wootton (2015), 57–109.

Chapter 5

1 A short biographical account of the life of Copernicus is available in Armitage (2004), 45–67, and Swerdlow and Neugebauer (2012), 3–32. A nonscholarly account can be found in Sobel (2011).

2 For a discussion of Copernicus' association with Novara, see Westman (2011), 87–99.

3 See Hetherington (2006), 80–82, and Swerdlow and Neugebauer (2012), 47–48, for a discussion of the connections between Copernicus and al-Ṭūsī and al-Shāṭir.

4 Barker (1999).

5 An English translation of the *Commentariolus* is available in Rosen (1939), 57–90.

6 Swerdlow and Neugebauer (2012), 9.

7 Swerdlow and Neugebauer (2012), 16.

8 Rosen (1975b), 535.

9 For a biography of Rheticus, see Danielson (2006).

10 An English translation of the *Narratio Prima* is available in Rosen (1939), 107–196.

11 See Copernicus (1992) for an English translation of *De revolutionibus*.

12 Swerdlow and Neugebauer (2012), 30.

13 See Malpangotto (2016).

14 This thesis is argued at length in Westman (2011).

15 Swerdlow and Neugebauer (2012), 36.

16 Copernicus laid out his general argument for the motions of Earth in Book I of *De revolutionibus* (Copernicus (1992), 7–27). The details of his theory for Earth's motion are given in Book III (Copernicus (1992), 119–172).

17 Copernicus (1992), 22.

18 Quoted from Copernicus (1992), 16.

19 Copernicus' argument for the immobility of the celestial sphere is given in Copernicus (1992), 15–17.

20 Copernicus (1992), 4.

21 Copernicus' lunar theory is presented in Book IV (Copernicus (1992), 173–226).

22 See, for example, Melanchthon's reaction in Westman (1975b), 173.

23 Copernicus' theory for planetary longitudes is given in Book V (Copernicus (1992), 227–306), while his theory of planetary latitudes is given in Book VI (Copernicus (1992), 307–330).

24 Useful comparisons of the Ptolemaic and Copernican models are given in Brehme (1976) and Martin (1984).

25 See, for example, his determination of the size of Mars' orbit in Copernicus (1992), 268–270.

26 Quoted from Copernicus (1992), 21–22.

27 On the role of coherence and symmetry in Copernicus' work see Hallyn and Leslie (1990), 73–103.

28 Quoted from Copernicus (1992), 5.

29 Quoted from Copernicus (1992), 4.

30 References to the "monstrousness" of the Ptolemaic system go back at least to Snecanus' preface to Peurbach's *Theroicae novae planetarum* (Hallyn and Leslie (1990), 47).

31 Clear but brief presentations of the technical details of Copernicus' models for planetary longitudes can be found in Linton (2004), 138–146, and Jacobsen (1999), 133–140. A detailed analysis is given in Swerdlow and Neugebauer (2012), 289–479.

32 A brief presentation of Copernicus' latitude theory is given in Jacobsen (1999), 141–149. A detailed analysis is presented in Swerdlow and Neugebauer (2012), 483–537.

33 On the religious response to Copernicus' theory, see Howell (2002); Kelter (1995); Kobe (1998); Rosen (1960); Christianson (1973).

34 Kelter (1995), 274.

35 See Copernicus (1992), 15–17.

36 Copernicus (1992), 17.

37 Quoted from Copernicus (1992), 18.

38 Copernicus (1992), 16.

39 Quoted from Copernicus (1992), 22.

40 Quoted from Copernicus (1992), XX.

41 See Swerdlow and Neugebauer (2012), 29, and Danielson (2006), 106–114.

42 Osiander also may have altered the title of the work (see Blumenberg (1987), 293–295).

43 For a discussion of instrumentalism (or fictionalism) and realism in 16th century astronomy, see the commentary by Peter Machamer in Westman (1975a), 346–353.

44 Quoted from Copernicus (1992), 22.

45 Quoted from Copernicus (1992), 4.

Chapter 6

1 Tredwell and Barker (2004), 151–153.

2 See Danielson (2006), 80–81.

3 Tredwell and Barker (2004), 146.

4 For a detailed discussion of the response to Copernicus, based on annotations in first editions of *De revolutionibus*, see Gingerich (2004).

5 For a discussion of this so-called "Melanchthon circle," see Westman (1975b).

6 A detailed account of Clavius' views is given in Lattis (2010).

7 See Kelter (1995) for the general Jesuit response to Copernicus.

8 Tredwell and Barker (2004), 147–150.

9 On Digges' role in advocating the Copernican model in England, see Johnson (1936).

10 For Bruno's views see Tredwell and Barker (2004), 153–154, Wootton (2015), 143–149, and Koyré (1957), 39–55.

11 The standard scholarly biography of Tycho is Thoren (1990), but see also Ferguson (2002).

12 Thoren (1990), 16.

13 Thoren (1990), 22–26.

14 Thoren (1990), 45–51.

15 Thoren (1990), 103–104.

16 Thoren (1990), 144–219.

17 Thoren (1990), 165.

18 Thoren (1990), 154–157.

19 Thoren (1990), 163–165.

20 Thoren (1990), 192–219.

21 Swerdlow and Neugebauer (2012), 20.

22 Quoted from Bartusiak (2004), 63–64.

23 Tycho's investigation of the nova of 1572 is discussed in Thoren (1990), 55–59.

24 Tycho's investigation of the comet of 1577 is described in Thoren (1990), 123–138, and in Gingerich (1977).

25 Thoren (1990), 249.

26 Schofield (1981), 73–74.

27 See Goldstein and Barker (1995) and Thoren (1990), 257–258.

28 For an analysis of the variety of views on the nature of celestial orbs, see Grant (1996), 324–370, and Donahue (1981).

29 Thoren (1990), 307.

30 Thoren (1990), 159–161 and 290–300.

31 On Brahe's critique of trepidation, see Evans (1998), 282, and Thoren (1990), 289–293.

32 Graney (2015), 36–37.

33 Gingerich and Voelkel (1998), 3.

34 Thoren (1990), 279.

35 On Tycho's examination of star sizes, see Graney (2015), 34–37.

36 Tycho had other objections to the Copernican system. See Blair (1990).

37 Graney (2015), 79.

38 Gingerich and Voelkel (1998), 23–24.

39 Thoren (1990), 91.

40 Schofield (1981), 74.

41 On Tycho's rejection of the motion of Earth, see Blair (1990), Thoren (1990), 275–280, and Graney (2015), 25–44.

42 Thoren (1990), 85–86.

43 For one possible route Tycho may have taken to reach his theory, and the role of the mathematician Paul Wittich in directing his thinking, see Thoren (1990), 239–247.

44 For a discussion of the Tychonic system, and other similar systems, see Schofield (1981).

45 On the Ursus controversy, see Schofield (1981), 108–136, and Thoren (1990), 255.

46 Details concerning Tycho's attempts to measure Martian parallax are given in Gingerich and Voelkel (1998).

47 Gingerich and Voelkel (1998), 5–9.

48 Gingerich and Voelkel (1998), 9–16.

49 Gingerich and Voelkel (1998), 21–23.

50 Thoren (1990), 231–235.

51 Gingerich and Voelkel (1998), 21–23, and Thoren (1990), 231–235.

52 Gingerich and Voelkel (1998), 16–21.

53 Gingerich and Voelkel (1998), 24.

54 Gingerich and Voelkel (1998), 25–28.

55 Gingerich and Voelkel (1998), 28.

56 Schofield (1981), 135, and Gingerich and Voelkel (1998), 28–29.

57 Thoren (1990), 370–373.

58 Thoren (1990), 376–415.

59 Wesley (1978).

Chapter 7

1 The standard scholarly biography of Kepler is Caspar (1993), but see also Connor (2004), Love (2015), and Ferguson (2002).

2 Caspar (1993), 37–38.

3 Caspar (1993), 36–47.

4 Methuen (1996).

5 Kepler mentions Maestlin's ideas about the Moon in Kepler (1981), 165.

6 An English translation of the *Somnium*, along with discussion of Kepler's initial 1593 disputation on the Moon, can be found in Kepler (2003).

7 See Caspar (1993), 48–52. Caspar suggests that Kepler was chosen for the position at Graz because he had the best qualification for that position. However, see Love (2015), 50, for an alternate view.

8 Caspar (1993), 53–60.

9 Kepler later described his moment of insight, and presented his diagram, in Kepler (1981), 65–66.

10 Kepler (1981), 67–68.

11 Kepler makes this analogy in Kepler (1981), 63.

12 See Kepler (1981), 199–201.

13 Kepler (1981), 19–20.

14 An English translation is available in Kepler (1981). For analysis see Field (1988), 30–95, and Martens (2000), 39–56.

15 Kepler (1981), 75–85.

16 Kepler (1981), 93–119. See also Field (1988), 30–72.

17 Adapted from Kepler (1981), 163.

18 Kepler (1981), 155–163.

19 His justification is based mostly on the oddities of Mercury's orbit in Copernicus. See Kepler (1981), 173–175.

20 Kepler (1981), 163–167.

21 Kepler (1981), 179.

22 Kepler (1981), 177–195.

23 Kepler (1981), 197–219.

24 Quoted from Kepler (1981), 199.

25 Kepler (1981), 217.

26 Kepler (1981), 219.

27 Kepler (1981), 221–223.

28 See, for example, Kepler's discussion in the dedication of the *Mysterium* (Kepler (1981), 55).

29 Caspar (1993), 93.

30 Caspar (1993), 64–69, and Kepler (1981), 19–22.

31 See Rosen (1975a) on Kepler's negotiations with the Tübingen faculty.

32 Caspar (1993), 69, and Drake (1978), 41.

33 Love (2015), 69, and Caspar (1993), 86–87.

34 Love (2015), 72–73.

35 Caspar (1993), 71–77.

36 Caspar (1993), 77–81.

37 Caspar (1993), 96–108.

38 Caspar (1993), 126.

39 Caspar (1993), 108–121.

40 A translation and analysis of this essay can be found in Jardine (1984).

41 Love (2015), 85–86.

42 Caspar (1993), 139–142.

43 Caspar (1993), 142–146, and Love (2015), 113–118.

44 Caspar (1993), 154–157, and Granada (2008), 479.

45 For a detailed analysis of how Kepler wrote *Astronomia Nova*, see Voelkel (2001).

46 Quoted from Kepler (2015), 16.

47 Kepler (2015), 3.

48 A detailed discussion of introduction of physical laws into astronomy is given in Stephenson (1994).

49 Kepler (2015), 17–36.

50 Quoted from Kepler (2015), 24.

51 Quoted from Kepler (2015), 25.

52 On Kepler's ideas about mass, gravity, and inertia, see Hecht (2017).

53 Kepler (2015), 75–130.

54 Kepler (2015), 80.

55 Kepler (2015), 149–173.

56 Quoted from Kepler (2015), 172.

57 Kepler (2015), 184–211.

58 Quoted from Kepler (2015), 211.

59 Kepler (2015), 225–230.

60 Kepler (2015), 231–240.

61 Kepler (2015), 241–253.

62 Kepler (2015), 275–277.

63 Kepler (2015), 278–289.

64 See Gilbert (1958).

65 Kepler (2015), 299–308.

66 Kepler (2015), 309–316.

67 Kepler (2015), 311n1.

68 Kepler (2015), 333.

69 Kepler (2015), 336–338.

70 Quoted from Kepler (2015), 338.

71 Kepler (2015), 339–369.

72 Quoted from Kepler (2015), 368.

73 Kepler (2015), 410–443.

74 Kepler (2015), 453–480.

75 Voelkel (2001), 205.

76 Caspar (1993), 198–202.

77 Caspar (1993), 202–208.

78 Caspar (1993), 209–227.

79 Caspar (1993), 240–258. See also Connor (2004), 231–306.

80 An English translation is available in Kepler (1997). For analysis of this work see Martens (2000), 112–141, and Field (1988), 96–166.

81 See Kepler (1997), 374–378.

82 Kepler (1997), 417–450.

83 Kepler (1997), 451–488.

84 See Gingerich (1975) on the origins of this law.

85 Quoted from Kepler (1997), 411–412.

86 A partial translation of Kepler's *Epitome* is available in Kepler (1995). For analysis see Martens (2000), 142–168.

87 Adapted from tables in Kepler (1997), pp. 420 and 486.

88 Kepler (1995), 141. See also Aiton (1969).

89 Kepler (1995), 35–36.

90 Kepler (1995), 42–43.

91 Russell (1964), 20.

92 Caspar (1993), 304–320.

93 See Gingerich (1971).

94 Caspar (1993), 321–361.

95 On the religious underpinnings of Kepler's approach to astronomy, see Barker and Goldstein (2001) and Holton (1956).

96 Quoted from Voelkel (2001), 165.

Chapter 8

1 Although several excellent biographies of Galileo are available, Drake (1978) provides a scholarly account that focuses on Galileo's scientific work. See also Wootton (2010); Heilbron (2012).

2 Heilbron (2012), 5.

3 Drake (1978), 4–7.

4 Drake (1978), 18–21.

5 Drake (1978), 33–49.

6 Drake (1978), 137–141, and Heilbron (2012), 147–150.

7 Drake (1978), 142–154.

8 Quoted from Galilei (1957), 31.

9 Galilei (1957), 32.

10 Galilei (1957), 40–41.

11 Galilei (1957), 38–40.

12 Galilei (1957), 42–45.

13 Quoted from Galilei (1957), 45.

14 Galilei (1957), 46.

15 Galilei (1957), 49–50.

16 This discovery is documented in Galilei (1957), 50–58. See also Gingerich and Van Helden (2011).

17 Drake (1978), 153–154, and Heilbron (2012), 160–162.

18 Lattis (2010), 180–195.

19 Kepler (1965), 14.

20 Quote from Kepler (1965), 39.

21 Galilei (1957), 59–65.

22 Van Helden (1974b).

23 On Galileo's observations of the phases of Venus, see Drake (1984); Gingerich (1984); Palmieri (2001).

24 Drake (1978), 166.

25 For a full account of Galileo's sunspot controversy, see Galilei and Scheiner (2010).

26 Galilei and Scheiner (2010), 9–34.

27 Galilei and Scheiner (2010), 76–79.

28 On Scheiner, see Engvold and Zirker (2016) and Galilei and Scheiner (2010), 37–57.

29 Galilei and Scheiner (2010), 59–73.

30 Translations of Galileo's letters can be found in Galilei and Scheiner (2010), 87–168.

31 Castelli's method is discussed in Galilei and Scheiner (2010), 80–83.

32 Galilei and Scheiner (2010), 109–110.

33 Galilei and Scheiner (2010), 111–112.

34 Galilei and Scheiner (2010), 99–103.

35 See Galilei and Scheiner (2010), 183–230.

36 Quoted from Galilei and Scheiner (2010), 199.

37 Galilei and Scheiner (2010), 251–304.

38 Quoted from Galilei and Scheiner (2010), 254.

39 Galilei and Scheiner (2010), 244–249.

40 Quoted from Galilei and Scheiner (2010), 377.

41 Galilei (1957), 145–153.

42 An English translation of the *Letter* is available in Galilei (1957), 173–216.

43 Galilei (1957), 155–156.

44 Quoted from Finocchiaro (1989), 146.

45 Galilei (1957), 217–220. See also Pedersen (1983).

46 Drake (1978), 252–256.

47 Drake (1978), 264–273.

48 Galilei and Scheiner (2010), 320.

49 Drake (1978), 275–278.

50 An English translation is available in Galilei (1957), 229–280.

51 Quoted from Galilei (1957), 258.

52 Quoted from Galilei (1957), 237.

53 Quoted from Galilei (1957), 274.

54 Drake (1978), 287–291.

55 On medieval studies of motion, see Grant (1971), 54–58, and Pederson and Pihl (1974), 217–228.

56 Grant (1971), 58–59.

57 Drake (1986), 303.

58 Drake (1986), 303.

59 Drake (1978), 72–73.

60 Drake (1978), 303.

61 Drake (1978), 103.

62 Drake (1978), 125–126.

63 See Galilei (2000), 162–164.

64 Galilei (2001), 25.

65 Galilei (2001), 169–172.

66 Drake (1978), 25–26.

67 For different views on how Galileo's concept of neutral circular motions relates to the modern law of inertia, see Drake (1964) and Losee (1966).

68 On Galileo's studies of projectile motion, see Drake (1973); Naylor (1974, 1980).

69 An English translation is available in Galilei (2001).

70 Galilei (2001), 164–173.

71 For Galileo's ideas on the motion of objects falling on a rotating Earth, see Naylor (2001), Martins (1998) and Galilei (2001), 191–194.

72 Galilei (2001), 220–230.

73 On Galileo's investigation of the size of fixed stars, see Brown (1985), Frankel (1978), and Graney (2008).

74 Galilei (2001), 400–412.

75 Galilei and Scheiner (2010), 315.

76 See Gingerich (2003).

77 Quoted from Galilei (2001), 516.

78 Quoted from Galilei (2001), 537–538.

79 Drake (1978), 338–352, and Heilbron (2012), 303–317. A documentary history of Galileo's trial, as well as the condemnation of 1616, is available in Finocchiaro (1989).

80 An English translation is available in Galilei (2000).

81 Drake (1978), 387.

82 Drake (1978), 436.

83 On Gassendi's transit observation, see Van Helden (1976).

84 On Gassendi's physics, see Pav (1966) and Pancheri (1978).

85 On Descartes' physics, see Blackwell (1966) and Arthur (2007).

86 An English translation of *Le Monde* is available in Descartes (1998).

87 Hall (1981), 117–120.

88 On Descartes' coordinate geometry, see Kline (1972), 308–317.

89 See Graney (2012a) and Graney (2015), 103–114.

90 See Graney (2010b); Graney and Grayson (2011); Graney (2010a) and Graney (2015), 129–139.

91 See Graney (2011) and Graney (2015), 115–128.

92 Graney (2015), 87–101.

93 On Borelli's theory of planetary motion, see Koyré (1992), 472–496, or the brief summary in Dreyer (1953), 422–423.

94 For a discussion of Huygens' work, see Yoder (2004), 9–70.

95 On Huygens' solution to the problem of Saturn's rings, see Van Helden (1974a).

96 Graney (2015), 151.

97 Chapman (1990), 345, and Graney (2015), 150–151.

98 On Horrocks, see Chapman (1990) and Plummer (1940).

99 On Gregory's reflecting telescope, see Simpson (1992).

100 On the introduction of the telescope into astronomy, see Brown (1985) and Van Helden (1974c). For a discussion of how Galileo's observations of Venus and Mars helped demonstrate the reliability of his telescope, see Chalmers (1985).

101 Quoted from Galilei (1957), 240.

102 Galilei (2000), 71.

103 Galilei (2000), 75.

104 On Galileo's scientific methodology see, for example, Seeger (1965).

Chapter 9

1 The standard scholarly biography of Newton is Westfall (1980), but see also Gleick (2003) for a brief account.

2 Westfall (1980), 40–55.

3 Westfall (1980), 55–65.

4 Westfall (1980), 66–76.

5 Guthke (1990), 144–158.

6 Westfall (1980), 94, Wilson (1970), Russell (1964), 19, but see also Thoren (1974).

7 Whiteside (1970) and Westfall (1980), 94.

8 Westfall (1980), 102–139.

9 A discussion of Newton's *anni mirabiles* can be found in Westfall (1980), 140–175.

10 See Herivel (1965), 129–130, for a transcription of Newton's work.

11 Newton's early work on centrifugal force, and his application of his results to the Moon and planets, can be found in Herivel (1965), 192–197.

12 Whiteside (1970), 13, and Herivel (1965), 58–60.

13 Westfall (1980), 156–175.

14 Mills and Turvey (1979) and Westfall (1980), 233.

15 Westfall (1980), 206–208.

16 Westfall (1980), 234–247.

17 Westfall (1980), 241–248.

18 Westfall (1980), 281–334.

19 For a biography of Hooke, see Inwood (2005), as well as Armitage (1951).

20 Nauenberg (1994), 335.

21 Quoted from Bacon and Rawley (1983), 11–12.

22 Quoted from Lohne (1960), 10.

23 Galilei (2001), 444–451.

24 Quoted from Hooke (1674), 25.

25 Hooke (1674), 25–26.

26 Quoted from Hooke (1674), 27–28.

27 Nauenberg (2005a), 521.

28 Westfall (1967), 248–249.

29 Quoted from Hooke (1674), 28.

30 See Nauenberg (1994).

31 Accounts of the exchange of letters between Hooke and Newton can be found in Lohne (1960), Ball (1893), 139–153, and Westfall (1980), 381–389.

32 Quoted from Lohne (1960), 23.

33 The investigation of the path of a falling body on a rotating Earth has a long history. See Koyré (1955) for a documentary history of early ideas on this topic.

34 Quotes from Lohne (1960), 24–25.

35 See Erlichson (1991b) for a plausible recontruction of Newton's method for generating his new sketch. That sketch has an error of its own, but it is generally of the correct shape for a constant magnitude force, as noted in Lohne (1960), 43–45.

36 Quotes from Lohne (1960), 29.

37 Quoted from Lohne (1960), 29–30.

38 Quotes from Ball (1893), 149.

39 For an analysis of Hooke's contribution to Newton's thinking about planetary motions and gravity, see Patterson (1949, 1950); Whiteside (1964); Rosenfeld (1965); Westfall (1967); Nauenberg (2005a,b).

40 Newton's proof that centripetal forces lead to Kepler's area law, and his proof that elliptical orbits require an inverse-square force, can be found in his *de Motu* of 1684. See Herivel (1965), 257–292.

41 On Newton's interest in the comets of 1680 and 1682, see Bork (1987) and Westfall (1980), 391–397.

42 Westfall (1980), 402–403, and Herivel (1965), 97.

43 Westfall (1980), 403–404.

44 Westfall (1980), 404–409.

45 Westfall (1980), 444–453.

46 Newton wrote the *Principia* in Latin. An English translation is available in Newton (1999). A useful summary, with comments, is available in Crowe (2007), 130–209.

47 Newton (1999), 403–415.
48 For a discussion of the different ways Newton used "force" in the *Principia*, see Erlichson (1991a).
49 Newton (1999), 416–430.
50 Quoted from Newton (1999), 416.
51 Quoted from Newton (1999), 416.
52 Quoted from Newton (1999), 417.
53 Quoted from Newton (1999), 417.
54 Newton (1999), 433–443.
55 Newton (1999), 444–448.
56 Newton (1999), 449–467.
57 See Weinstock (1989) for criticism of Newton's handling of this issue.
58 Newton (1999), 561–565.
59 Newton (1999), 567–588.
60 Newton (1999), 590–595.
61 Newton (1999), 779–790.
62 Newton (1999), 787–788.
63 Newton (1999), 797–801.
64 Newton (1999), 802–805.
65 Newton and Cohen (2004), 6.
66 Newton (1999), 806–811.
67 Newton (1999), 811–815.
68 Newton (1999), 821–826.
69 Newton (1999), 835–839.
70 Newton (1999), 839–874.
71 Newton (1999), 885–888.
72 Newton (1999), 888–938.
73 On Halley's prediction and its ultimate success, see Broughton (1985).
74 Westfall (1980), 469–473.
75 Westfall (1980), 730–732.
76 Quoted from Newton (1999), 943.
77 Quoted from Newton (1999), 940.
78 Westfall (1980), 504–511.
79 On Voltaire and du Châtelet's role in spreading Newtonian physics to France, see Johnson and Chandrasekar (1990).
80 Quoted from Newton (1999), 794–795.
81 Quoted from Newton (1999), 795–796.
82 See Cushing (1982) and Crowe (2007), 209–220, for a discussion of Newton's methodology.
83 See Hoskin (1977).

Chapter 10

1 On Flamsteed's alleged parallax measurement, see Williams (1979).
2 Williams (1979), 107–109.

3 Bradley's account of this work was published in Bradley (1727), but see also Blackwell (1963); Fisher (2010).

4 An apocryphal account of how Bradley came up with his idea is given in Thomson (2011), 346–347, but see Fisher (2010).

5 Fisher (2010), 38–39.

6 On Manfredi's observations, see Gualandi and Bònoli (2009) and Fisher (2010), 42–45.

7 Bradley (1727), 659–660.

8 Fernie (1975b), 225.

9 Galilei (2001), 444.

10 On Galileo's parallax method and its early implementations, see Siebert (2005); Hoskin (1966).

11 On Herschel's plan to use double stars to detect parallax, see Herschel and Banks (1782).

12 Herschel (1802), 480–486.

13 Hirshfeld (2002), 220–221, and Williams (1982), 93–94.

14 Bessel reported his parallax measurement in Bessel (1838).

15 Henderson reported his results in Henderson (1840).

16 See Airy (1837).

17 For an engaging account of the long struggle to detect and measure annual stellar parallax, see Hirshfeld (2002).

18 Newton (1999), 821–832. See also Greenberg (1996).

19 Terrall (1992), 218–221.

20 Terrall (1992), 221–226.

21 Terrall (1992), 226–237.

22 A detailed discussion of the theory of dropped objects on the rotating earth is given in Tiersten and Soodak (2000), 134–138. An account of the history of measurement attempts is in Hall (1903), 182–186.

23 Hall (1903), 189.

24 Tiersten and Soodak (2000), 137–138.

25 For a full account of Foucault's famous demonstration, see Aczel (2003).

26 Conlin (1999), 183–184.

27 On the spread of Foucault pendulums throughout Europe and the USA, see Conlin (1999).

28 In the first edition Newton had accepted a solar parallax of $20''$. See Newton (1999), 812–813.

29 On the Richer–Cassini parallax measurements, see Olmsted (1942), 125.

30 On using the transit of Venus to find the Astronomical Unit, see Teets (2003).

31 Teets (2003), 337–338.

32 Short's analysis is given in Short (1761).

33 Hornsby's analysis is given in Hornsby (1771).

34 Graney and Grayson (2011), 7–10.

35 Graney and Grayson (2011), 11–12.

36 Michelson and Pease (1921), 146. See also Devorkin (1975).

37 Broughton (1985), 123–126.

38 Broughton (1985), 126–130.

39 On the discovery of Uranus, see Schaffer (1981).

40 Smith (1989), 397–398.

41 Smith (1989), 400–402.

42 Hanson (1962), 359–363.

43 Smith (1989), 405–406.

44 Hanson (1962), 365–367.

45 Hanson (1962), 367–374.

46 Hanson (1962), 375–376, and Harper (2007), 937–938.

47 Harper (2007), 938–939.

Appendix

1 Evans (1998), 68–69.

2 Stephenson (1994), 166–167.

3 Voelkel (2001), 205.

4 Kepler (1995), 38–42.

5 Kepler (1995), 65–66.

6 See Galilei (1957), 40–41.

7 Drake (1986), 304.

8 See Drake (1973).

9 Herivel (1965), 192–198.

10 Quoted from Herivel (1965), 196.

11 Newton (1999), 804.

12 Herivel (1965), 197.

13 Newton (1999), 810–811.

14 Newton (1999), 813.

15 Bradley (1727), 648.

16 Bradley (1727), 653.

References

Aczel, A. D. (2003). *Pendulum: Léon Foucault and the Triumph of Science*. Atria Books, New York.

Airy, G. B. (1837). On the parallax of alpha Lyrae. *Monthly Notices of the Royal Astronomical Society*, 4:91.

Aiton, E. J. (1969). Kepler's second law of planetary motion. *Isis*, 60(1):75–90.

Aiton, E. J. (1987). Peurbach's *Theoricae novae planetarum*: A translation with commentary. *Osiris*, 3:4–43.

Ariew, R. (1987). The phases of Venus before 1610. *Studies in History and Philosophy of Science Part A*, 18(1):81–92.

Aristotle (2009). *The Basic Works of Aristotle*, translated by R. Mckeon. Modern Library, New York.

Aristotle (n.d.). *On the Heavens*, translated by J. L. Stock. Generic NL Freebook Publisher.

Armitage, A. (1951). Robert Hooke as an astronomer. *Popular Astronomy*, 59:287.

Armitage, A. (2004). *Copernicus and Modern Astronomy*. Dover Publications, New York.

Arthur, R. (2007). Beeckman, Descartes and the force of motion. *Journal of the History of Philosophy*, 45(1):1–28.

Bacon, F. and Rawley, W. (1983). *Sylva Sylvarum, Or, A Naturall Historie in Ten Centuries*. John Haviland, London.

Ball, W. W. R. (1893). *An Essay on Newton's Principia*. Macmillan and Company, London.

Barker, P. (1999). Copernicus and the critics of Ptolemy. *Journal for the History of Astronomy*, 30(4):343–358.

Barker, P. and Goldstein, B. R. (2001). Theological foundations of Kepler's astronomy. *Osiris*, 16:88–113.

Bartusiak, M. (2004). *Archives of the Universe, 100 Discoveries that Transformed Our Understanding of the Cosmos*. Vintage Books, New York.

Bessel, F.-W. (1838). On the parallax of 61 Cygni. *Monthly Notices of the Royal Astronomical Society*, 4:152–161.

Blackwell, D. (1963). The discovery of stellar aberration. *Quarterly Journal of the Royal Astronomical Society*, 4:44.

Blackwell, R. J. (1966). Descartes' laws of motion. *Isis*, 57(2):220–234.

Blair, A. (1990). Tycho Brahe's critique of Copernicus and the Copernican system. *Journal of the History of Ideas*, 51(3):355–377.

Blumenberg, H. (1987). *The Genesis of the Copernican World*. MIT Press, Cambridge, MA.

Bork, A. (1987). Newton and comets. *American Journal of Physics*, 55(12):1089–1095.

Bowen, A. C. (2002). Simplicius and the early history of Greek planetary theory. *Perspectives on Science*, 10(2):155–167.

Bradley, J. (1727). An account of a new discovered motion of the fix'd stars. *Philosophical Transactions*, 35(399–406):637–661.

Brehme, R. W. (1976). New look at the Ptolemaic system. *American Journal of Physics*, 44(6):506–514.

Broughton, P. (1985). The first predicted return of comet Halley. *Journal for the History of Astronomy*, 16(2):123–133.

Brown, H. I. (1985). Galileo on the telescope and the eye. *Journal of the History of Ideas*, 46(4):487–501.

Carman, C. C. (2015). The planetary increase of brightness during retrograde motion: An explanandum constructed ad explanantem. *Studies in History and Philosophy of Science Part A*, 54:90–101.

Carman, C. C. and Evans, J. (2015). The two earths of Eratosthenes. *Isis*, 106(1):1–16.

Caspar, M. (1993). *Kepler*, translated by C. D. Hellman. Dover Publications, New York.

Chalmers, A. (1985). Galileo's telescopic observations of Venus and Mars. *The British Journal for the Philosophy of Science*, 36(2):175–184.

Chapman, A. (1990). Jeremiah Horrocks, the transit of Venus, and the 'New Astronomy' in early Seventeenth-Century England. *Quarterly Journal of the Royal Astronomical Society*, 31:333.

Christianson, J. R. (1973). Copernicus and the Lutherans. *The Sixteenth Century Journal*, 1–10.

Cohen, I. B. (1985). *The Birth of a New Physics*. WW Norton & Company, New York.

Conlin, M. F. (1999). The popular and scientific reception of the Foucault pendulum in the United States. *Isis*, 90(2):181–204.

Connor, J. A. (2004). *Kepler's Witch*. Harper Collins, New York.

Copernicus, N. (1992). *Nicholas Copernicus: On the Revolutions*, translated by E. Rosen. Johns Hopkins Press, Baltimore, MD.

Crowe, M. J. (1990). *Theories of the World from Antiquity to the Copernican Revolution*. Dover Publications, New York.

Crowe, M. J. (2007). *Mechanics from Aristotle to Einstein*. Green Lion Press, Santa Fe, NM.

Cushing, J. T. (1982). Kepler's laws and universal gravitation in Newton's *Principia*. *American Journal of Physics*, 50(7):617–628.

Danielson, D. and Graney, C. M. (2014). The case against Copernicus. *Scientific American*, 310(1):72–77.

Danielson, D. R. (2001). The great Copernican cliché. *American Journal of Physics*, 69(10):1029–1035.

Danielson, D. R. (2006). *The First Copernican: Georg Joachim Rheticus and the Rise of the Copernican Revolution*. Walker Publishing, New York.

Descartes, R. (1998). *Descartes: The World and Other Writings*. Cambridge University Press, Cambridge, UK.

Devorkin, D. H. (1975). Michelson and the problem of stellar diameters. *Journal for the History of Astronomy*, 6(1):1–18.

Dijksterhuis, E. J. (1950). *The Mechanization of the World Picture*, translated by C. Dikshoorn. Princeton University Press, Princeton, NJ.

Donahue, W. H. (1981). *The Dissolution of the Celestial Spheres*. Arno Press, New York.

Drabkin, I. E. (1938). Notes on the laws of motion in Aristotle. *The American Journal of Philology*, 59(1):60–84.

Drake, S. (1964). Galileo and the law of inertia. *American Journal of Physics*, 32(8):601–608.

Drake, S. (1973). Galileo's experimental confirmation of horizontal inertia. *Isis*, 64(3):291–305.

Drake, S. (1978). *Galileo at Work: His Scientific Biography*. Dover Publications, New York.

Drake, S. (1984). Galileo, Kepler, and phases of Venus. *Journal for the History of Astronomy*, 15(3):198–208.

Drake, S. (1986). Galileo's physical measurements. *American Journal of Physics*, 54(4):302–306.

Dreyer, J. L. E. (1953). *A History of Astronomy from Thales to Kepler*. Dover Publications, New York.

Duhem, P. (1987). *Medieval Cosmology: Theories of Infinity, Place, Time, Void, and the Plurality of Worlds*, translated by R. Ariew. University of Chicago Press, Chicago, IL.

Engvold, O. and Zirker, J. B. (2016). The parallel worlds of Christoph Scheiner and Galileo Galilei. *Journal for the History of Astronomy*, 47(3):332–345.

Erlichson, H. (1991a). Motive force and centripetal force in Newton's mechanics. *American Journal of Physics*, 59(9):842–849.

Erlichson, H. (1991b). Newton's 1679/80 solution of the constant gravity problem. *American Journal of Physics*, 59(8):728–733.

Evans, J. (1984). On the function and the probable origin of Ptolemy's equant. *American Journal of Physics*, 52(12):1080–1089.

Evans, J. (1998). *The History and Practice of Ancient Astronomy*. Oxford University Press, Oxford, UK.

Ferguson, K. (2002). *Tycho and Kepler: The Unlikely Partnership that Forever Changed Our Understanding of the Heavens*. Walker Publishing, New York.

Fernie, J. (1975a). The historical search for stellar parallax, part I. *Journal of the Royal Astronomical Society of Canada*, 69:153.

Fernie, J. (1975b). The historical search for stellar parallax, part II. *Journal of the Royal Astronomical Society of Canada*, 69:222.

Field, J. V. (1988). *Kepler's Geometrical Cosmology*. Athlone Press, London.

Finocchiaro, M. A. (1989). *The Galileo Affair: A Documentary History*. University of California Press, Berkeley, CA.

Fisher, J. (2010). Conjectures and reputations: The composition and reception of James Bradley's paper on the aberration of light with some reference to a third unpublished version. *The British Journal for the History of Science*, 43(1):19–48.

Frankel, H. R. (1978). The importance of Galileo's nontelescopic observations concerning the size of the fixed stars. *Isis*, 69(1):77–82.

Franklin, A. (1976). Principle of inertia in the middle ages. *American Journal of Physics*, 44(6):529–545.

Galilei, G. (1957). *Discoveries and Opinions of Galileo*, translation and commentary by S. Drake. Anchor Books, New York.

Galilei, G. (2000). *Two New Sciences*, translated by S. Drake. Wall and Emerson, Toronto, Canada.

Galilei, G. (2001). *Dialogue Concerning the Two Chief World Systems*, translated by S. Drake. Modern Library, New York.

Galilei, G. and Scheiner, C. (2010). *On Sunspots*, translated by E. Reeves and A. Van Helden. University of Chicago Press, Chicago, IL.

Gilbert, W. (1958). *De Magnete*, translated by P. F. Mottelay. Dover Publications, New York.

Gingerich, O. (1971). Johannes Kepler and the Rudolphine Tables. *Sky and Telescope*, 42.

Gingerich, O. (1975). The origins of Kepler's third law. *Vistas in Astronomy*, 18:595–601.

Gingerich, O. (1977). Tycho Brahe and the Great Comet of 1577. *Sky and Telescope*, 54.

Gingerich, O. (1984). Phases of Venus in 1610. *Journal for the History of Astronomy*, 15(3):209.

Gingerich, O. (1997). *The Eye of Heaven: Ptolemy, Copernicus, Kepler*, volume 7 of Masters of Modern Physics. Springer, New York.

Gingerich, O. (2003). The Galileo sunspot controversy: Proof and persuasion. *Journal for the History of Astronomy*, 34(1):77–78.

Gingerich, O. (2004). *The Book Nobody Read: Chasing the Revolutions of Nicolaus Copernicus*. Penguin Books, New York.

Gingerich, O. and Van Helden, A. (2011). How Galileo constructed the moons of Jupiter. *Journal for the History of Astronomy*, 42(2):259–264.

Gingerich, O. and Voelkel, J. R. (1998). Tycho Brahe's Copernican campaign. *Journal for the History of Astronomy*, 29(1):1–34.

Gleick, J. (2003). *Isaac Newton*. Pantheon Books, New York.

Goldstein, B. R. (1967). The Arabic version of Ptolemy's *Planetary hypotheses*. *Transactions of the American Philosophical Society*, 57(4):3–55.

Goldstein, B. R. (1996a). Levi ben Gerson and the brightness of Mars. *Journal for the History of Astronomy*, 27(4):297–300.

Goldstein, B. R. (1996b). The pre-telescopic treatment of the phases and apparent size of Venus. *Journal for the History of Astronomy*, 27(1):1–12.

Goldstein, B. R. (1997). Saving the phenomena: The background to Ptolemy's planetary theory. *Journal for the History of Astronomy*, 28(1):1–12.

Goldstein, B. R. (2012). *The Astronomy of Levi ben Gerson (1288–1344): A Critical Edition of Chapters 1–20 with Translation and Commentary*, volume 11. Springer Science & Business Media.

Goldstein, B. R. and Barker, P. (1995). The role of Rothmann in the dissolution of the celestial spheres. *The British Journal for the History of Science*, 28(4):385–403.

Goldstein, B. R. and Bowen, A. C. (1983). A new view of early Greek astronomy. *Isis*, 74(3):330–340.

Granada, M. A. (2008). Kepler and Bruno on the infinity of the universe and of solar systems. *Journal for the History of Astronomy*, 39(4):469–495.

Graney, C. M. (2008). But still, it moves: Tides, stellar parallax, and Galileo's commitment to the Copernican theory. *Physics in Perspective*, 10(3):258–268.

Graney, C. M. (2010a). Seeds of a Tychonic revolution: Telescopic observations of the stars by Galileo Galilei and Simon Marius. *Physics in Perspective*, 12(1):4–24.

Graney, C. M. (2010b). The telescope against Copernicus: Star observations by Riccioli supporting a geocentric universe. *Journal for the History of Astronomy*, 41(4):453–467.

Graney, C. M. (2011). Contra Galileo: Riccioli's "Coriolis-force" argument on the Earth's diurnal rotation. *Physics in Perspective*, 13(4):387–400.

Graney, C. M. (2012a). Science rather than God: Riccioli's review of the case for and against the Copernican hypothesis. *Journal for the History of Astronomy*, 43(2):215–225.

Graney, C. M. (2012b). The work of the Best and Greatest Artist: a forgotten story of religion, science, and stars in the Copernican Revolution. *Logos: A Journal of Catholic Thought and Culture*, 15(4):97–124.

Graney, C. M. (2015). *Setting Aside All Authority: Giovanni Battista Riccioli and the Science Against Copernicus in the Age of Galileo*. University of Notre Dame Press, Notre Dame, IN.

Graney, C. M. and Grayson, T. P. (2011). On the telescopic disks of stars: A review and analysis of stellar observations from the early seventeenth through the middle nineteenth centuries. *Annals of Science*, 68(3):351–373.

Grant, E. (1964). Motion in the void and the principle of inertia in the middle ages. *Isis*, 55(3):265–292.

Grant, E. (1971). *Physical Science in the Middle Ages*. Cambridge University Press, Cambridge, UK.

Grant, E. (1996). *Planets, Stars, and Orbs: The Medieval Cosmos, 1200-1687*. Cambridge University Press, Cambridge, UK.

Greenberg, J. L. (1996). Isaac Newton and the problem of the Earth's shape. *Archive for History of Exact Sciences*, 49(4):371–391.

Gualandi, A. and Bònoli, F. (2009). The search for stellar parallaxes and the discovery of the aberration of light: The observational proofs of the Earth's revolution, Eustachio Manfredi, and the 'Bologna case'. *Journal for the History of Astronomy*, 40(2):155–172.

Guthke, K. (1990). *The Last Frontier: Imagining Other Worlds, from the Copernican Revolution to Modern Science Fiction*. Cornell University Press, Ithaca, NY.

Hall, A. R. (1981). *From Galileo to Newton*. Dover Publications, New York.

Hall, E. H. (1903). Do falling bodies move south? *Physical Review (Series I)*, 17(3):179.

Hallyn, F. and Leslie, D. M. (1990). *The Poetic Structure of the World: Copernicus and Kepler*. Zone Books, New York.

Hannah, R. (2008). *Time in Antiquity*. Routledge, London.

Hanson, N. R. (1962). Leverrier: The zenith and nadir of Newtonian mechanics. *Isis*, 53(3):359–378.

Harper, W. (2007). Newton's methodology and Mercury's perihelion before and after Einstein. *Philosophy of Science*, 74(5):932–942.

Hecht, E. (2017). Kepler and the origins of pre-Newtonian mass. *American Journal of Physics*, 85(2):115–123.

Heilbron, J. L. (2012). *Galileo*. Oxford University Press, Oxford, UK.

Henderson, T. (1840). On the parallax of alpha Centauri. *Memoirs of the Royal Astronomical Society*, 11:61.

Herivel, J. (1965). *The Background to Newton's Principia: A Study of Newton's Dynamical Researches in the Years 1664-84*. Oxford University Press, Oxford, UK.

Herschel, M. and Banks, J. (1782). On the parallax of the fixed stars. *Philosophical Transactions of the Royal Society of London*, 72:82–111.

Herschel, W. (1802). Catalogue of 500 new nebulous stars, planetary nebula, and clusters of stars; with remarks on the construction of the heavens. *Philosophical Transactions of the Royal Society of London*, 92:477–528.

Hetherington, N. S. (2006). *Planetary Motions: A Historical Perspective*. Greenwood Press, Westport, CT.

Hirshfeld, A. W. (2002). *Parallax: The Race to Measure the Cosmos*. Macmillan, London.

Holton, G. (1956). Johannes Kepler's universe: Its physics and metaphysics. *American Journal of Physics*, 24(5):340–351.

Hooke, R. (1674). *An Attempt to Prove the Motion of the Earth from Observations*. Printed by TR for John Martyn Printer to the Royal Society at the Bell in St. Pauls Church-yard, London.

Hornsby, T. (1771). The quantity of the Sun's parallax as deduced from the observations of the transit of Venus, on June 3, 1769. *Philosophical Transactions*, 61:574–579.

Hoskin, M. (1966). Stellar distances: Galileo's method and its subsequent history. *Indian Journal of History of Science*, 1:22–29.

Hoskin, M. (1977). Newton, providence and the universe of stars. *Journal for the History of Astronomy*, 8(2):77–101.

Howell, K. J. (2002). *God's Two Books: Copernican Cosmology and Biblical Interpretation in Early Modern Science*. University of Notre Dame Press, Notre Dame, IN.

Inwood, S. (2005). *The Forgotten Genius: The Biography of Robert Hooke, 1635–1703*. MacAdam/Cage Publishing, San Francisco, CA.

Jacobsen, T. S. (1999). *Planetary Systems from the Ancient Greeks to Kepler*. University of Washington Press, Seattle, WA.

Jardine, N. (1984). *The Birth of History and Philosophy of Science*. Cambridge University Press, Cambridge, UK.

Johnson, F. R. (1936). The influence of Thomas Digges on the progress of modern astronomy in sixteenth-century England. *Osiris*, 1:390–410.

Johnson, W. and Chandrasekar, S. (1990). Voltaire's contribution to the spread of Newtonianism – ii. Élemens de la philosophie de Neuton: The elements of the philosophy of Sir Isaac Newton. *International Journal of Mechanical Sciences*, 32(6):521–546.

Kelter, I. A. (1995). The refusal to accommodate: Jesuit exegetes and the Copernican system. *The Sixteenth Century Journal*, 273–283.

Kepler, J. (1965). *Kepler's Conversation with Galileo's Sidereal Messenger*, translated by E. Rosen. Johnson Reprint Corp., New York.

Kepler, J. (1981). *Mysterium Cosmographicum: The Secret of the Universe*, translated by A. M. Duncan. Abaris Books, New York.

Kepler, J. (1995). *Epitome of Copernican Astrononomy and Harmonies of the Worlds*, translated by C. G. Wallis. Prometheus Books, New York.

Kepler, J. (1997). *The Harmony of the World*, translated by E. J. Aiton, A. M. Duncan, and J. V. Field, volume 209. American Philosophical Society, Philadelphia, PA.

Kepler, J. (2003). *Kepler's Somnium: The Dream, or Posthumous Work on Lunar Astronomy*, translated by E. Rosen. Dover Publications, Mineola, NY.

Kepler, J. (2015). *Astronomia Nova*, translated by W. H. Donahue. Green Lion Press, Santa Fe, NM.

Kline, M. (1972). *Mathematical Thought From Ancient to Modern Times*, volume 1. Oxford University Press, Oxford, UK.

Knorr, W. R. (1990). Plato and Eudoxus on the planetary motions. *Journal for the History of Astronomy*, 21(4):313–329.

Kobe, D. H. (1998). Copernicus and Martin Luther: An encounter between science and religion. *American Journal of Physics*, 66(3):190–196.

Koestler, A. (1968). *The Sleepwalkers: A History of Man's Changing Vision of the Universe*. Penguin Books, Aylesbury, UK.

Koyré, A. (1955). A documentary history of the problem of fall from Kepler to Newton. *Transactions of the American Philosophical Society*, 45(4):329–395.

Koyré, A. (1957). *From the Closed World to the Infinite Universe*. Johns Hopkins Press, Baltimore, MD.

Koyré, A. (1992). *The Astronomical Revolution: Copernicus, Kepler, Borelli*. Dover Publications, New York.

Kuhn, T. S. (1957). *The Copernican Revolution: Planetary Astronomy in the Development of Western Thought*, volume 16. Harvard University Press, Cambridge, MA.

Lattis, J. M. (2010). *Between Copernicus and Galileo: Christoph Clavius and the Collapse of Ptolemaic Cosmology*. University of Chicago Press, Chicago, IL.

Linton, C. M. (2004). *From Eudoxus to Einstein: A History of Mathematical Astronomy*. Cambridge University Press, Cambridge, UK.

Lohne, J. (1960). Hooke versus Newton. *Centaurus*, 7(1):6–52.

Lohne, J. A. (1967). The increasing corruption of Newton's diagrams. *History of Science*, 6(1):69–89.

Losee, J. (1966). Drake, Galileo, and the law of inertia. *American Journal of Physics*, 34(5):430–432.

Love, D. (2015). *Kepler and the Universe: How One Man Revolutionized Astronomy*. Prometheus Books, Amherst, NY.

Malpangotto, M. (2016). The original motivation for Copernicus's research: Albert of Brudzewo's *Commentariolum super Theoricas novas Georgii Purbachii*. *Archive for History of Exact Sciences*, 70(4):361–411.

Martens, R. (2000). *Kepler's Philosophy and the New Astronomy*. Princeton University Press, Princeton, NJ.

Martin, D. R. (1984). Status of the Copernican theory before Kepler, Galileo, and Newton. *American Journal of Physics*, 52(11):982–986.

Martins, R. d. A. (1998). Natural or violent motion: Galileo's conjectures on the fall of heavy bodies. *Dialoghi-Rivista di Studi Italici*, 2(1/2):45–67.

Mendell, H. (1998). Reflections on Eudoxus, Callippus and their curves: Hippopedes and Callippopedes. *Centaurus*, 40(3–4):177–275.

Methuen, C. (1996). Maestlin's teaching of Copernicus: The evidence of his university textbook and disputations. *Isis*, 87(2):230–247.

Michelson, A. A. and Pease, F. G. (1921). Measurement of the diameter of Alpha-Orionis by the interferometer. *Proceedings of the National Academy of Sciences*, 7(5):143–146.

Millevolte, A. (2014). *The Copernican Revolution: Putting the Earth in Motion*. Tuscobia Press.

Mills, A. and Turvey, P. (1979). Newton's telescope, an examination of the reflecting telescope attributed to Sir Isaac Newton in the possession of the Royal Society. *Notes and Records of the Royal Society of London*, 33(2):133–155.

Murschel, A. (1995). The structure and function of Ptolemy's physical hypotheses of planetary motion. *Journal for the History of Astronomy*, 26(1):33–61.

Nauenberg, M. (1994). Hooke, orbital motion, and Newton's *Principia*. *American Journal of Physics*, 62(4):331–350.

Nauenberg, M. (2005a). Hooke's and Newton's contributions to the early development of orbital dynamics and the theory of universal gravitation. *Early Science and Medicine*, 10(4):518–528.

Nauenberg, M. (2005b). Robert Hooke's seminal contribution to orbital dynamics. *Physics in Perspective*, 7(1):4–34.

Naylor, R. (2001). Galileo's physics for a rotating Earth. *Largo campo di filosofare, Eurosyposium Galileo, Capitulo,* 3:337–355.

Naylor, R. H. (1974). Galileo and the problem of free fall. *The British Journal for the History of Science,* 7(2):105–134.

Naylor, R. H. (1980). Galileo's theory of projectile motion. *Isis,* 71(4):550–570.

Newton, I. (1999). *The Principia, Mathematical Principles of Natural Philosophy,* translated by I. B. Cohen and A. Whitman. University of California Press, Berkeley, CA.

Newton, I. and Cohen, I. (2004). *A Treatise of the System of the World.* Dover Publications, New York.

Newton, R. R. (1980). The sources of Eratosthenes measurement of the Earth. *Quarterly Journal of the Royal Astronomical Society,* 21:379.

Olmsted, J. W. (1942). The scientific expedition of Jean Richer to Cayenne (1672–1673). *Isis,* 34(2):117–128.

Palmieri, P. (2001). Galileo and the discovery of the phases of Venus. *Journal for the History of Astronomy,* 32(2):109–129.

Pancheri, L. U. (1978). Pierre Gassendi, a forgotten but important man in the history of physics. *American Journal of Physics,* 46(5):455–463.

Patterson, L. D. (1949). Hooke's gravitation theory and its influence on Newton. I: Hooke's gravitation theory. *Isis,* 40(4):327–341.

Patterson, L. D. (1950). Hooke's gravitation theory and its influence on Newton. II: The insufficiency of the traditional estimate. *Isis,* 41(1):32–45.

Pav, P. A. (1966). Gassendi's statement of the principle of inertia. *Isis,* 57(1):24–34.

Pedersen, O. (1983). Galileo and the council of Trent: the Galileo affair revisited. *Journal for the History of Astronomy,* 14(1):1–29.

Pederson, O. and Pihl, M. (1974). *Early Physics and Astronomy: A Historical Introduction.* Macdonald and Janes, London.

Plato (1888). *The Timaeus of Plato,* translated by R. D. Archer-Hind. Macmillan, London.

Plato (1989). *The Republic and Other Works,* translated by B. Jowett. Anchor Books, New York.

Plummer, H. (1940). Jeremiah Horrocks and his *Opera posthuma. Notes and Records of the Royal Society of London,* 3(1):39–52.

Ptolemy, C. (1998). *Ptolemy's Almagest,* translated by G. J. Toomer. Princeton University Press, Princeton, NJ.

Rawlins, D. (1987). Ancient heliocentrists, Ptolemy, and the equant. *American Journal of Physics,* 55(3):235–239.

Rosen, E. (1939). *Three Copernican Treatises: The Commentariolus of Copernicus; The Letter against Werner; The Narratio Prima of Rheticus* (revised edn.). Dover Publications, New York.

Rosen, E. (1960). Calvin's attitude toward Copernicus. *Journal of the History of Ideas,* 431–441.

Rosen, E. (1975a). Kepler and the Lutheran attitude towards Copernicanism in the context of the struggle between science and religion. *Vistas in Astronomy,* 18:317–338.

Rosen, E. (1975b). Was Copernicus' Revolutions approved by the Pope? *Journal of the History of Ideas,* 36(3):531–542.

Rosenfeld, L. (1965). Newton and the law of gravitation. *Archive for History of Exact Sciences,* 2(5):365–386.

Russell, J. L. (1964). Kepler's laws of planetary motion: 1609–1666. *The British Journal for the History of Science,* 2(1):1–24.

Saliba, G. (1995). *A History of Arabic Astronomy: Planetary Theories During the Golden Age of Islam*. NYU Press, New York.

Schaffer, S. (1981). Uranus and the establishment of Herschel's astronomy. *Journal for the History of Astronomy*, 12(1):11–26.

Schofield, C. (1981). *The Tychonic and Semi-Tychonic World Systems*. Arno Press, New York.

Seeger, R. J. (1965). Galileo, yesterday and today. *American Journal of Physics*, 33(9):680–698.

Short, J. (1761). The observations of the internal contact of Venus with the sun's limb, in the late transit, made in different places of Europe, compared with the time of the same contact observed at the Cape of Good Hope, and the parallax of the Sun from thence determined. *Philosophical Transactions (1683–1775)*, 52:611–628.

Siebert, H. (2005). The early search for stellar parallax: Galileo, Castelli, and Ramponi. *Journal for the History of Astronomy*, 36(3):251–271.

Simpson, A. (1992). James Gregory and the reflecting telescope. *Journal for the History of Astronomy*, 23(2):77–92.

Singham, M. (2007). The Copernican myths. *Physics Today*, 60(12):48.

Smith, R. W. (1989). The Cambridge network in action: The discovery of Neptune. *Isis*, 80(3):395–422.

Sobel, D. (2011). *A More Perfect Heaven: How Copernicus Revolutionized the Cosmos*. Walker Publishing, New York.

Stephenson, B. (1994). *Kepler's Physical Astronomy*. Princeton University Press, Princeton, NJ.

Swerdlow, N. (1972). Aristotelian planetary theory in the Renaissance: Giovanni Battista Amico's homocentric spheres. *Journal for the History of Astronomy*, 3(1):36–48.

Swerdlow, N. M. (2005). Ptolemy's theories of the latitude of the planets in the *Almagest*, *Handy Tables*, and *Planetary Hypotheses*. In J. Z. Buchwald and A. Franklin (eds.), *Wrong for the Right Reasons*, pages 41–71. Springer, Amsterdam.

Swerdlow, N. M. and Neugebauer, O. (2012). *Mathematical Astronomy in Copernicus' De Revolutionibus: In Two Parts*, volume 10. Springer Science & Business Media, New York.

Taub, L. C. (1993). *Ptolemy's Universe: The Natural Philosophical and Ethical Foundations of Ptolemy's Astronomy*. Open Court, Chicago, IL.

Teets, D. A. (2003). Transits of Venus and the astronomical unit. *Mathematics Magazine*, 76(5):335–348.

Terrall, M. (1992). Representing the Earth's shape: the polemics surrounding Maupertuis's expedition to Lapland. *Isis*, 83(2):218–237.

Thomson, T. (2011). *History of the Royal Society: From Its Institution to the End of the Eighteenth Century*. Cambridge University Press, Cambridge, UK.

Thoren, V. E. (1974). Kepler's second law in England. *The British Journal for the History of Science*, 7(3):243–256.

Thoren, V. E. (1990). *The Lord of Uraniborg: A Biography of Tycho Brahe*. Cambridge University Press, Cambridge, UK.

Thorndike, L. (1949). *The Sphere of Sacrobosco and its commentators*. University of Chicago Press, Chicago, IL.

Tiersten, M. S. and Soodak, H. (2000). Dropped objects and other motions relative to the noninertial earth. *American Journal of Physics*, 68(2):129–142.

Timberlake, T. K. (2013). Seeing earth's orbit in the stars: Parallax and aberration. *The Physics Teacher*, 51(8):478–481.

Toulmin, S. and Goodfield, J. (1961). *The Fabric of the Heavens: The Development of Astronomy and Dynamics*. Harper & Row, New York.

Tredwell, K. and Barker, P. (2004). Copernicus' first friends: Physical Copernicanism from 1543 to 1610. *Filozofski vestnik*, 25(2):143–166.

Van Helden, A. (1974a). "annulo cingitur": The solution of the problem of Saturn. *Journal for the History of Astronomy*, 5(3):155–174.

Van Helden, A. (1974b). Saturn and his anses. *Journal for the History of Astronomy*, 5(2):105–121.

Van Helden, A. (1974c). The telescope in the seventeenth century. *Isis*, 65(1):38–58.

Van Helden, A. (1976). The importance of the transit of Mercury of 1631. *Journal for the History of Astronomy*, 7(1):1–10.

Voelkel, J. R. (2001). *The Composition of Kepler's Astronomia Nova*. Princeton University Press, Princeton, NJ.

Weinstock, R. (1989). Long-buried dismantling of a centuries-old myth: Newton's *Principia* and inverse-square orbits. *American Journal of Physics*, 57(9):846–849.

Wesley, W. G. (1978). The accuracy of Tycho Brahe's instruments. *Journal for the History of Astronomy*, 9(1):42–53.

Westfall, R. S. (1967). Hooke and the law of universal gravitation: A reappraisal of a reappraisal. *The British Journal for the History of Science*, 3(3):245–261.

Westfall, R. S. (1980). *Never at Rest: A Biography of Isaac Newton*. Cambridge University Press, Cambridge, UK.

Westman, R. S. (1975a). *The Copernican Achievement*. University of California Press, Berkeley, CA.

Westman, R. S. (1975b). The Melanchthon circle, Rheticus, and the Wittenberg interpretation of the Copernican theory. *Isis*, 66(2):165–193.

Westman, R. S. (2011). *The Copernican Question: Prognostication, Skepticism, and Celestial Order*. University of California Press, Berkeley, CA.

Whiteside, D. T. (1964). Newton's early thoughts on planetary motion: a fresh look. *The British Journal for the History of Science*, 2(2):117–137.

Whiteside, D. T. (1970). Before the *Principia*: The maturing of Newton's thoughts on dynamical astronomy, 1664–1684. *Journal for the History of Astronomy*, 1(1):5–19.

Williams, M. (1979). Flamsteed's alleged measurement of annual parallax for the Pole Star. *Journal for the History of Astronomy*, 10(2):102–116.

Williams, M. (1982). James Bradley and the eighteenth century 'gap' in attempts to measure annual stellar parallax. *Notes and Records of the Royal Society of London*, 37(1):83–100.

Wilson, C. A. (1970). From Kepler's laws, so-called, to universal gravitation: empirical factors. *Archive for History of Exact Sciences*, 6(2):89–170.

Wootton, D. (2010). *Galileo: Watcher of the Skies*. Yale University Press.

Wootton, D. (2015). *The Invention of Science: A New History of the Scientific Revolution*. Harper Collins, New York.

Yavetz, I. (1998). On the homocentric spheres of Eudoxus. *Archive for History of Exact Sciences*, 52(3):221–278.

Yoder, J. G. (2004). *Unrolling Time: Christiaan Huygens and the Mathematization of Nature*. Cambridge University Press, Cambridge, UK.

Index